Complexity in Physics and Technology

Complexity in Physics and Technology

edited by

Manuel S. Garrido
Instituto Nacional de Investigação Científica (Lisbon)
and Université Catholique de Louvain

R. Vilela Mendes
European Center for Nuclear Research (Geneva)
and Instituto Nacional de Investigação Científica (Lisbon)

World Scientific
Singapore • New Jersey • London • Hong Kong

Published by

World Scientific Publishing Co. Pte. Ltd.
P O Box 128, Farrer Road, Singapore 9128
USA office: Suite 1B, 1060 Main Street, River Edge, NJ 07661
UK office: 73 Lynton Mead, Totteridge, London N20 8DH

COMPLEXITY IN PHYSICS AND TECHNOLOGY

ISBN 981-02-1016-7

Printed in Singapore.

PREFACE

Informally we may define *complex system* as one with many parts and a collective dynamical behaviour qualitatively different from the dynamics of its components. The dynamical behaviour of a complex system may either be *simple* (i. e. attracted to fixed points or periodic solutions) or *complex* in the sense of being sensitively dependent on the initial conditions. In this latter case the exponentially fast separation of nearby trajectories amplifies small errors in the initial conditions and the system becomes unpredictable beyond a small time interval.

In many situations complex behaviour arises as soon as elementary systems with regular behaviour are allowed to interact (see for example the contributions by T. S. Krasnopolskaya et al. and P. Santini et al. in this volume). Although complex chaotic behaviour is generic for large classes of non-linear systems, this behaviour may somehow be controlled or suppressed by time-dependent parameter modulation (cf. L. Fronzoni). The exponential instabilities in non-linear dynamical systems find also useful applications in control engineering because the fast response time of unstable systems leads to faster more efficient control devices.

Simple systems may display complex behaviour, but on the other hand simple systems which, when isolated, have complex behaviour may, when allowed to interact, develop regular patterns of collective behaviour. The simplest example of such properties is probably the coupled map lattices and coupled oscillator chains. Despite their simplicity these systems display a wide range of behaviour from coherent structures to computational behaviour to models of equipartition and relaxation to thermal equilibrium (cf. R. Livi and S. Ruffo). Propagation in a non-linear medium is another paradigmatic example of the emergence of coherent structures from non-linearity (cf. Kivshar).

The *science of complexity* covers both the study of the *complex behaviour of simple systems* and the study of the *collective behaviour of complex systems*.

The general notions of complex system and complex behaviour are becoming familiar concepts, fairly well understood at least in a qualitative way. However a quantitative characterization of complexity is still a matter of controversy. Many different notions like algorithm complexity, computational complexity, depth, sophistication, etc. have been proposed (for a review see Ref. 1) . These notions are not equivalent and probably an absolute definition of complexity is not possible because the notion depends on which particular features of the system one wants to emphasize. Some of the issues arising when inferring the complexity or modelling complex systems or systems with incomplete information are taken up in the contributions by J. P.Crutchfield, A. Kaufmann, P.R. Medina-Martins and L. Rocha.

Among the emergent collective properties of complex systems are those of learning and self-organization. Very interesting properties of this type have been found in systems composed of a large number of highly connected simple processing units. In this so-called *connectionist systems* the local processing units (*nodes*) obey simple uniform rules and the relevant information, coded or learned by the system, is carried by the pattern of *connection strengths.* When the right dictionary is established[2] a large class of connectionist systems (classifiers, autocatalytic, etc.) are equivalent to *neural networks.* For this reason and for its technological and biological relevance, neural networks is a very active topic in the field of complex systems. Some aspects of the current research in this field are illustrated in the contributions by B. Appoloni et al., K. E. Kürten and B. Söderberg. Probably at the origin of the usefulness of the connectionist approach is the fact that most of the relevant information, that we have to deal with, is of a relational type. For example in a written text the information is carried by the spatial relationship between the characters, not by the isolated characters themselves. Also in visual perception an object is recognized from the structured spatial arrangement of its parts. Incidentally (human or machine) vision carries some of the most challenging problems in the field of complex systems. On the one hand vision contains local problems like removal of noise, deblurring and contour determination in low-level or early vision (cf. R. Marques Pereira). On the other hand the high-level vision problem of object recognition is clearly a global problem involving long range correlations. The situation is aggravated by the fact that the correlations are domain dependent, in the sense that they involve the knowledge of shapes and properties of classes of objects. Some of these problems cannot be dealt with by rigid deduction algorithms and the *weak constraint satisfaction* embodied in the minimization of *cost functions* holds some promise.

Not only in vision problems but also in other kinds of signal processing the extraction of correlations is a useful tool. Normal mode decomposition and its multidimensional generalizations (cf. J. Dente and R. Lima) provide in some cases a way to extract, from the correlation information, models for the underlying signal-generating system. Actually, when the complexity of the system is such that any direct modelling becomes impossible, a signal processing approach to the observational data is the only sensible method. Some of the problems and techniques used in the analysis and extraction of useful information from the data are discussed in the contributions by J. M. N. Leitão, A. C. Rosa, R. T. Duarte, J. Santana and R. Vilela Mendes.

As is apparent from the contributions in this book, complex systems and complex behaviour, in the sense stated above, appear in many fields of natural science and technology, production lines (M. O. Hongler), etc. Economy and social sciences provide further examples. For this reason many of the same phenomena are rediscovered again and again in all these fields. The same happens for the tools that are used to deal with the phenomena. This situation calls for some interdisciplinar effort to develop a common language and to exchange tools and experiences. This book is an outgrowth of an attempt to combine the efforts and experiences of physical scientists and technologists which, although sometimes seem to be using very different languages, are in fact dealing most of the time with the same paradigms of our complex world.

[1] P.Grassberger, "Problems in quantifying self-generated complexity", Helvetica Physica Acta, 62 (1989) 489.

[2] J. Doyne Farmer, "A Rosetta stone for connectionism", Physica D 42 (1990) 153.

[illegible]

CONTENTS

VISION

COMPLEXITY, DATA ANALYSIS AND PREDICTION

PRODUCTION LINES

Complexity, Uncertainty and Modelling

DISCOVERING COHERENT STRUCTURES IN NONLINEAR SPATIAL SYSTEMS*

JAMES P. CRUTCHFIELD

Physics Department, University of California
Berkeley, California 94720 USA
chaos@gojira.berkeley.edu

ABSTRACT

A synthesis of elementary computation and dynamical system theories leads to a constructive approach to discovering coherent structures in spatial systems and to quantifying a pattern's complexity. The basic technique reviewed here builds probabilistic automata from temporal and spatial data series generated by a simple nonlinear spatial system. In this way, a given pattern's unpredictability and structure are measured by the entropy rate and complexity, respectively, of the "machine" reconstructed from the pattern data. Ancillary remarks indicate how the analysis gives a global view of the high-dimensional state space structures associated with spatial systems and, in particular, the geometry of coherent structure interactions. The bulk of the review, though, emphasizes practical results on inferring coherent space-time structures and on building detectors to track particle-like objects.

1. Coherent Structures

1.1 Between a Dynamical System and a Hard Turbulence

It is a well-worn dilemma that contrasts the apparent unpredictability of chaotic few-degree-of-freedom dynamical systems with the orderly and structured solitons appearing in nonlinear many-degree-of-freedom systems. Progress in mechanics over the last half-century has frustrated the intuitive view of "molecular chaos", that the complication observed in a system's behavior should be proportional to the number of subsystems that it contains. The mathematical description of nature is continually forced to the middle ground, the domain of complexity, from its simplifying extremes. The problem of finding "significant" and useful structure in nonlinear spatial systems is particularly important, but difficult since it too focuses on this intermediate regime. Coherent structures fall in the dynamical spectrum between the low-dimensional, but geometrically-tractable unpredictability studied so widely in recent applications of dynamical systems theory and the statistical simplicity of homogeneous, hard turbulence.

* This also appears under the same title in **Nonlinear Ocean Waves**, A. Brandt, S. Ramberg, and M. Shlesinger, editors, World Scientific Publishers, Singapore (1992). It is reproduced here with permission of the publisher.

To date there is a veritable zoology of nonlinear waves, solitons, dislocations, solitary waves, vortices, and the like. These structures appear to play a very large role in organizing the range of behavior that spatial systems can exhibit. But how are such coherent structures discovered? Often this is simply by visual inspection of experimental systems or of numerical simulations. A process that is on the whole not objective and, given the robust nonlinearity of the governing equations, not amenable to first-principles analysis.

Similar analytic difficulties are typically encountered in nonlinear dynamical systems. Due to this, efforts in this area have taken a different, "qualitative dynamics" tack. The emphasis in this is a geometric view of solution sets and their identification with behavior classes such as basins, basin separatrices, and fixed point, limit cycle, and chaotic attractors. Decomposing a given system's behavior into this list of objects, the attractor-basin portrait, yields a picture of the global organization of its solutions. Important information is extracted in this way without recourse to exact solution of the equations of motion.[4]

This geometric view when coupled with the realization that apparent unpredictability can be generated by simple, but nonlinear dynamical systems has lead in recent years to the development of "nonlinear modeling".[3] This endeavor has many aspects, but speaking broadly the goal is to discover from given data the underlying geometric structures and possibly infer the equations of motion. It is a marked change from existing exploratory data analysis[21] in that it incorporates qualitative dynamics.

The question arises then of whether the insights into unpredictability gleaned using dynamical systems theory and the more recent results from nonlinear modeling can be adapted to the study of patterns, coherent structures, and the like found in nonlinear spatial systems. The basic observation underlying this is simply that spatial time-dependent systems are, in fact, dynamical systems. So dynamical systems theory applies to them too; in principle, at least.

1.2 Nonlinear Spatial Systems as Dynamical Systems

There are some very basic problems, though, with a literal dynamical system view of spatial systems.

First, a spatial system is seen as a very high- or infinite-dimensional system. The geometric and visualization techniques and many of the analytic methods of dynamical systems theory (DST), however, are useful largely in the situation where one performs a modal decomposition or a normal-form analysis. In doing so, the analyst imposes a host of restrictions on the range of behavior that can be explained; viz., that behavior supported by the space spanned by the assumed basis patterns.

Second, in DST the system's state is a structureless point. There is, for example, no indication of the internal spatial symmetry that constrains possible states.

Third, DST does not, in and of itself, give a notion of pattern, let alone a way to detect and quantify the structure in a pattern. There are, of course, several measures that are used to classify dynamical systems. The most widely used is the dimension of "the" attractor. This is something like the number of "active degrees of freedom". Unfortunately, it is often not clear to what structures these degrees of freedom correspond. The Kolmogorov-Sinai entropy, somewhat more usefully, measures the degree of unpredictability of the dynamical system. It quantifies the diversity of state-space trajectories. It does not, however, indicate the development of structure during spatial pattern evolution. Simply stated, high and low entropy and high and low dimension processes can generate more or less complicated spatial structures.

Finally, the bulk of DST concentrates on stationary processes. There is a particular emphasis on attractors and other invariant sets. In contrast to this, it has been argued that transients, both nonstationary and quasistationary, can dominate a spatial system's behavior for extremely long times. Complicated and unpredictable spatial patterns can give way, for example, after universe-lifetimes to simple, periodic behavior.[9] (Nonstationary transients will be the subject of investigation later on.) Thus, to be more useful in the study of spatial systems DST needs to focus more on transients and the state space structures that control them.

1.3 An Approach

The computational approach that is the subject here addresses these problems directly. In particular, it provides a way to geometrize the high-dimensional state space of a spatial system. The resulting global analysis is expressed in terms of the attractor-basin portrait; as just noted, this is one of the most basic behavioral classifications of DST. It also leads to an explicit representation of "patterns" in spatial data and to a quantitative measure of their complexity in terms of the statistical computation required to produce them. A constructive procedure is given that builds from given space-time data the effective equations of motion. Properties of coherent structures in the data are captured by and can be analyzed via the estimated model. The analysis allows for one to detect and track "particles" and other coherent structures in space-time.

The following is an overview of applying these techniques, "computational mechanics" for short, to nonlinear spatial systems. The outline is as follows. The key idea turns on the ability to build models directly from a data series. Before that can be detailed, the character of the available data and the limitations imposed by the measuring instrument are mentioned. Then the model estimation procedure, machine reconstruction, is reviewed. It consists of two steps. The first is the discovery of generalized states in the data series and the second is the estimation of the dynamic, that is, the state-to-state transitions. Several useful statistics are then defined: the entropy rate, which is a measure of the diversity of patterns in the data stream, and the (finitary) complexity, which is a measure of the amount of memory in the generating process.

With the computation analytic tools laid out, they are then applied to the complicated space-time patterns produced by a nonlinear cellular automaton. Spatial structure, pattern, and domain are defined in terms of the reconstructed machine. In this way, the machine represents the type of computation that the spatial system has embedded in the pattern over time. The global state space geometry is summarized with a few remarks on a formal analysis of the high-dimensional state space in terms of the system's attractor-basin portrait. The emphasis, though, is on the domain and dislocation structures found by reconstruction. The associated machines are used to detect and track domains and dislocations in a chaotic sea. The concluding remarks indicate how the computational analysis could be automated.

1.4 Nonlinearity versus Discovery

Before embarking on the main topics a cautionary reminder is in order concerning the problem of "discovering" structure in a nonlinear system. In the case of spatial systems this discovery problem is particularly acute. On the one hand, this is due to the large number of degrees of freedom and so, also, the rather large amount of data that must be handled. On the other hand, this is due to the human propensity to see order in spatial images where there may be none.

In simple terms, the "discovery" problem of nonlinear modeling is: Have we discovered something in our data or have we projected the newly-found structure onto it by selecting an inappropriate or biased representation? This was the main lesson of attempting to reconstruct equations of motion from a time series:[10] When it works, it works; when it doesn't, you don't know what to do; and in both cases it is ambiguous what you have learned. Even though data was generated by well-behaved, smooth dynamical systems, there was an extreme sensitivity to the assumed model class that completely swamped "model order estimation". Worse still there was no *a priori* way to select the class appropriate to the generating process. This should be contrasted with what is probably one of the more important practical results in statistical modeling: within a model class a procedure exists to find, given a finite amount of data, an optimal model that balances prediction error against model complexity.[19] Despite allusions to the contrary, this "model order estimation" procedure does not address issues of class inappropriateness and what to do when confronted with failure.

One way to address the model class discovery dilemma is hierarchical machine reconstruction. This approach starts at the lowest level of representation, the given discrete data, and builds adaptively a series of models within a series of model classes of increasing computational capability until a finite causal model is found.[6] It has been useful in pointing out some general principles for detecting hierarchies of structure and connections with phase transitions. Both of these undoubtedly will have important implications in the future for spatially-extended systems. The following will take a substantially less abstract view of the modeling process, viz. modeling focussed on a single level of representation. But it will not be forgotten altogether, since when analyzing spatial systems, hierarchical reconstruction suggests a way to "factor out" layers of structure from very complicated pattern data.

With this attempt to mollify the muse of objectivity out of the way, the particular modeling methodology can be introduced.

2. Computational Mechanics

The overall goal is to infer from a series of measurements of a process a model of the generating mechanism. Additionally, the model is to indicate the process's computational structure. This refers not only to its statistical properties, such as the decay of correlation, but also to the amount of memory it contains and, for example, whether or not it is capable of producing the digit string of (say) $\sqrt{3}$. The extraction of these properties from the model determines the utility of the model class beyond mere temporal prediction of the process's behavior.

The first subsection starts off by defining the character of the data from which the model is built. The next two subsections then review machine reconstruction and the statistical mechanics of the inferred stochastic machines. The section closes with some comments on complexity and causality.

2.1 The Measurement Channel

The universe of discourse for nonlinear modeling consists of a process P, the measuring apparatus $\mathcal{I}$, and the modeler itself. The goal is for the modeler, by taking advantage of its available resources, to make the "best" representation of the nonlinear process.

The process P, the object of the modeler's ultimate attention, is the unknown, but hopefully knowable, variable in this picture. And so there is little to say, except that it can be viewed as governed by stochastic evolution equations

$$\vec{X}_{t+\Delta t} = \vec{F}\left(\vec{X}_t, \vec{\xi}_t, t\right) \tag{1}$$

where $\vec{X}_t$ is the configuration, e.g. a finite dimensional state or a spatial pattern, at time t, $\vec{\xi}_t$ some noise process, and $\vec{F}$ the governing equations of motion. The following discussion also will have occasion to refer to the process's probability distribution $\mu\left(\vec{X}\right)$ on its configuration space and the entropy rate $h_\mu\left(\vec{X}\right)$ at which it produces information.

The measuring apparatus is a transducer that maps $\vec{X}_t$ to some accessible states of an instrument $\mathcal{I}$. This instrument has a number of characteristics, most of which should be under the modeler's control. The primary interaction between the instrument and the process is through the measurement space $\mathcal{R}^{\mathrm{D}}$ which is a projection $\mathcal{P}$ of $\vec{X}_t$ onto (say) a Euclidean space whose dimension is given by the number D of experimental probes.

The instrument, which distinguishes the projected states to within resolution ϵ, partitions the measurement space into a set $\Pi_\epsilon(\mathrm{D}) = \{\pi_i : \pi_i \subset \mathcal{R}^{\mathrm{D}}, i = 0, \ldots \epsilon^{-\mathrm{D}}\}$ of cells. Each cell π_i is the equivalence class of projected states that are indistinguishable using that instrument. The instrument represents the event of finding $\mathcal{P}\left(\vec{X}_t\right) \in \pi_i$ by the cell's label i. With neither loss of generality nor information, these indices are then encoded into a time-serial binary code. As each measurement is made its code is output into the data stream. In this way, a time series of measurements made by the instrument becomes a binary string, the data stream s available to the modeler. This is a discretized set of symbols $\mathrm{s} = \ldots s_{-4}s_{-3}s_{-2}s_{-1}s_0s_1s_2s_3s_4\ldots$ where in a single measurement made by the modeler the instrument returns a symbol $s_t \in \mathcal{A}$ in an alphabet $\mathcal{A}$ at time index $t \in \mathbf{Z}$. Here we take a binary alphabet $\mathcal{A} = \{0, 1\}$.

The modeler is presented with s, the bit string, some properties of which were just given. The modeler's concern is to go from it to a useful representation, which will be called a model. To do this the modeler needs a notion of the process's effective state and its effective equations of motion. Having built a model consisting of these two components, any residual error or deviation from the behavior described by the model can be used to estimate the process's effective noise level. This level is determined by the amount of data "unexplained" by the inferred model. It should be clear when said this way that the noise level and the model's sophistication depend directly on the data and on the modeler's resources. Finally, the modeler may have access to experimental control parameters. And these can be used to aid in obtaining different data streams useful in improving the model by (say) concentrating on behavior where the effective noise level is highest.

The central problem of nonlinear modeling now can be stated. Given an instrument, some number of measurements, and fixed *finite* inference resources, how much regularity and computational structure in the underlying process can be extracted?

2.2 Computation from a Time Series

On what sort of structure in the data stream should the models be based? If part of the goal is prediction, then a natural object to reconstruct from the data series is a representation of the process's instantaneous state. Unfortunately, as already noted, individual measurements are only indirect representations of the state. Indeed, the instrument simply may not supply data of a quality sufficient to discover the true states, independent of the amount of data. So how can the process's "effective" states be accessed?

For discrete time series a state is defined to be the set of subsequences that render the future conditionally independent of the past.[11] This notion of state is widespread; appearing in various guises in early symbolic dynamics and ergodic and automata theories. It is close to the basic notion of state in Markov process theory. And so, here, an observer

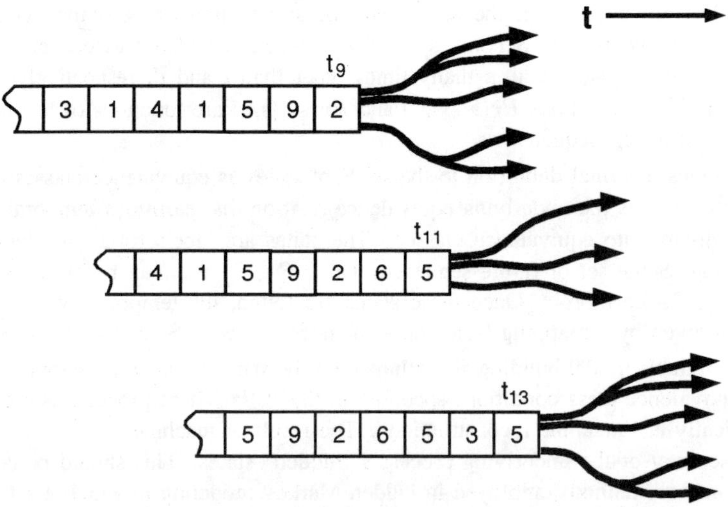

Fig. 1 Morph-equivalence induces conditionally independent states. When the template of future possibilities, i.e. allowed future subsequences and their past-conditioned probabilities, has the same structure then the process is in the same generalized state. At t_9 and at t_{13}, the process is in the same state; at t_{11} it is in another different state.

identifies a state at different times in the data stream as its being in identical conditions of ignorance about the future. (See Fig. 1.)

Consider two parts of the data stream $\mathrm{s} = \ldots s_{-2}s_{-1}s_0s_1s_2\ldots$. The one-sided forward sequence $\mathrm{s}_t^{\rightarrow} = s_ts_{t+1}s_{t+2}s_{t+3}\ldots$ and one-sided reverse sequence $\mathrm{s}_t^{\leftarrow} = \ldots s_{t-3}s_{t-2}s_{t-1}s_t$ are obtained from s by splitting it at time t into the forward- and reverse-time semi-infinite subsequences. Consider the joint distribution of possible forward sequences $\{\mathrm{s}^{\rightarrow}\}$ and reverse sequences $\{\mathrm{s}^{\leftarrow}\}$ over all times t

$$Pr(\mathrm{s}^{\rightarrow}, \mathrm{s}^{\leftarrow}) = Pr(\mathrm{s}^{\rightarrow}|\mathrm{s}^{\leftarrow})Pr(\mathrm{s}^{\leftarrow}) \tag{2}$$

The conditional distribution $Pr(\mathrm{s}^{\rightarrow}|\omega)$ is to be understood as a function over all possible forward sequences $\{\mathrm{s}^{\rightarrow}\}$ that can follow the particular sequence ω where ever ω occurs in s.

Then the same state $S \in \mathbf{S}$ is associated with all those times $t, t' \in \{t_{i_1}, t_{i_2}, t_{i_3} \ldots : i_k \in \mathbf{Z}\}$ such that past-conditioned future distributions are the same. That is,

$$t \sim t' \text{ if and only if } Pr(\mathrm{s}^{\rightarrow}|\mathrm{s}_t^{\leftarrow}) = Pr(\mathrm{s}^{\rightarrow}|\mathrm{s}_{t'}^{\leftarrow}) \tag{3}$$

If the source generating the data stream is ergodic, then there are several comments that serve to clarify how this relation defines states. First, the sequences $\mathrm{s}_t^{\leftarrow}$ and $\mathrm{s}_{t'}^{\leftarrow}$ are typically distinct. If $t \sim t'$, Eq. 3 means that upon having seen different histories one can be, nonetheless, in the same state of anticipation or ignorance about what will happen in the future. Second, $s_t^{\leftarrow}$ and $s_{t'}^{\leftarrow}$, when considered as particular symbol sequences, will each occur in s many times other than t and t', respectively. Finally, the conditional distributions $Pr(s^{\rightarrow}|s_t^{\leftarrow})$ and $Pr(s^{\rightarrow}|s_{t'}^{\leftarrow})$ are functions over a nontrivial range of "follower" sequences $s^{\rightarrow}$.

This gives a formal definition to the set $\mathbf{S}$ of states as equivalence classes of future predictability: $\sim$ is the underlying equivalence relation that partitions temporal shifts of the data stream into equivalence classes. The states are labels for these classes. For a given state S the set of future sequences $\{s_S^{\rightarrow} : S \in \mathbf{S}\}$ that can be observed from it is called its future morph. Once these states are found, the temporal evolution of the process is given by a mapping from states to states, $T : \mathbf{S} \rightarrow \mathbf{S}$; that is, $S_{t+1} = TS_t$.

The available model building algorithms infer the states $\mathbf{S}$ via various approximations of the equivalence class conditions specified in (3).[6,11,12] These procedures are referred to generically as "machine reconstruction". The result of machine reconstruction, then, is the discovery of the underlying process's "hidden" states. This should be contrasted with the *ad hoc* methods employed in hidden Markov modeling in which a set of states and a transition structure are imposed by the modeler at the outset[18] and common in "transform" analysis using Fourier, Laplace, or wavelet basis functions.

The overall procedure has two steps. This first is to identify the states $\mathbf{S}$ and the second is to infer the transformation T. In the following I review the simplest implementation since this affords the most direct means of commenting on certain properties of the resulting models and derivative statistics.

Initially, a parse tree is built from the data stream s. A window of width D is advanced through s one symbol at a time. If $s = \ldots 1010101110101 \ldots$ then at some time t the $D = 5$ subsequence $s_t^5 = 10101$ is seen, followed by $s_{t+1}^5 = 01010$. Each such subsequence is represented in the parse tree as a path. The tree has depth $D = 5$. (See Fig. 2).

Counts are accumulated at each tree node as each subsequence is put into to the parse tree. If the associated path is not present in the tree, it is added; new nodes each begin with a count of 1 and counts in existing nodes are incremented. If s has length N, then a node probability, which is also the probability of the sequence leading to it, is estimated by its relative frequency

$$Pr(n) = \frac{c_n}{N - D} \tag{4}$$

where c_n is the count accumulated at node n. The node-to-node transition probabilities $Pr(n \rightarrow n')$ are estimated by

$$Pr(n \rightarrow n') = Pr\left(s|s^k\right) = \begin{cases} \frac{Pr(ss^k)}{Pr(s^k)} \approx \frac{c_{n'}}{c_n} & c_n > 0 \\ 0 & \text{otherwise} \end{cases} \tag{5}$$

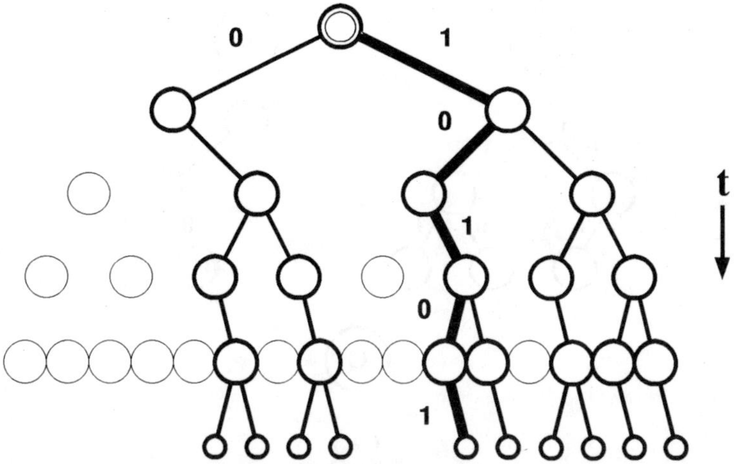

Fig. 2 The parse tree for the process "every other symbol is a 1". The tree is a hierarchical representation of the subsequences produced by the process. Here the tree depth is $D = 5$. If s = $\ldots 1010101110101 \ldots$, then the subsequence $s_i^5 = 10101$ will be seen. This is put into the parse tree, as shown by the bold line. Note that time goes down the tree. When counts of the subsequence that lead to a given tree node are accumulated, the tree gives a hierarchical representation of the infinite-sequence probability distribution.

where the length k sequence s^k leads to node n and the length $k+1$ sequence $s^{k+1} = ss^k$ leads to node n' on symbol $s \in \mathcal{A}$.

To infer the set of states one looks for a set of similar future-morphs. On the parse tree the latter are distinct-shape subtrees. By way of incremental approximation, the morph similarity is limited to future subsequences of length L. On the parse tree, equivalently, one looks for subtrees of morph depth L. Naturally, $L \leq D$ and typically one takes $2L = D$. Investigating the parse tree of Fig. 2 using depth $L = 2$ subtrees one finds the three morphs shown in Fig. 3. In this way, three "machine" states $\mathbf{S} = \{\mathrm{A}, \mathrm{B}, \mathrm{C}\}$ have been discovered for the "every other symbol is a 1" process of Fig. 2 up to the approximation implied by setting $D = 5$ and $L = 2$. A moment's reflection shows that the latter parameter settings are sufficient to infer the process's correct and minimal model.

The state-to-state transition structure is obtained by looking at how the morphs change into one another upon moving down the parse tree, i.e. upon reading $s = 0$ or $s = 1$. The resulting transformation T is represented graphically by the machine shown in Fig. 4. (Ignore for the moment the transition probabilities mentioned there.) This should be compared to the parse tree (Fig. 2). There the morph below the top node, which is associated with machine state A, makes a transition on $s = 1$ to a tree node with a morph below it in the same equivalence class (machine state A). On $s = 0$, though, the top tree node makes a transition to a tree node with a morph associated with machine state B. In just this way the machine of Fig. 4 summarizes the morph to morph transition structure T on the parse tree.

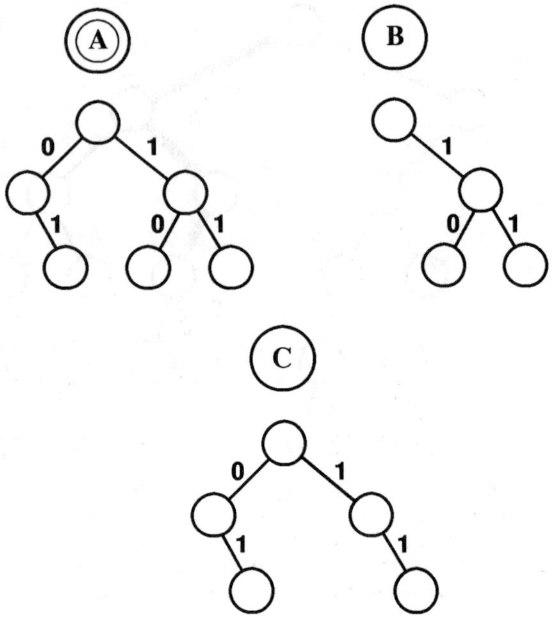

Fig. 3 The morph depth $L = 2$ subtrees found below tree nodes down to depth 3 in Fig. 2's parse tree. Each morph, or subtree, has been labeled by its associated machine state (cf. Fig. 4).

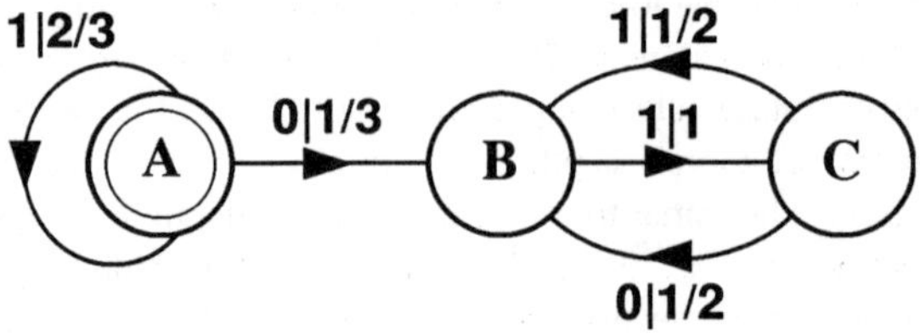

Fig. 4 The reconstructed stochastic machine for the process "every other symbol is a 1". The state with an inscribed circle is the start state, in which the machine begins reading or generating symbols in a string. Each edge is labeled $s|p$ where $s \in \mathcal{A}$ and $p \in [0, 1]$ is the probability of making the transition between the states connected by the edge with label s.

This exposition only considered topological reconstruction: subtrees were compared only up to the subsequences $\{s_t^{\rightarrow}\}$ which were observed. Of course, what is needed is a set of states that are not only topologically distinct, but also distinct in probability as indicated by (3). Details for this and for the transition probabilities can be found elsewhere.[11]

2.3 Statistical Mechanics

Machine reconstruction gives then a set of states, that will be associated with a set $\mathbf{V} = \{v\}$ of vertices, and a set of transitions, that will be associated with a set

$$\mathbf{E} = \left\{ e : e \sim v \underset{s}{\rightarrow} v', \quad v, v' \in \mathbf{V}, s \in \mathcal{A} \right\} \tag{6}$$

of labeled edges. The graphical depiction of $\mathrm{M} = \{\mathbf{V}, \mathbf{E}, \mathcal{A}\}$ is a labeled directed graph as seen above in Fig. 4. The full probabilistic structure is described by a set of input-alphabet labeled transition matrices

$$\left\{ T^{(s)} : \left(T^{(s)} \right)_{vv'} = p_{v \underset{s}{\rightarrow} v'}, \quad v, v' \in \mathbf{V}, s \in \mathcal{A} \right\} \tag{7}$$

where $p_{v \underset{s}{\rightarrow} v'}$ denotes the conditional probability to make a transition to state v' from state v on observing symbol s.

Given the current state $v_t \in \mathbf{V}$ at time t the future is conditionally independent of the past

$$\begin{aligned} Pr(\mathrm{s}) &= Pr\left(\overleftarrow{\mathrm{s}_t} \overrightarrow{\mathrm{s}_{t+1}}\right) \\ &= Pr\left(\overrightarrow{\mathrm{s}_{t+1}} | \overleftarrow{\mathrm{s}_t}\right) Pr\left(\overleftarrow{\mathrm{s}_t}\right) \\ &= Pr\left(\overrightarrow{\mathrm{s}_{t+1}} | v_t\right) Pr(v_t) \end{aligned} \tag{8}$$

where $\overleftarrow{\mathrm{s}_t}$ is the historical sequence leading to v_t. By factoring the joint distribution $Pr(\mathrm{s})$ over the observed subsequences, the discovered states vastly reduce the dimensionality of the process's representation. In the case of strings of length N, the representation changes from requiring an N-dimensional probability vector for the joint distribution to a set of $\|\mathbf{E}\|$ transition probabilities.

Rather than simply looking up the probability of a given sequence $\mathrm{s}^N = s_0 s_1 s_2 \ldots s_{N-1}, \quad s_i \in \mathbf{A}$, in the table of joint probabilities, its probability is recovered from the machine by computing the telescoping product of conditional transition probabilities

$$Pr\left(\mathrm{s}^N\right) = p_{v_0} p_{v_0 \underset{s_0}{\rightarrow} v_1} p_{v_1 \underset{s_1}{\rightarrow} v_2} \cdots p_{v_{N-1} \underset{s_{N-1}}{\rightarrow} v_N} \tag{9}$$

Here v_0 is the unique start state. By the nature of machine reconstruction it is the state of total ignorance: before the first measurement is made $p_{v_0} = 1$. The sequence $v_0, v_1, v_2, \ldots, v_{N-1}, v_N$ are those states through which the sequence s^N drives the machine.

A machine is a compact and informative representation of the underlying source. In order to appreciate the properties that it captures, there are several statistics that can be computed from a given machine. Each of these is, naturally, a simplification in one way or another of the structure captured by the machine. Up to this point there has been no restriction on the number $\|\mathbf{V}\|$ of states. For the sake of simplicity in the following the number of states will be assumed to be finite. This restricts the general model class to that of finitary machines or, more commonly, stochastic finite automata.

One useful reduction of a machine M is to ask for its equivalent Markov process. This is described by the stochastic connection matrix

$$T = \sum_{s \in \mathcal{A}} T^{(s)} \tag{10}$$

where $(T)_{vv'} = p_{v \to v'}$ is the state to state transition probability, unconditioned by the measurement symbols. By construction every state has an outgoing transition. This is reflected in the fact that T is a stochastic matrix: $\sum_{v' \in \mathbf{V}} p_{vv'} = 1$. It should be clear from dropping the input-alphabet transition labels from the machine that the detailed, I call it "computational", structure of the input data stream has been lost. All that is retained in T is the state transition structure and this is a Markov chain. The interesting fact is that Markov chains are a proper subset of finitary machines.

The stationary state probabilities $\vec{p}_{\mathbf{V}} = \left\{ p_v \in [0,1] : \sum_{v \in \mathbf{V}} p_v = 1, v \in \mathbf{V} \right\}$ are given by the left eigenvector of T

$$\vec{p}_{\mathbf{V}} T = \vec{p}_{\mathbf{V}} \tag{11}$$

The entropy rate of the Markov chain is then[20]

$$h_\mu(T) = - \sum_{v \in \mathbf{V}} p_v \sum_{v' \in \mathbf{V}} p_{v \to v'} \log_2 p_{v \to v'} \tag{12}$$

Being the transition uncertainty averaged over all the states, it measures the information production rate in bits per time step. It is the entropy rate $h_\mu(\mathrm{M})$ of the stochastic machine M. The entropy rate gives the growth rate of the Shannon information in subsequences

$$h_\mu = \lim_{L \to \infty} \frac{H(L)}{L} \tag{13}$$

where $H(L) = - \sum_{\omega \in \{s^L\}} Pr(\omega) \log_2 Pr(\omega)$. That is, h_μ is a scaling exponent: $H(L) \underset{L \to \infty}{\propto} h_\mu L$. Once a machine is reconstructed from a data stream, its entropy is an estimate of the underlying process's entropy rate.

The complexity† quantifies the information in the state-alphabet sequences

$$C_\mu(\mathrm{M}) = H(\vec{p}_{\mathbf{V}}) = - \sum_{v \in \mathbf{V}} p_v \log_2 p_v \tag{14}$$

† Within the reconstruction hierarchy this is actually the finitary complexity, since discussion is restricted to processes with a finite number of states.

It measures the amount of memory in the generating process.

The entropy h_μ, as a measure of the diversity of patterns, and the complexity C_μ, as a measure of memory, have been taken as the two elementary "information processing" coordinates with which to analyze a range of nonlinear dynamical systems.[12]

Although it falls somewhat outside of the present discussion, it is worthwhile noting that there is a general thermodynamics of ϵ-machines. This describes the process of optimal model estimation as one in which the observer comes to "inferential" equilibrium with the data stream. Additionally, it indicates the detailed structure of invariant subsequences in terms of generalized thermodynamic potentials.[13]

2.4 Complexity and Causality

It is useful at this stage to stop and reflect on some properties of the models whose reconstruction and structure have just been described. Consider two extreme data sources. The first, highly predictable, produces a stream of 1s; the second, highly unpredictable, is an ideal random source of binary symbols. The parse tree of the predictable source is a single path of 1s. And there is a single distinct subtree, at any depth. As a result the machine has a single state and a single transition on $s = 1$: a simple model of a simple source. For the ideal random source the parse tree, again to any depth, is the full binary tree. All paths appear in the parse tree since all binary subsequences are produced by the source. There is a single subtree, of any morph depth, at all parse tree depths: the full binary subtree. And the machine has a single state with two transitions: one on $s = 1$ and one on $s = 0$. The result is a simple machine, even though the source produces the widest diversity of binary sequences. Thus, these two zero entropy and maximal entropy sources have zero complexity.

A simple gedanken experiment serves to illustrate how complexity is a measure of a machine's memory capacity. Consider two observers **A** and **B**, each with the same model M of some process. **A** is allowed to start machine M in any state and uses it to generate binary strings that are determined by the edge labels of the transitions taken. These strings are passed to observer **B** which traces their effect through its own copy of M. On average how much information about M's state can **A** communicate to **B** via the binary strings? If the machine describes (say) a period three process, e.g. it outputs strings like $101101101\ldots$, $011011011\ldots$, and $110110110\ldots$; it has $\|\mathbf{V}\| = 3$ states. Since **A** starts M in any state, **B** can learn only the information of the process's phase in the period 3 cycle. This is $\log_2 \|\mathbf{V}\| = 1.584\ldots$ bits of information on average about the process's state, if **A** chooses the initial states with equal probability. However, if the machine describes an ideal random binary process, by definition **A** can communicate no information to **B**, since there is no structure in the sequences to use for this purpose. This is reflected in the fact, as already noted above, that the corresponding machine has a single state and its complexity is $C_\mu(\mathrm{M}) = \log_2 1 = 0$. In this way, a process's complexity is the amount of information that someone controlling its start state can communicate to another.

These examples serve to highlight one of the most basic properties of complexity, as I use the term. Both predictable and random sources are simple in the sense that their models are small. Complex processes in this view have large models. In computational terms, complex processes have, as a minimum requirement, a large amount of memory as revealed by many internal states in the reconstructed machine. Most importantly, this memory is structured in particular ways that support different types of computation.

There are two additional points to be brought out concerning what reconstructed machines represent. First, by the definition of future-equivalent states, they give the minimal information dependency between the morphs. In this respect, they represent the causality of the morphs considered as events. If state B follows state A then A is a cause of B and B is one effect of A. Second, the machines capture the information flow within the given data stream. Machine reconstruction produces minimal models up to the given approximation level; that is, up to the amount of data available for the estimation and up to the setting of the parameters (D, L). This minimality guarantees that there are no other events (morphs) that intervene between successive states, at the given error level, to render A and B independent. In this and only this case, can one unambiguously say that information flows from A to B, under the chosen data stream parsing direction. The amount of information that flows is given by the mutual information $I(\mathrm{A};\mathrm{B}) = H(\mathrm{B}) - H(\mathrm{B}|\mathrm{A})$ of observing the state-event A followed by state-event B.[20] This criterion for information flow also extends to spatial systems for which it gives a way to trace out iso-causality contours.

3. Cellular Automata: Prototype Spatial Systems

The preceding sections reviewed a model estimation method that reconstructs a stochastic automaton from a data series. The explanations were facilitated by thinking of the individual measurements as ordered by time. This, of course, need not be the case. The remainder will show, in fact, an application to building models of spatial patterns using machine reconstruction. Along the way, a number of interpretations and uses for the resulting models will be pointed out.

This analysis will be done for what are probably the simplest deterministic models of spatial dynamical systems: cellular automata (CA). They have the notable feature of being as discrete as possible: in space (cells), in site value (e.g. binary), and in time.[24] The particular example selected, rule 18 to be defined below, is itself the simplest nonlinear CA. It is specified by a very simple rule, but obeys no linear superposition principle as do other periodic and chaotic CA.

CA might seem to be too simple to be a useful indicator of nonlinear spatial modeling for continuum, nonlinear wave systems. However, despite their simplicity, or perhaps because of it, CA have been used recently as a simulation technique for Navier-Stokes fluid flows and for nonlinear reaction-diffusion partial differential equations, such as the shock-wave forming Burghers equation.[2] Though these lattice-gas CA will not be further discussed here, the techniques that follow apply in principle to them and so form an interesting area for their application.

The following reviews a synthesis of the dynamical systems view of a CA as a mapping of a state space and the computation theoretic view of a CA as a parallel algorithm that produces complicated bit strings.[15] CA dynamics will be analyzed in terms of local space-time and state-space "coherent" structures that organize the system's evolution. The approach introduces a rather literal analysis of CA as dynamical systems. Using it a wide range of related CA, those dominated by domains and walls, can be analyzed in terms of the analogs of attractors, basins, and separatrix structures; the centrals object of study in DST. The remainder of the discussion briefly reviews results presented elsewhere that bear on the question of detecting coherent structures.[15]

3.1 Definitions

A cellular automaton consists of a countable array of discrete sites or cells i and a discrete-time update rule ϕ operating in parallel on local neighborhoods of a given radius r.[22] At each time the sites take on values in a finite alphabet $\mathcal{A}$ of primitive symbols: $\sigma_t^i \in \{0, 1, ...k-1\} \equiv \mathcal{A}$. The local site-update function is written

$$\sigma_{t+1}^i = \phi(\sigma_t^{i-r}, \ldots, \sigma_t^{i+r}) \tag{15}$$

The state s_t of the CA at time t is the configuration of the finite or infinite spatial array: $\mathrm{s}_t \in \mathcal{A}^N$, where $\mathcal{A}^N$ is the set of all possible site value configurations on a lattice of N cells. The "extended state space", denoted $\mathcal{A}^*$, is the union of all states of any N:

$$\mathcal{A}^* = \bigcup_{N \geq 0} \mathcal{A}^N \tag{16}$$

with $\mathcal{A}^0 = \emptyset$.

The CA global update rule $\Phi : \mathcal{A}^N \to \mathcal{A}^N$ applies ϕ in parallel to all sites in the lattice: $\mathrm{s}_t = \Phi \mathrm{s}_{t-1}$. For finite N it is also necessary to specify a boundary condition. The following will use periodic boundary conditions.

A pattern ensemble $\Lambda \subseteq \mathcal{A}^*$ is a collection of strings each of which is an observable configuration at a given time. The machine representation of a pattern ensemble Λ will be denoted $\mathrm{M}(\Lambda)$.

3.2 Rule 18

The automaton of interest here is an "elementary" CA for which $(k, r) = (2, 1)$; that is, nearest-neighbor interactions and binary cell values, giving a total of $k^{2r+1} = 8$

possible neighborhood patterns. Following the conventional numbering scheme,[22] one arranges these patterns in the order below, writes the value obtained by applying ϕ to each pattern, and interprets the resulting string of 8 symbols as a binary number giving the rule index. The following analyzes the CA having rule index 18, for which the rule table is as follows:

$$\begin{array}{lcccccccc} \text{Neighborhood} & \underline{111} & \underline{110} & \underline{101} & \underline{100} & \underline{011} & \underline{010} & \underline{001} & \underline{000} \\ \text{Next site value} & 0 & 0 & 0 & 1 & 0 & 0 & 1 & 0 \end{array} \qquad \frac{\sigma_t^{i-1}\sigma_t^i \quad \sigma_t^{i+1}}{\sigma_{t+1}^i} \tag{17}$$

As is readily seen from the rule table this CA is dissipative in the sense that the global update rule is many-to-one: $\Phi\mathcal{A}^N \subset \mathcal{A}^N$.

The basic phenomenology of rule 18 has been known for some time.[14] It organizes itself into local regions in which every other cell in space and time is 0 and the remaining cells are unrestricted. The boundaries between adjacent "phase-locked" regions are dislocations, where the spatial phase slips a cell. These are the coherent structures in rule 18. (The reader may wish jump ahead to the space-time diagram of Fig. 7. This shows rule 18 evolving from a random initial pattern.) When evolved from a random initial condition, numerical results indicate that the number density d of dislocations decays in time as $d \approx (8\pi Dt)^{-\frac{1}{2}}$, with $D \approx \frac{1}{2}$. Furthermore, the excursion x of a "plume" of dislocations grows as $\langle x^2 \rangle \approx 2Dt$. The interpretation of this is that the dislocations move as if following a space-time random walk with diffusion constant D, annihilating in pairs when their paths cross. In short, starting from a random configuration the dislocations behave like a diffusive gas of annihilating particles.

3.3 Transducers

The ϵ-machines defined above can be used either as input devices for classifying existing strings or as output devices that generate strings belonging to a given language. These two modes of operation are easily combined by modifying the machine so that on each transition a symbol is emitted as well as read. The resulting "transducer" machines are functions from strings to strings. Transducers give a very compact notation for functions that map complicated sets, e.g. Cantor sets, to complicated sets. Of equal importance, they are constructive in the sense that they indicate literally how the function is to be implemented and so computed.

A transducer $\mathrm{T} = \{\mathbf{V}, \Sigma_{in}, \mathbf{E}, \Sigma_{out}, \beta\}$ is an automaton where $\{\mathbf{V}, \mathbf{E}\}$ are as defined above and where Σ_{in} is the previous input alphabet $\mathcal{A}$. Σ_{out} is an output alphabet. The observation function β maps $\mathbf{V} \times \Sigma_{in}$ to Σ_{out}. A transducer operates as an ϵ-machine with the extra feature of writing a symbol in Σ_{out} to an output string upon reading a symbol in Σ_{in} from the input string. In the context of spatial systems, the effect of running the transducer over a pattern is to recode the data by replacing original symbols with those emitted by the machine.

4. Coherent Structures

With these definitions in hand, this section begins to address the problems of spatially-extended systems alluded to in the introduction. The first task is to motivate several definitions of coherent structures. The following section then applies these to structure detection.

The global machine $M_t = M(\Omega_t)$, when $\Omega_t = \mathbf{\Phi}_t(\mathcal{A}^*)$, describes the time-evolution of the ensemble of all possible initial conditions up to some time t. It necessarily captures all of the information processing accessible to the CA, which includes correlations and information transmission, as well as computation. Previous investigations of CA patterns have focused on the global machine M_t.[23,17]

In terms of dynamical systems theory, though, the global machine describes the entire attractor-basin portrait: the collection of all invariant sets and transients, including attractors and separatrices.[4] In many cases, this description is seen to be prohibitively difficult to construct because, for example, M_t grows too large with time. In fact, such complete global descriptions are rarely pursued in dynamical systems theory. Even for elementary CA, there are many cases in which the size of M_t increases extremely rapidly in time. This is true in particular of rule 18: $M(\Omega_3)$ has $\|\mathbf{V}\| = 143$ states; $M(\Omega_4)$ is estimated to have $\|\mathbf{V}\| > 20,000$.[23] Furthermore, there are other CA rules for which the growth is even faster.[17]

An alternative approach is to identify the *structures* that dominate and organize the spatial system's behavior. The most basic of these for dissipative systems are attractors, basins, and separatrices.[4] The state space is then understood in terms of those structures. In the case of CA, the goal is to find the dominant, dynamically-homogeneous space-time patterns and their associated machines.

The introductory comments noted that a state of a dynamical system is a structureless point. In contrast, it is exactly the internal, now spatial, structure of a CA configuration s_t that is of interest. This deficit can be addressed by interpolating between the literal notion of a CA state as a position-dependent configuration of cells and the intuitive notion of a translation-independent pattern. Generally speaking, a pattern is (i) a position-independent configuration and (ii) a typical sequence generated by the process, where "typical" is used in the sense of the Shannon-McMillan theorem of information theory.[1] The use of a finitary machine description of processes captures exactly these features.

For CA rule 18, the dominant structure is the domain already described: every other site has value 0, while the remaining sites can be either 0 or 1. The iterate of a domain consists of a domain shifted by one site. This observation is consistent with the intuitive notion of a domain as a dynamically-homogeneous region of the spatial array.

A domain Λ of a CA is an ensemble representing a set of spatial regions with the following properties:
(i) *Temporal invariance:* Λ is mapped onto itself by the dynamic: $\mathbf{\Phi}\Lambda = \Lambda$; and

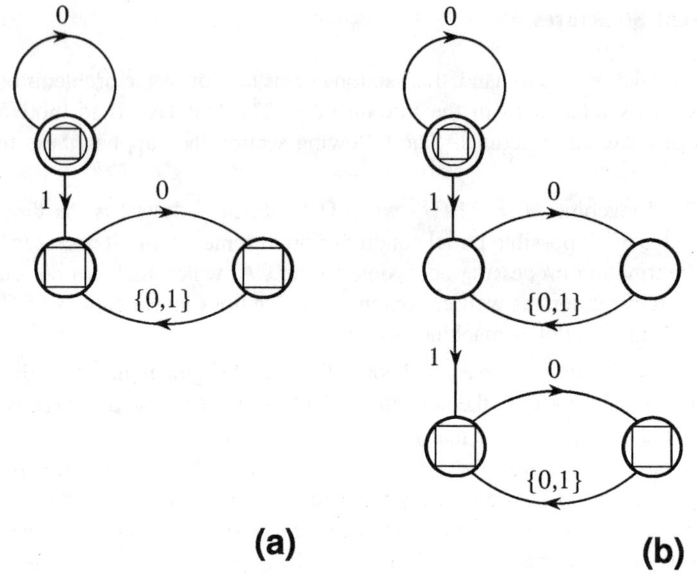

Fig. 5 $M(\Lambda^0)$ for single domain language Λ^0 and (b) $M(\Lambda^{0,0})$ for domain-wall-domain language $\Lambda^{0,0}$. Inscribed squares indicate the accept states: finite strings are accepted if and only if they leave the machine in one of the states. Edges labeled $\{0,1\}$ indicate two transitions, one on $s = 0$ and one on $s = 1$, are allowed. Transition probabilities are uniform and are not indicated.

(ii) *Spatial homogeneity:* The recurrent states of $M(\Lambda)$ are strongly connected. This is similar to statistical stationarity of the spatial pattern.

A domain describes a computationally-homogeneous *region* of the spatial array. In this way we take "domain" to mean the domain language Λ or, alternatively, any configuration $s \in \Lambda$. For general CA there can be any number of different domains Λ^i, each with its own computational structure. Additional particle-like coherent structures can now be defined with respect to domains.

A wall is a boundary between domains. Two domains Λ^i and Λ^j separated by a wall will be denoted $\Lambda^{i,j}$. Walls may consist of any number of cells, including zero. A dislocation is a wall separating two identical domains and whose minimum width is zero.

The phase of site i in a domain Λ^j is the state $v \in \mathbf{V}$ of the machine $M(\Lambda^j)$ when s^i is read in a chosen direction.

In the case of rule 18, the observed behavior described above defines a domain language Λ^0. All spatial states are made up of concatenations of words $\omega \in \Lambda^0$ separated by dislocations. In particular, the two-domain language $\Lambda^{0,0}$ consists of all states in which there is a single dislocation. The (minimal) machines for both Λ^0 and $\Lambda^{0,0}$ are shown in Figure 5.

Starting from the above definition of a domain and a dislocation, definitions can be developed for neighborhoods, attractors, basins, and separatrices. The global structure of rule 18's state space can be laid out in terms of an attractor-basin portrait, almost wholly independent of the lattice size or, equivalently, the dimension of the state space. With the identification of the coherent structures, even more geometry can be extracted concerning the organization of the basins into subbasins connected by portals. The number of dislocations is constant in a subbasin. The portals are determined by the annihilation conditions operating when dislocations collide; that is, when the number of dislocations decreases. From the view of average properties, it is then seen that there are two statistically significant attractors, viz. those given in Fig. 5. Formally, it can be shown that (i) Λ^0 is a domain; (ii) $\Lambda^{0,0}$ is a time-invariant (fixed-point) ensemble. Additionally, it can be shown that on an infinite lattice with finite initial conditions and on an overwhelmingly large set of finite lattices, Λ^0 and $\Lambda^{0,0}$ are attractors. These results established the geometric picture that the attracting domain Λ^0 organizes the state space. $\Lambda^{0,0}$ appears as the lowest "energy" fluctuation above the vacuum state Λ^0. Like analogous situations with topological solitons the dislocation in $\Lambda^{0,0}$ is stabilized by the global topology of either finite odd-size or infinite lattices. And so $\Lambda^{0,0}$ is an attractor, although not a domain.

5. Structure Detection

An essential issue that arises here is the question of discovering domains in the first place. Even with unlimited space-time data, the domain language(s) for a given CA may be sufficiently complex as to defy visual identification. For rule 18, the dominant structure had already been found; but this is not the case for arbitrary CA. One consequence of the foregoing analysis is that a general approach to identifying these structures, once developed, would provide a basis for automated structure discovery and detection, as will now be shown. The contention here is that machine reconstruction is the central technique facilitating this type of analysis.[7]

As a starting point in the analysis of rule 18 machine reconstruction was applied to the state reached by 10^5 iterations on a periodic array of 10^5 sites, starting from a random initial condition. The results of a series of reconstructions were trivial variants of the two-domain machine $M(\Lambda^{0,0})$ shown in Fig. 5. With machine reconstruction it was quite straightforward to identify the domains, dislocations, and attractors with their "vicinity" neighborhoods and proceed with the dynamical systems analysis. Current plans include an interactive expert-system-like environment to facilitate the discovery of domains and to perform automated construction of vicinities and basins and proofs of invariance and attraction. The knowledge base for such a system consists of the properties, rules of manipulation and theorems for finite automata.[16] It also incorporates basic notions from dynamical systems, such as those defining the attractor-basin portrait.

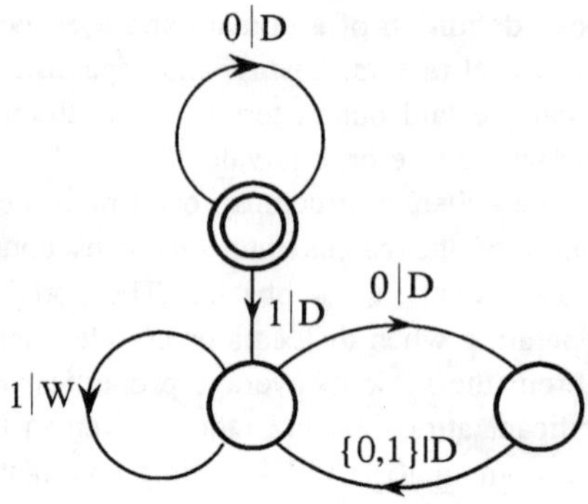

Fig. 6 Transducer $T_{domain}(\Lambda^0)$ for recognizing dislocations in rule 18 domain structure Λ^0.

Given the coherent structures for rule 18, space-time data can be recoded to isolate them. Dislocation trajectories, for example, are detected by modifying the domain-machine $M(\Lambda^0)$ to be a transducer $T_{domain}(\Lambda^0)$ that emits a (domain) symbol D on allowed transitions of Λ^0, and a (wall) symbol W on disallowed transitions. This dislocation recognizer is shown in Fig. 6. For general CA with some set $\{\Lambda^i\}$ of domain types, it is straightforward to construct a transducer that will encode the data to show each of the domain and wall structures.

Figure 7 shows a typical evolution of patterns generated by rule 18. The space-time diagram consists of 251 sites taken out of a 1000 site simulation over 251 time steps. The initial pattern was a random binary string. In the space-time diagram sites whose state is a 1 are black; those with 0, white.

Figure 8 illustrates the effect of recoding the preceding figure with the transducer $T_{domain}(\Lambda^0)$. As seen there, domains and dislocations are easily detected within the chaotic domain background. There all sites in an "even"-site domain are colored white; those in an "odd"-site domain, gray. The dislocations are shown as black cells on the domain borders. This type of display indicates in a transparent way the temporal evolution of the coherent domain and dislocation structures. This should be compared to the original space-time plot of the preceding figure.

There are two additional benefits of using $T_{domain}(\Lambda^0)$ worth noting. The first is a marked increase in display speed of one or two orders of magnitude, which would be even more significant for patterns of higher spatial dimension or a wider dynamic range of site values than binary. The second is the very substantial data compression that can be achieved by storing only the trajectories of the coherent structures. For example, Fig. 7 can be stored in 63 kilobits as a raw pattern; using the domain detector to factor out the "every other cell a zero" regularity this can be reduced to 32 kbits; and, giving up exact retrieval of the space-time pattern and storing only the initial condition and the relative moves along the dislocation trajectories, Fig. 8 can be stored in about 3 kbits.

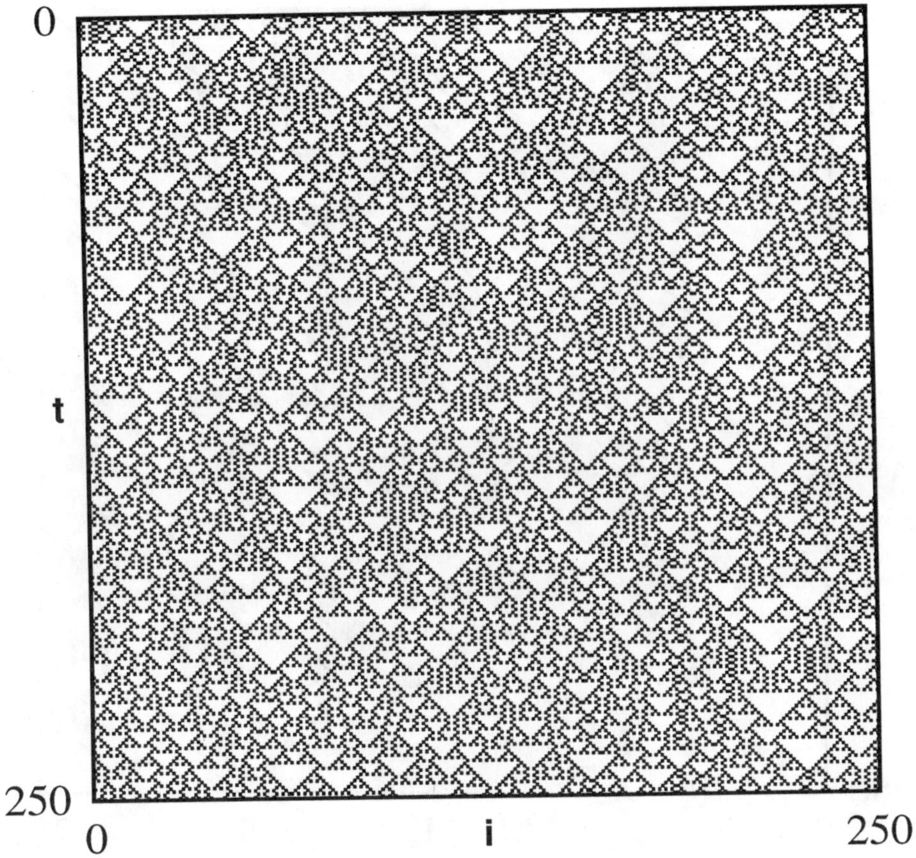

Fig. 7 CA rule 18 space-time diagram showing 251 sites over 251 time steps starting with a random initial condition. The 251 sites are a window into a 1000 lattice. Sites with a 1 are black; with 0, white.

The transducer is run over the spatial data in a particular direction: in the present example, from left to right. This introduces a spatial asymmetry into the recoded picture. For example, the visible asymmetry in the trajectory at the top of a triangle of white space (Figure 8) is due to the fact that as $T_{\mathrm{domain}}(\Lambda^0)$ is moved to the right over the data, it identifies a dislocation only by reading 1 in the wrong place. Thus, a 1 followed by an even number of consecutive 0's shows a dislocation at the right side, when the

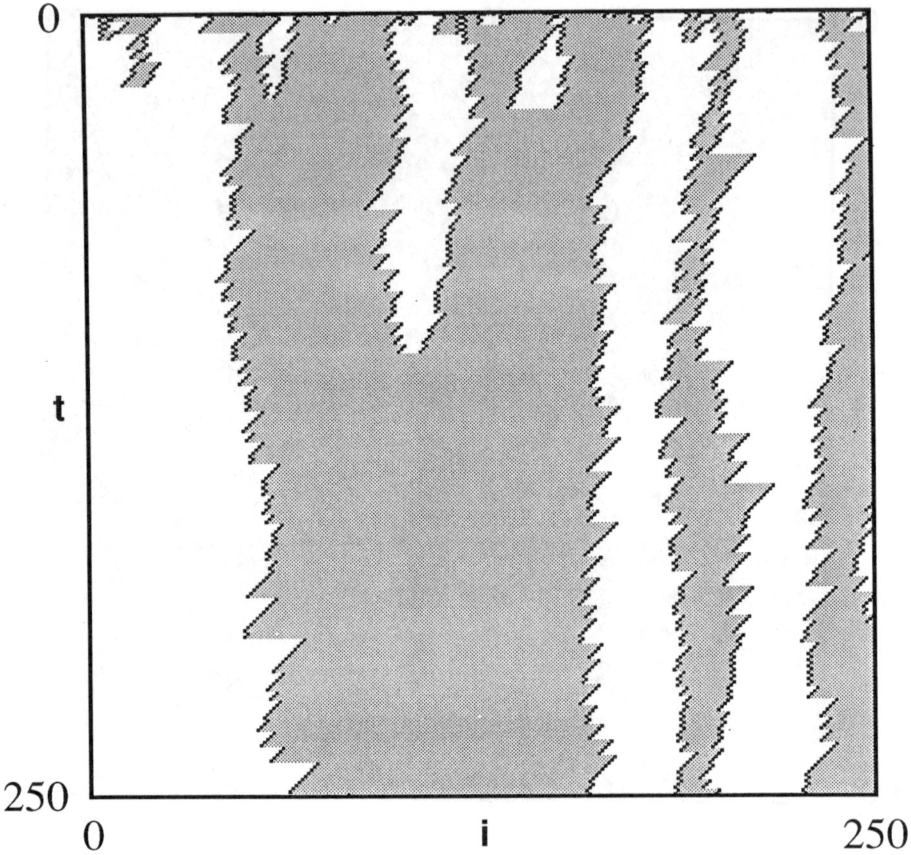

Fig. 8 Domains and dislocations detected in rule 18. Space-time diagram as in preceding figure and starting from the same initial condition. Here all sites in an "even"-site domain are colored white; those in an "odd"-site domain, gray. The dislocations are shown as black cells on the domain borders.

next 1 is encountered. This has the effect of a constraint on the motion of the observed dislocation: all moves to the left are a single cell, but moves to the right may be any odd-valued distance. Further, a move left may be followed by a move either to the right or left, but a move right must be followed by one to the left. More complicated transducers can be defined to evade this directionality. For example, the transducer could

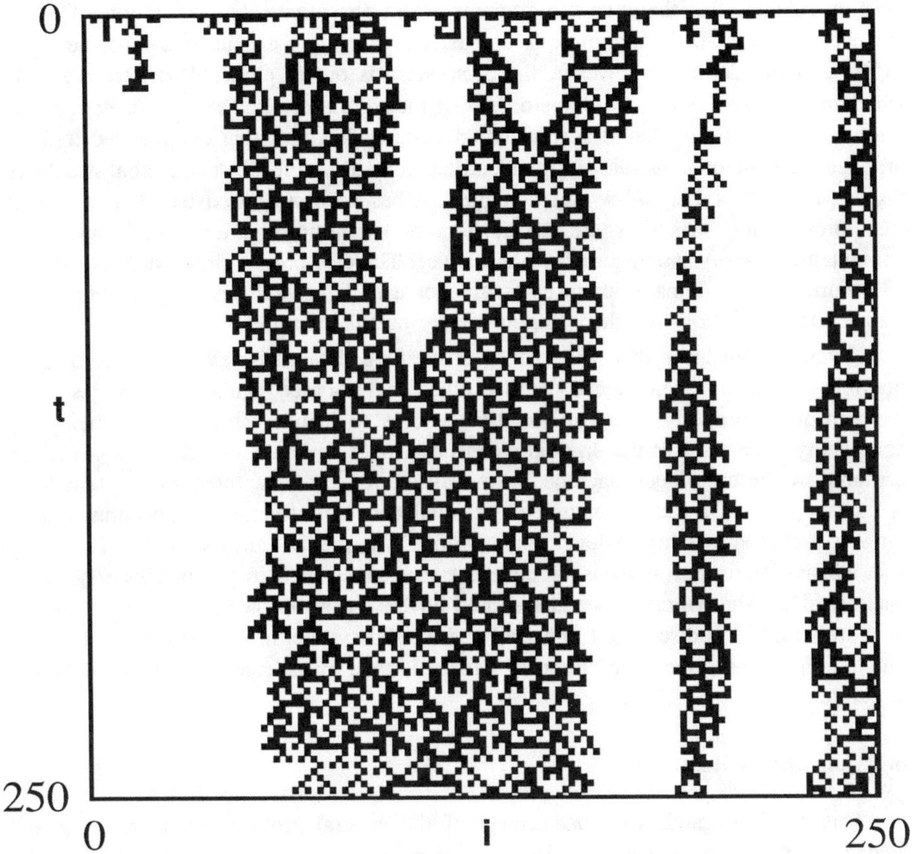

Fig. 9 Misleading structure detection using a blocking transformation that decimates every other site. The space-time diagram is the same as in preceding figure; again including the same initial condition.

label all blocks $1(00)^n1$, $n = 0, 1, 2, \ldots$, as walls. With this detector, the dislocation becomes delocalized as it enters a triangle of 0's, as seen in Fig. 7.

Figure 9 demonstrates that some detectors can suggest misleading interpretations of coherent structures. A common CA recoding method uses block transformations in which every other cell is removed from the display. Notice in Fig. 9, where this recoding is used on rule 18, that the even-site domains are in phase with the decimation. That is, the even-site domains' random cells are deleted, leaving the cells that must be zero. In the odd-site domains, however, the decimation is out of phase. It deletes the cells which must be zero in the odd-domain, leaving the arbitrary cells. Previous investigations often concluded from block-transformed space-time diagrams that complicated-looking propagating structures, such as the misidentified domains in Fig. 9, are coherent structures distinct from the other (all-white) domains. It has been suggested based on this that such propagating structures are the carriers of useful information processing. The computational analysis suggests an alternative. There is only a single domain in rule 18. Furthermore, domains themselves are not useful for high levels of computation. Rather the dislocations are the computationally useful structures.

Given a transducer that will recode space-time data to isolate dislocations, it is possible to examine the encoded data for other, higher-order regularities. By designing the transducer properly, one can use this process to identify structures of increasing complexity. The result is the original pattern dynamics "modulo" the coherent structures captured by the transducer machine. In this all the regularity captured by the transducer is excised, giving a clearer picture of residual coherent structures supporting longer-range correlation and higher-level computation. The method allows for more accurate and incisive statistical analysis of coherent structures, even in the presence of noisy backgrounds. The overall procedure should be viewed as an incremental one in which as more structure is discovered in a nonlinear spatial system, more regularity is factored out. The remaining machine "filtered" pattern is then investigated for further coherent structure.

6. Concluding Remarks

This overview addressed one aspect of the general problem of discovering and analyzing space-time structures. By way of contrast, consider the class of coherent structures called solitons. Historically, as in the case of a soliton in a shallow water channel, the recognition of a propagating structure was done intuitively via visual inspection. In some cases, such as with solitons, this identification eventually develops a formal expression. The propagating entity has a mathematically defined shape expressed in closed form. With it one can search for them by convolving the space-time pattern with the soliton kernel. In a literal sense, the shapes form a basis for the space of recognizable patterns. The method of transducer machine recognition introduced above gives a different, but complementary, computational approach to space-time pattern recognition, as we have just described it, that is both general and constructive. It, along with machine reconstruction, suggests an automated method of identifying coherent structures and tracking them in space-time. Although the present approach starts with discretized data, note that rather than a liability this is the domain of many important problems in time series and image processing. The study of domain-wall dominated CA strongly suggests similar analyses for a wider class of spatially-extended systems, such as map lattices, where each site has a continuum of states, and soliton-bearing PDEs.

The explosive complexity found in the global machine representation of rule 18's temporal evolution was circumvented by studying subspaces of initial configurations. These were represented by taking initial pattern ensembles Ω_0 to be various ensembles other than $\mathcal{A}^*$. Implicit in this is a set of "designer" ensemble methods in which pattern ensembles are constructed to facilitate the study of specific geometric properties in high-dimensional spatially-extended systems. This analysis for CA was motivated, in fact, by techniques used in earlier work on continuous-state lattice dynamical systems[8] and their extremely long transients.[5,9] And these too will be helpful in analyzing a wide range of similar problems posed by nonlinear spatial systems.

7. Acknowledgements

Thanks to Jim Hanson for making Figs. 7, 8, and 9. Many thanks to the Santa Fe Institute, where the author was supported by a Robert Maxwell Foundation Visiting Professorship, for the warm hospitality during the writing of the present essay. Funds from ONR contract N00014-90-J-1774, AFOSR-91-0293, and NASA-Ames University Interchange NCA2-488 contributed to this work.

8. References

1. Blahut, R. E. *Principles and Practice of Information Theory*. Addison-Wesley, Reading, Massachusetts, 1987.

2. Boghosian, B. M. Lattice gases. In *1989 Lectures in Complex Systems* (Reading, Massachusetts, 1990), E. Jen, Ed., vol. Lect. Vol. II of *SFI Studies in the Sciences of Complexity*, Addison-Wesley.

3. Casdagli, M., and Eubank, S., Eds. *Nonlinear Modeling* (Reading, Massachusetts, 1992), SFI Studies in the Sciences of Complexity, Addison-Wesley.

4. Chillingworth, D. R. J. *Differential Topology with a View to Applications*. Pitman, London, 1976.

5. Crutchfield, J. P. Subbasins, portals, and mazes: Transients in high dimensions. *J. Nucl. Phys. B 5A* (1988), 287.

6. Crutchfield, J. P. Reconstructing language hierarchies. In *Information Dynamics* (New York, 1991), H. A. Atmanspracher and H. Scheingraber, Eds., Plenum, p. 45.

7. Crutchfield, J. P., and Hanson, J. E. Computational mechanics of cellular automata: Space-time machines. in preparation, 1992.

8. Crutchfield, J. P., and Kaneko, K. Phenomenology of spatio-temporal chaos. In *Directions in Chaos*, H. Bai-lin, Ed. World Scientific Publishers, Singapore, 1987, p. 272.

9. Crutchfield, J. P., and Kaneko, K. Are attractors relevant to turbulence? *Phys. Rev. Let. 60* (1988), 2715.

10. Crutchfield, J. P., and McNamara, B. S. Equations of motion from a data series. *Complex Systems 1* (1987), 417.

11. Crutchfield, J. P., and Young, K. Inferring statistical complexity. *Phys. Rev. Let. 63* (1989), 105.

12. Crutchfield, J. P., and Young, K. Computation at the onset of chaos. In *Entropy, Complexity, and the Physics of Information* (Reading, Massachusetts, 1990), W. Zurek, Ed., vol. VIII of *SFI Studies in the Sciences of Complexity*, Addison-Wesley, p. 223.

13. Crutchfield, J. P., and Young, K. ϵ-machine spectroscopy. preprint, 1992.

14. Grassberger, P. New mechanism for deterministic diffusion. *Phys. Rev. A 28* (1983), 3666.

15. Hanson, J. E., and Crutchfield, J. P. The attractor-basin portrait of a cellular automaton. *J. Stat. Phys. 66* (1992), 1415.

16. Hopcroft, J. E., and Ullman, J. D. *Introduction to Automata Theory, Languages, and Computation*. Addison-Wesley, Reading, 1979.

17. Hurd, L. P. Appendix. table 10. regular language complexities. In *Theory and Applications of Cellular Automata*, S. Wolfram, Ed. World Scientific, Singapore, 1986.

18. Rabiner, L. R. A tutorial on hidden markov models and selected applications. *IEEE Proc. 77* (1989), 257.

19. Rissanen, J. *Stochastic Complexity in Statistical Inquiry*. World Scientific, Singapore, 1989.

20. Shannon, C. E., and Weaver, W. *The Mathematical Theory of Communication*. University of Illinois Press, Champaign-Urbana, 1962.

21. Tukey, J. W. *Exploratory Data Analysis*. Addison-Wesley, Reading, Massachusetts, 1977.

22. Wolfram, S. Statistical mechanics of cellular automata. *Rev. Mod. Phys. 55* (1983), 601.

23. Wolfram, S. Computation theory of cellular automata. *Comm. Math. Phys. 96* (1984), 15.

24. Wolfram, S. *Theory and Applications of Cellular Automata*. World Scientific Publishers, Singapore, 1986.

UNCERTAINTY, SUBJECTIVITY, FUZZINESS

ARNOLD KAUFMANN
Formerly Professor
at
Institut National
Polytechnique de Grenoble

ABSTRACT

Introduction - Uncertainty and hazard. When experts replace instruments - To begin with uncertainty : the structure - The theory of fuzzy subsets. Intervals and triplets of confidence - Several experts : the theory of expertons - Fuzzy inferences - A general view about fuzzification of models in engineering and human sciences - A very short bibliography

1 - INTRODUCTION

It is obvious that the world, the whole world, becomes more and more complicated not only for any citizen but also for every scientists whatever their discipline. This complexity concerns not only interrelations and systems but also the facts themselves and their data, semantically or numerically data, measurable or not, becomes less precise specially in soft sciences like, for instance, human sciences. One of the major reasons is the fact that these scientists are confronted with many more interactions, unknown or disregarded yesterday and taken in account today. Not only these interactions are numerous but also variable and unstable. Of course, many data are mesurable owing to many kinds of hardware progresses and novel gains of theoretical research but in fact, perfect objectivity becomes rare.

When a scientist is able to do a measurement he must do it otherwise he is not a scientist but if measurement is doubtful or even impossible, and when some information is necessary, he must use his own subjectivity and experts opinion. A lot of models in natural and human sciences contain non measurable data. For all that, it is almost necessary to use this data. It is the reason why since several decades new sectors of mathematics are spreading out rapidly, specially if logic binary concepts appear insufficients in many models. One or zero, even used sequencially are not sufficient to describe more realistic facts. A multi-valent logic, also called fuzzy logic seems closed to the real world. Beside predictive models need more and more often expert opinions (one or several), and often subjective judgement replace exact or mesurable information. Of course, mathematics of uncertain, subjective and fuzzy data must be exact mathematics with convenient axioms, theorems and structures like non fuzzy mathematics. Fuzzy in the modern mathematical sense do not mean incorrect theory, but a non Aristotelian or Chrysippian mathematics using shade between true and false. If you agree with this kind of shape, you open infinite ways in the modelling, some of them better adapted to a system than others; the diversity and the possibility of some choices is sometimes to complicated to reach a solution, but the main objective of the human operator is the realism.

For more than two centuries statisticians had the same intentions; statistician point of view and fuzzy concepts are not in competition but the second can be used when statistics are not acceptable and even impossible. Often associated, at the best, they allow to introduce and use uncertainty in mathematical modellization in the most adequate ways available.

2 - UNCERTAINTY AND HAZARD

Too often, uncertainty and hazard are confuse together. For a scientist hazard or randomness is acquainted with the concept of probability, that is today essentially measurable and exactly computable by some reasonings. At the opposite, uncertainty is not measurable, and is taken into account by our thinking after some observations and analyses in a soft sense. If you meet your friend after a long time, don't tell him : I meet you by hazard. It is false, you have no information about the fact to meet or not to meet. The true sentence would be : I am uncertain to have the opportunity to meet you! In fact, in the current life every one use hazard instead uncertainty. Myself very often and all dictionnaries, in all languages, all over the world, make the same confusion. But for a scientist the difference is very wide and this difference is essential. Another question occurs about these difference.

Probability theory has its foundations from Borel-Kolmogorov axioms which give a boolean lattice, meanwhile uncertainty needs some tools like valuation confidence segments, fuzzy sets, expertons, etc, only a segment [0, 1] like reference set and uncertainty theories (segments, fuzzy sets, expertons) use the same segment for their number, it is only a human habit and is not indispensable. Besides, another point. Probability theory uses the word "probability" admitted with the same meaning by anyone because it is an objective theory. On the contrary uncertainty theories use the word "valuation" which is a subjective term. A probability is a valuation but a valuation is not a probability. Step by step, in this paper, this difference will appear more and more precisely.

3 - WHEN EXPERTS REPLACE INSTRUMENTS

When an objective data cannot be obtained by correct experiments, measures, repetitives observations and when the scientist needs some data necessarily for the model he is building, he will request the knowledge of qualified people that means experts.

These experts will propose valuations from their own knowledge and personal experience. Of course, a subjective data has not in several ways the quality of an objective one but in so many cases the perfect objectivity is a way to deceive oneself. A good expert translates semantically or numerically his own opinion, his valuation. The word valuation is coming from the theory of lattices and is perfectly convenient for the data given by an expert. A valuation can be given in the segment [0, 1], sometimes in [-1, 1] and likewise in [- ∞, ∞] or in a semantic correspondence between the concept and the valuation. When an expert gives his valuation in a numerical scale he is able to use a lot (theoretically an infinity) of scales in such a way to translate at the best his own perception. Among this lot of scale we have the habit to use "endecadery scale" (from the greek which means eleven). In this way with any qualifying instead true or false the following correspondence is propounded. If the list is to small, specificities are not sufficient, if the list is to large, the human operator has too much choices and can be confused.

Let us show a correspondence we consider not too small and not too large with advantage of the use of one decimal numbers.

Valuation	**Semantic meaning**	
1	true	
0.9	almost true	
0.8	nearly true	
0.7	enough true	
0.6	more true than false	
0.5	neither true nor false	(3.1)
0.4	more false than true	
0.3	enough false	
0.2	nearly false	
0.1	almost false	
0	false	

Of course, such a correspondence is arbitrary and a different correspondence can be prefered. To enlarge the possibility to translate his own opinion an expert is permitted to give a segment for his opinion, for instance [.7, .9] that means: between enough true and almost true. In fact in the litterature which concerns the uncertainty many others correspondences are useful. The main question concerns agreements between users.

Before to go beyond, it is important to recall that chaos and uncertainty concern two different fields : chaos is an unknown structure, uncertainty is reported to a field where a domain is almost known but not a point on the field. Information about uncertainty is subjective and for the chaos even subjectivity seems away.

4 - TO BEGIN WITH UNCERTAINTY : THE STRUCTURE

It is of major importance to recall to each reader some properties and a general structure which will be found for several kinds of variables used in the treatment of the uncertainty. This structure is the distributive lattice.

Let us consider a referential set E and two internal composition laws. Let us suppose we have the following properties.

$\forall$ x, y, $\in$ E:

$x * y = y * x$ commutativity (4.1)

$x *' y = y *' x$ (4.2)

$x * (y * z) = (x * y) * z$ associativity (4.3)

$x *' (y *' z) = (x *' y) *' z$ (4.4)

$x * x = x$ idempotency, (4.5)

$x *' x = x$ (4.6)

$x * (x *' y) = x$ absorption, (4.7)

$x *' (x * y) = x$ (4.8)

A such algebraic structures is a "lattice".

If, in addition we have also :

$(x * y) *' z = (x *' y) * (y *' z)$ distributivity (4.9)

$(x *' y) * z = (x * z) *' (y * z)$ (4.10)

Such lattice is called "distributive lattice".

It is easy to prove from these 8 properties these others where $\overline{x}$ is the "complement" of x reporting to the referential set :

$(\overline{x}) = x$ Involutive property, (4.11)

$x * \overline{x} \neq \varnothing,$ ($\varnothing$ is the empty set) (4.12)

$x *' \overline{x} \neq E,$ (4.13)

$x * \varnothing = \varnothing,$ (4.14)

$x *' \varnothing = x,$ (4.15)

$x * E = x$ (4.16)

$x *' E = E$ (4.17)

$\overline{x * y} = \overline{x} *' \overline{y},$ De Morgan's theorems (4.18)

$\overline{x *' y} = \overline{x} * \overline{y},$ (4.19)

The main concepts used for the treatment of uncertainty have the distributive lattice structure.

We note a fundamental difference between distributive lattive and boolean lattice. In a boolean lattice we find the Aristotelian properties:

$$x * \overline{x} = \varnothing, \tag{4.20}$$

$$x *' \overline{x} = E. \tag{4.21}$$

Boolean algebra or boolean lattice is the usual structure used with deterministic concepts like Boolean variables 0 or 1 in computers, and also in probability theory.

In computers (binary computers) the operators * and *' are respectively ∩ intersection, ∪ union for sets and ∧ (minimum) and v (maximum) for elements of sets. Usualy ∧ is called and noted (.) product and v is called and noted (+) sum. But as each one knows this is a confusing notation.

With variables and subsets in the theory used for the uncertainty the Aristotelian properties (4.20) and (4.21) are not available. This point is fundamental for each one who intends to understand the theory of fuzzy mathematics which is, in our epoque, convenient for uncertain data. For uncertain data and uncertain subsets * will be ∩ intersection (for subsets) and ∧ minimum (for variables), *' will be ∪ union (for subsets) and v maximum (for variables).

In boolean algebra $\overline{a} = 1 - a$, it is also the case in fuzzy algebra but in boolean algebra $\overline{a}$ is the negation of a, meanwhile in fuzzy algebra $\overline{a}$ is the complementation of a. This is not a game of words but a conceptual distinction.

What we explained above will be examinated in the following paragraphs. In this paragraph 4, we only intend, at the beginning, to move aside the thinking of the reader that boolean algebra and probabilities are available in the case of uncertainty (except in very particular cases). Our brain is not deterministic, even probabilistic, it is fuzzy in the mathematical sense.

5 - THE THEORY OF FUZZY SUBSETS

In the classical theory of sets a subset is defined by the "characteristic function" which specify if an element of a given referential belongs or not to the subset. If E is a set and A is a subset $A \subset E$,

$$(\mu_A(x) = 1) \Leftrightarrow (x \in A), \tag{5.1}$$

$$(\mu_A(x) = 0) \Leftrightarrow (x \notin A), \tag{5.2}$$

But, in a larger theory, the fuzzy (1) set theory, we consider that μ_A (x) instead of being an element of 0,1 can be an element of [0, 1], that is to say, a fuzzy subset A of E is defined by

$$\forall x \in E \qquad \mu_A(x) \in [0,1] \tag{5.3}$$

and A is called a "fuzzy subset" of E, which E is always a non fuzzy set (ordinary set).

We show some figures to explicit these formulations.

(1) Fuzzy like "flou" in french, "borrosos" in spanish, and also in other languages all over the world, is a pejorative word but not in mathematics.

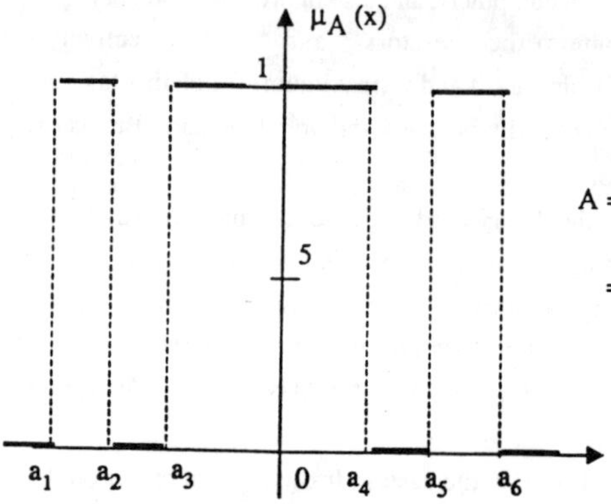

An ordinary subset of R,
figure 5.1

$E = \{a, b, c, d, e, f, g\}$

	a	b	c	d	e	f	g
A =	1	0	0	1	1	0	1

$= \{a, d, e, g\}$

An ordinary subset of E, a finite set.
figure 5.2

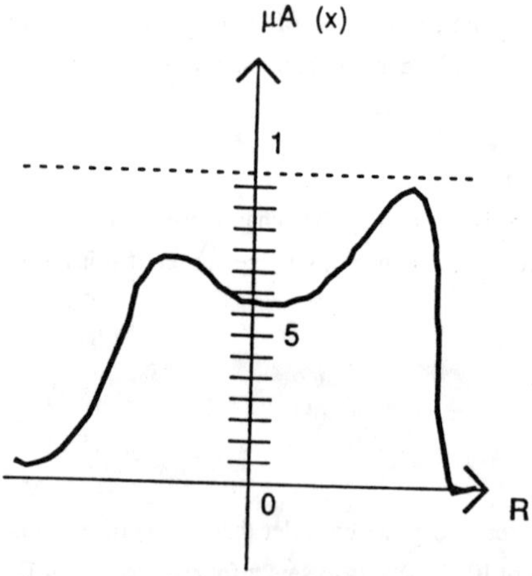

A fuzzy subset of R
figure 5.3

$E = \{a, b, c, d, e, f, g\}$

	a	b	c	d	e	f	g
A =	.4	.9	1	0	.6	.5	.2

An fuzzy subset of E
figure 5.4

The algebraic structure of ordinary subsets of a set is a boolean lattice meanwhile the algebraic structure of fuzzy subsets of a set is a distributive lattice (non Aristotelian).

Intersection of two fuzzy subsets is given by :

$$\mu_{A \cap B}(x) = \mu_A(x) \wedge \mu_B(x) \quad (5.4)$$

the union by :

$$\mu_{A \cup B}(x) = \mu_A(x) \vee \mu_B(x) \quad (5.5)$$

the complementation:

$$\mu_{\bar{A}}(x) = 1 - \mu_A(x) \quad (5.6)$$

But, in the classical theory fo sets there is only three basic operators $\cap$, $\cup$, $\bar{}$, in fuzzy sets theory there is an infinity which keeps all results in [0, 1].

- T - norms and T - conorms like :

$$\forall\ a, b \in [0,1] \quad (5.7)$$

T - norms	T - conorms	
$a \wedge b,$	$a \vee b,$	(5.8)
$0 \vee (a + b - 1),$	$1 \wedge (a + b),$	(5.9)
$a\ .\ b,$	$a \hat{+} b,$	(5.10)
$\overline{1 + \bar{a}\ .\ \bar{b}}$	$\overline{\overline{1 + \bar{a}.\ \bar{b}}}$	(5.11)
$a.\ b$	$a \hat{+} b = (a + b - ab),$	(5.12)

etc.

And also an infinity of semantic operators maintaining results in [0, 1] which corresponds to some qualificatives like :

very, smaller, larger, very very, never..or, etc.

The algebra of fuzzy subsets is closer to the human thinking that ordinary algebra of sets.

In the ordinary theory of set, relations and graphs are defined and extensively used. It is also true with fuzzy theory. We have not enough rooms to describe all visual properties. We show by figures relations beetween fuzzy subsets. Two referential sets :

$$E_1 = \{a, b, c, d, e\}, \quad (5.13)$$

$$E_2 = \{m, n, p, q\} \quad (5.14)$$

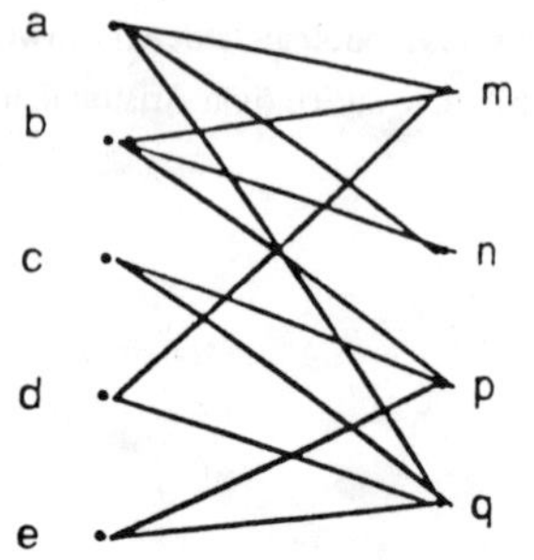

Figure 5.5

	m	n	r	q
a	.0	.1	0	.9
b	.6	1	.4	0
c	0	0	.2	.8
d	.3	.1	0	.4
e	0	0	.6	1

Figure 5.6

A fuzzy relation is characterized by its "membership function". If $R \subset E_1 \times E_2$:

$$\forall\, x, y \in E_1 \times E_2 : \mu_R(x, y) \qquad (5.15)$$

is the membership of (x, y) e R.

The convolution of two fuzzy relations R and S is given by :

$\forall\, x, y \in R,$

$$y, z \in S \qquad \mu_{ROS}(x, z) = \vee_y \big(\mu_R(x, y) \wedge \mu_S(y, z)\big) \qquad (5.16)$$

Such an operation is represented symbolically by :

R 0 S

which is not, of course, commutative.

When $E_1 = E_2$ the relation is more commonly called "graph". Fuzzy graphs have some operations common with ordinary graphs such as: reflexivity, symetry, transitivity, transitive closure, pre-order, order, similarity, antisymetry, similitude, ressemblance, etc, and some others special for the fuzzy case. Very large chapters are given in the adequate litterature about these properties.

Before passing to another chapter, it is important to recall what is an α - cut of a fuzzy subset.

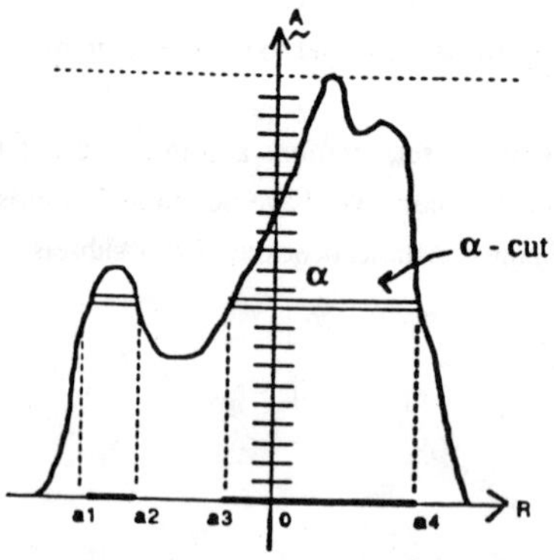

Figure 5.7

The figure 5.7 shows an α - cut of level α of the fuzzy subset A. A fuzzy subset is perfectly defined by the set of its α - cuts. An α - cut is an ordinary subset of the referential set (for instance R continum from $-\infty$ to ∞, E a finite set, etc). For more information we advise the reader to see the corresponding litterature which is very very rich.

Before leaving this paragraph we have to describe an important class of fuzzy sets: fuzzy numbers. A fuzzy number is a fuzzy set which is normal and convex. Normal means that the membership function has at least one point with level 1. Convex means that all the α - cut are segments where each one is embodied in the preceding one from the level 1 to the 0.

Fuzzy numbers generalize in the fuzzy sense ordinary numbers. The more usual fuzzy numbers are triangular fuzzy numbers (T.F.N.) with shapes which are linear lines and trapezoidal fuzzy numbers (T_p.F.N). A lot of simplifications are then available.

6 - INTERVALS AND TRIPLETS OF CONFIDENCE

Some ways to take in to account uncertainty in a simpler way that fuzzy numbers are the concepts of intervals of confidence(*) and triplets of confidence.

Beginning with intervals we explain the nature of this concept. Before we consider intervals of confidence in R^+, where we recall that R^+ represents the continuous $[0, \infty]$.

Let us consider the interval (in pure mathematics a segment) $[a_1, a_2] \subset R^+$. The lower bound a_1 is included in the interval likewise the higher bound a_2. This means that exists an unknow number a such $a_1 \leq a \leq a_2$ but we dont know a, only $[a_1, a_2]$. Bounds a_1 and a_2 can be known objectivily or subjectively depending on the informations available.

The algebra of $[a_1, a_2] \subset R^+$ is a little bit more complicated that those of ordinary elements of R^+.

$$[a_1, a_2] (+) [b_1, b_2] = [a_1+b_1,\ a_2+ b_2], \tag{6.1}$$

$$[a_1, a_2] (.) [b_1, b_2] = [a_1 . b_1,\ a_2 . b_2], \tag{6.2}$$

$$[a_1, a_2] (-) [b_1, b_2] = [a_1 - b_2,\ a_2 - b_1], \tag{6.3}$$

and ordinary subtractions

$$[a_1, a_2] \textcircled{m} [b_1, b_2] = [a_1 - b_1, a_2 - b_2] \quad \text{with } a_1 - b_1 \leq a_2 - b_2 \tag{6.4}$$

which concerns the solutions of equations such as:

(*) Interval of confidence here means a different concept from the one known in statistical theory.

$$[b_1, b_2](+)[c_1, c_2] = [a_1, a_2]$$

Also

$$[a_1, a_2](:)[b_1, b_2] = \left[\frac{a_1}{b_2}, \frac{a_2}{b_1}\right] \tag{6.5}$$

and ordinary division.

$$[a_1, a_2](\ddot{m})[b_1, b_2] = \left[\frac{a_1}{b_1}, \frac{a_2}{b_2}\right] \quad \text{with} \quad \frac{a_1}{b_1} < \frac{a_2}{b_2} \tag{6.6}$$

what concern the solutions of equations such :

$[b_1, b_2]$ (.) $[c_1, c_2] = [a_1, a_2]$.

The neutral for (+) is $[0, 0] = 0$ and for (.) is $[1, 1] = 1$.

We have not enough space to show the case of intervals in $R = (-\infty, \infty)$ which are more complicated.

Now let us see what is a triplet* of confidence using $a_1 \in R^+$, $a_2 \in R^+$, $a_3 \in R^+$, $a_4 \in R^+$, with $a_1 \leq a_2 \leq a_3 \leq a_4$. We represent this concept by :

$$(a_1, [a_2, a_3], a_4) \tag{6.7}$$

where a_1 is the lower bound, a_4 the higher and $[a_2, a_3]$ the "maximum of presumption". Of course, particular cases like $a_2 = a_3$, $a_1 = a_2$, $a_1 = a_2 = a_3$ ... etc can be found. The maximum of presumption can be very useful if an expert (or several) is able to give it. The case $a_2 < a_3$ corresponds to trapezoïdal fuzzy number and the case $a_2 = a_3$ to triangular fuzzy number.

The reader will find easily the algebra of the triplets of confidence. For instance :

$$(a_1, [a_2, a_3], a_4)\ (+)\ (b_1, [b_2, b_3], b_4) = [a_1 + b_1, [a_2 + b_2, a_3 + b_3], a_4 + b_4) \tag{6.8}$$

The case of intervals and triplets of confidence in [0, 1] is of major importance when the uncertainty can be quantified by experts. The algebra is the same as algebra in R^+ with, of course, the adaptation to the referential valuation $a \in [0, 1]$. Intervals $[a_1, a_2] \subset [0, 1]$, triplets $(a_1, [a_2, a_3], a_4) \subset [0, 1]$ have the same configuration : distributive lattice which are discribed in paragraph 4. We advise to use the endecadery scale given in (3.1), it is not too small and not too large for expertise and an expert can also use intervals and triplets if his vagueness needs more than a number in [0, 1] for its valuation.

We apologize for not giving extensive details. The theory of uncertainty needs a lot of books. This paper is an elementary introduction in a field generally not well known.

---

* - In fact this triplet is a quadruplet but we consider the maximum of presumption like a datum

7 - SEVERAL EXPERTS - THE THEORY OF EXPERTONS

In many and various situations the subjectivity is extended to several experts. The agregation of several estimates is a problem well solved by a new fuzzy concept: the experton.

To explain the theory we shall present an example, the general theory will be simpler to understand.

Suppose seven experts are consulted to give their own opinion to a property, using an endecadery scale like (3.1) in a specified problem. Experts are authorised to give a valuation by an endecadery number or an interval. Let us suppose that the following collect has been optained and a statistic of bounds is computed:

(7.1)

Expert	
1	[.3, .5]
2	.6
3	[.1, .2]
4	[.1, .4]
5	[.8, 1]
6	[.5 .6]
7	.4

(7.2)

Level		
0		
.1	2	
.2		1
.3	1	
.4	1	2
.5	1	1
.6	1	2
.7		
.8	1	
.9		
1		1

This statistis is cumulated from 1 to 0 (complemented cumulation).

Now, the cumulated statistic is normalized (divided by 7), that gives :

(7.3)

Level		
0	7	7
.1	7	7
.2	5	7
.3	5	6
.4	4	6
.5	3	4
.6	2	3
.7	1	1
.8	1	1
.9	0	1
1	0	1

(7.4)

0	1	1
.1	1	1
.2	.714	1
.3	.714	.857
.4	.571	.857
.5	.428	.571
.6	.285	.428
.7	.142	.142
.8	.142	.142
.9	0	.142
1	0	.142

This is the corresponding experton. Its algebraic configuration is the same as valuation interval in [0, 1], triplets in [0, 1] and also fuzzy subsets : distributive lattice. And T - norms, T - conorms, semantic operators used for valuations are convenient for expertons when the monotonicity from the left to the right and the bottom to the high is respected (that is necessarily the case).

You can do with expertons what you are allow to do with valuations, intervals, triplets, fuzzy subsets. To find the mathematical expectation of an experton is very simple : you make the additions of each column without taking in account the level 0. The result is divided by 10. That gives for (7.3) or (7.4).

$$\varepsilon\,(7.4) = \frac{1}{10}.\frac{1}{7}(1+1+2+3+4+5+5+7) \tag{7.5}$$

left $= \frac{1}{70} \times 28 = 0.400$

$$\varepsilon\,(7.4) = \frac{1}{10}.\frac{1}{7}(1+1+\ 1+1+3+4+6+6+7+7) \tag{7.6}$$

right $= \frac{1}{70} \times 37 = 0.528$

The mathematical expectancy is the interval :

$$[0.400,\ 0.528] \tag{7.7}$$

But with expertons, it exists a strong rule : take the mathematical expectancy the later possible. The computation with expertons preserves all properties of statistical distributions with all operators used in [0, 1]. The mean and others moments don't preserve strictly the statistical properties. So, only after all computations have been made, you compute the expectancy, for instance if you are in position for a decision.

But a question arises : how to compute the complementation of an interval, a triplet, a fuzzy set, an experton.

For a valuation :

$$\bar{a} = 1 - a \qquad a \in [0,\ 1] \tag{7.8}$$

For an interval :

$$\overline{[a_1,\ a_2]} = [1 - a_2,\ 1 - a_1] \tag{7.9}$$

For a triplet :

$$\overline{(a_1,\ [a_2,\ a_3],\ a_4)} = (1 - a_4,\ [1 - a_3,\ 1 - a_2],\ 1 - a_1) \tag{7.10}$$

For a fuzzy set :

see (3.6) (7.11)

For an experton. We use an example. It is very simple.

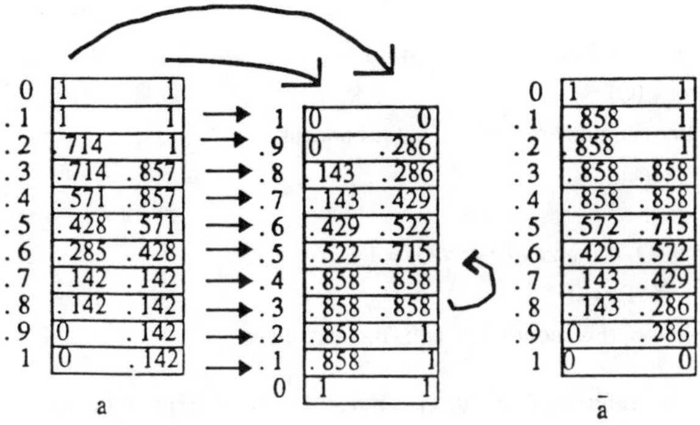

(7.12)

Expertons are one of the useful tools for expert systems, artificial intelligence, operation research, etc, when several appraisements are necessary.

8 - FUZZY INFERENCES

In the binary or boolean logic a proposition P is true or false and the value is :

P is true : v (P) = 1, (8.1)

P is false : v (P) = 0

(8.2)

Of course, the correspondence is arbitrary but usual.

An implication or inference using the notation :

P ---> Q (8.3)

is unique in this logic and it represented by :

$\bar{a} \vee b$, (8.4)

where a = v (P), b = v (Q), $\bar{a}$ = 1 - a = 1 - v (P) and the following tableau gives the true table :

$\bar{a} \vee b$, a \ b	0	1
0	1	1
1	0	1

(8.5)

With the boolean inference (8.5) only two modes are available :

Mode 1 :
premises : v (P) = 1 and v (P ---> Q) = 1
conclusion : v (Q) = 1 (8.6)
This well known mode is called : "modus ponens"

Mode 2 :
premises : v (Q) = 0 and v (P ---> Q) = 1
conclusion : v (P) = 0 (8.7)
Well known too, this mode is called : "modus tollens"

But the referential set {0, 1} is not sufficient to describe the natural mechanisms of the human brain which are more fuzzy than binary. It is one of the main reasons that mathematicians in our epoque propound a "multivalent" or "fuzzy" logic. In this logic the value of a proposition P is:

$$v(P) = a, \qquad a \in [0,1] \tag{8.9}$$

so, semantically, a proposition can be true, false, but also have a position between true and false.

For instance :
v (P) = 0.6 using the arbitrary semantic correspondence given in (3.1). (8.10)
P is more true than false. (8.11)

Any n - ary scale can be used with the referential set [0, 1] discrete or continuous.

But, if with binary logic it exists one and only one inference like (8.4), with the fuzzy logic an infinity of inferences can be proposed. The following axioms are given to define a fuzzy inference ; these axioms, of course, are also available for the binary case :

$\forall a, b \in [0, 1]$ and the operator * represent an inference

1) $$a * b \in [0,1] \tag{8.12}$$

2) $$0 * 0 = 1,\ 0 * 1 = 1,\ 1 * 0 = 0,\ 1 * 1 = 1 \tag{8.13}$$

3) $\forall_a, b, a', b' \in [0, 1]$:

$$(a < a') \Rightarrow ((a * b) \geq (a' * b)) \tag{8.14}$$

$$(b < b') \Rightarrow ((a * b) \leq (a * b')) \tag{8.15}$$

monotone inference

The following figure gives the fuzzy extension of (8.5).

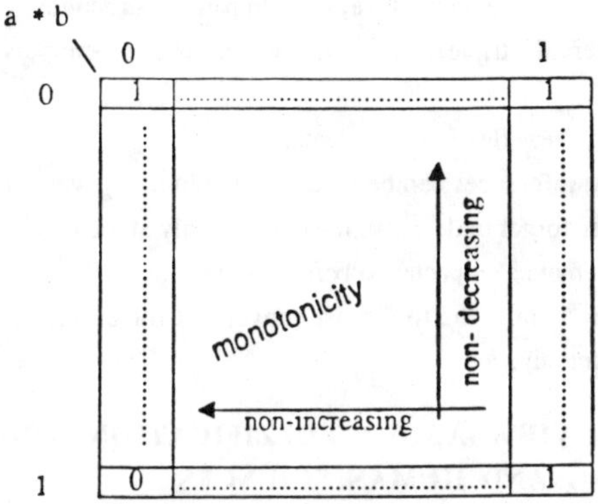

(8.16)

Among the infinity of possible inferences some are more serviceable than others.

$a * b = \bar{a} \vee b$, Lee's inference (8.17)

$a * b = 1 \wedge (\bar{a} + b)$, Lukaciewicz's inference **(8.18)**

$a * b = 1,\ a \leq b$ Godel-.Mizumoto's inference

$= b,\ a > b$ (8.19)

$a * b = 1 \wedge \frac{b}{a}$, Goguen's inference (8.20)

This is a small sample.

Like an explicit example, we give a tableau with endecadery scale of the inference of Lukaciewicz, one of the most useful.

$1 \wedge (\bar{a} + b)$

a \ b	0	.1	.2	.3	.4	.5	.6	.7	.8	.9	1
0	1										
.1	.9	1									
.2	.8	.9	1								
.3	.7	.8	.9	1							
.4	.6	.7	.8	.9	1						
.5	.5	.6	.7	.8	.9	1					
.6	.4	.5	.6	.7	.8	.9	1				
.7	.3	.4	.5	.6	.7	.8	.9	1			
.8	.2	.3	.4	.5	.6	.7	.8	.9	1		
.9	.1	.2	.3	.4	.5	.6	.7	.8	.9	1	
1	0	.1	.2	.3	.4	.5	.6	.7	.8	.9	1

(8.21)

Inference of Lukaciewicz

All types of fuzzy inference can be inversed forward and backward for their use in expert systems, some with more difficulties than others.

They can be used in a sequential way or with parallel structures.

Valuations, intervals, triplets, fuzzy sets, expertons can be inputs or outputs with fuzzy inferences.

The utilization is very flexible and adaptable.

Inputs, outputs, inferences can be parametrized in many ways if necessary. In a complex systems never forget to decrease the entropy only at the end of the process; do not use means or mathematical expectation before the end except if it is necessary.

The fuzzy logic is now the tool most convenient for expert systems and more generally for artificial intelligence.

9 - A GENERAL VIEW ABOUT FUZZIFICATION OF MODELS IN ENGINEERING AND HUMAN SCIENCES

An incomplete but enough rich list of applications is given below:

- Economics,	- Finance,	- Decision theory,	- Earthquakes,
- Operation research,	- Cost analysis,	- Appraisements	- Geology,
- Production control,	- Marketing,	- Pattern recognition,	- Archeology,
- Reliability,	- Expert systems,	- Physics,	- Psychology,
- Quality control,	- Artificial intelligence,	- Chemistery,	- Medical science,
- Complex systems,	- Neural networks,	- Biology,	- Pharmacology,
- Communications,	- Computers,	- Methods of creativity,	- Social sciences,
- Forecasting,	- Selection of personal,	- Meteoroly,	

and many others fields ; in fact all hard and soft sciences without forgetting arts, litterature, history, geography, ...

The theory of fuzzy subsets have been presented for the first time with their main properties by L. A. Zadeh, Professor at the University of California in Berkeley, some predecessors also investigated this domain : Moïsil, Lukaciewicz, Post, and several others.

At the present time, in 1992, more than 15 000 papers have been published all over the world on this subject, more than 120 books, more than 150 P.H.D. Thesis. The author himself like author or co-author has published 12 books in english, french, spanish, japanese, russian. He advises you to be attentive about this part of the scientific progress.

10 - ACKNOWLEGEMENTS

The author thanks Dr L.PIERRAT for his help and Mrs Catherine PENELON for her excellent typewriting.

11 - A VERY SHORT BIBLIOGRAPHY

1 - Dubois D. et Prade H - Theory of possibility - Ed. Masson (Paris).

2 - Dubois D. et Prade H - Fuzzy Sets and Systems Theory and applications - Publ.Academic Press.

3 - Kandel A. - Fuzzy Techniques and Pattern Recognition - Publ. Willey

4 - Kaufmann A. - Theory of Fuzzy Subsets - Publ. Academic Press.

5 - Kaufmann A. and Gupta M.M - Introduction to fuzzy arithmetic - Publ. Van Nostrand Reinhold.

6 - Kaufmann A and Gupta M.M - Fuzzy mathematical models in engineering and Management - Publ. North Holland.

7 - Mamdani E. and Gaines B.A. - Fuzzy reasoning and its applications - Publ. Academic Press.

8 - Negoita C.V. and Ralescu D.A. - Applications of fuzzy sets to system analysis - Publ. Birhauser.

9 - Zimmermann H.J. - Fuzzy sets and applications - Publ. Kluwer Nighoff. UNCERTAINTY,

THE 'IN' AND THE 'OUT'
An Evolutionary Approach

P.R. MEDINA-MARTINS
Instituto Superior Tecnico, Av. Rovisco Pais, 1
1096, Lisbon, Portugal

and

L. ROCHA
State University of New York, Binghamton
U S A

ABSTRACT

It is claimed that a great deal of the problems which the mechanist approaches to the emulation/modelling of the human mind are presently facing are due to a host of canons so 'readily' accepted and acquainted that some of their underlying processes have not yet been the objective of intensive research. The essay proposes a (possible) solution for some of these problems introducing the tenets of a new paradigm which, based on a reappraisal of the concepts of purposiveness and teleology, lays emphasis on the evolutionary aspects (biological and mental) of alive beings. A complex neuro-fuzzy system which works as the supporting realization of this paradigm is briefly described in the essay.

1. INTRODUCTION

This essay has been written as long as a seldom, albeit crucial, moment of the history of Europe and, no doubt, of the whole world, is being witnessed by all of us. A moment of decisive change in which ancient frames of mind seem to have collapsed definitively and the global, massifying concept of a 'scientific' state they unfold is being substituted by a new paradigm in which not only the individual becomes thenceforth regarded as a thorough *sentient being* (no longer a simple cog in a gigantic economic machine) but also moral, ethical and social *values* - as those of liberty, will, dignity, the right and the respect for the difference (both of the individual and of minor ethnical groups or communities embedded into larger countries), of shared feelings of togetherness, of old nationalisms (founded on common traditions, myths, religions, language) long ago repressed by the institutionalized powers of that state - seem to blossom again all around.

Being not a sociologist, economist or a social historian I am not entitled to venture what further fate and what further transformations such new countries, societies or communities will undergo. Indeed, I would like to forecast (if the task should be realizable) a future of mutual comprehension for what their citizens have in common and for so much tolerance for their differences that 'foreigner' and 'ennemy' could no longer be merged, as they were in classical antiquity, into a single concept. I doubt however that, unlikely to what is *presently* happening amongst the 'older' nations, this stage is easily attainable. Human beings are creatures among whose instinctual endowment (together with their desire for love) is also reckoned a powerful share of agressiveness which - as a rule - waits for some provocation to be externalized. And since, furthermore, the longer its repression holds the stronger its bearings are - then some of the tragic results we are already witnessing in such new countries are so not altogether surprising. Perhaps - as Aristotle has already asserted about happiness and the whole social history of mankind has evinced - such a desirable comprehension, dialogue and agreement amidst people and nations need *time* to be reached. If this is so, then my hopes from a personal but historical perspective, are that such an agreement (which does not preclude the agreement for disagreeing) is early attained in this dawn of their new world.

As a cybernetician whose main interests are related both to Pask's Conversation Theory and to the emulation of the depth psychological processes which general interaction among actors entail, these historical events are, thus, of manifold importance. I am not speaking, of course, of my own attitude as a citizen of a commonly shared world towards such events. But, from a scientific and (in so far as possible) impartial perspective, there are two factors directly related to the objectives of the essay that I propose to pick out as regards such changes. Firstly, for they enhance the strength of the most archaic, irrational and/or affective structures of the human mind upon our behaviours in situations against which no logical reasons or rational explanations offer relieve to distress or to instinctual needs. Secondly, because I cannot forbear to set up a parallelism between such a massifying, political, economical, social and even psychological paradigm and another 'frame of mind', now a scientic one, that deals with the poverty of the (supposedely all-embrancing) 'logistic-computational approaches' according to which their deffenders aim at the *full* modelling of the human mind.

For the objectives of the essay a survey of the major tendencies, currents, problems, provisional solutions, technological improvements, new mathematical techniques, etc., that these approaches have entailed since their foundations is, obviously, meaningless. None the less, I would like to point out that, fifty years later, the immense and yet unfinished debate that such an emulation/modelling has begotten amongst philosophers, psychologists, physiologists, biologists, cyberneticians, mathematicians, etc., has already adduced duly reasons for believing that the greatest expectations and the deepest problems stemmed from the 'golden era' of cybernetics still hold nowadays. Some major questions posed in the sixties like "How would we recognize an 'entity' as having a mind?", "Can machines think?", "Can robots have feelings?" and so forth (all of which, with more or less variations, are titles of published articles, papers or books) still remain to be solved. Asserting this I am not, of course, debasing everything already achieved in the area, particularly in terms of *application* (without which, indeed, this essay would be meaningless). But from a *research perspective* (which *ab initio* excludes *immediate* designs) there are two omnipresent bottom-questions I pose about the present-day status of cybernetics and computer science that, concisely, may be stated as follows:

A) Can we go farther? Can we, like the ancient Greek demiurgs, impart to a system made up of electromechanical/electronic bits, some of the features which human beings are endowed with, like the ability for creating its 'own' (not pre-programmed) goals, for desiring or for acting upon its (external and internal) environments according to its 'own' (also not pre-programmed) purposes?

B) In spite of that undeniable technological progress, which sort of hindrances have prevented us from bestowing machines with the features previously referred to?

In the sequel I propose to tackle these questions from two different, albeit cognate, perspectives. Firstly, I shall survey some of the psychological, historical and philosophical problems which they both underly and from which the standpoints of a new paradigm providing (possible) solutions for them will then be brought to light. Secondly, based on such standpoints, the co-author of the essay will introduce the major features of an evolutionary neuromorphic system (not properly a 'machine' in the traditional sense) which exhibits some of the characteristics referred to Question A).

2. THE MECHANIST PARADIGM

I shall approach the question of wheter or not a machine may be endowed with human characteristcs from a *historical* perspective which, at the outset, will lay emphasis on Question B).

Since its very beginnings that two major approaches to the modelling of the human being were, in fact, entailed in the cybernetic movement: one, dealing with which, for brevity, I shall name 'biological' whose main emphasis was laid on the *physiological* consequences that could be extracted from the concepts of 'feedback' and 'purposiveness' (Rosenblueth, Bigellows, Wiener, Ashby, etc.), the other, more specifically devoted to the analysis and emulation of what, again for brevity, can be named 'mental' (Mc Cullochs, Pitts, Von Neumann, etc.) and from which the currents of the so-called 'Artificial

Intelligence' (AI) were born. But, however surprising the following assertion may be, I claim that the option made by the cyberneticians (thanks especially to Von Neumann's influence) prevailing the latter in detriment of the former (especially after the computational explosion verified in the 70's) has been a real trapp of which the present-day AI's crisis and the overall problems previously referred to are bare reflexes.

That, in the one hand, such an option was *in some sense* inevitable I have no doubt. I remember, for instance, Chapman's self - organizing classification machine which uses threads of cotton moistened with lime water or, in a different area, Pask's work (Pask, 1958) on what may be thought of as a chemical-colloidal machine and how easier the *simulation* of these systems becomes by means of serial computers. But, on the other hand, how high was, has been and still is the price paid by this facilitation. And, in order to justify this contention, I propose to lay some emphasis on six canons or 'dogmas' which, until quite recently, have framed not only the kingdom of artificial systems but also - and moreover - their designers' minds. I shall name them the *dogma of the logical mind*, the *dogma of the disembodied mind*, the *dogma of the timeless mind*, the *dogma of the adult observer*, the *dogma of the objectivity* and the *dogma of the 'in' and the 'out'*.

The first three of these dogmas proceeded, essentially from the works of Mc Culloch, Pitts, Von Neumann, Turing and Putnam. Of Mc Cullochs and Pitts because they showed that the operations of real, neural networks seemmed to be modelled in terms of (classical) logic. Nerves could be thought of as logical statements, and the all-or-none property of nerves firing or not firing could be compared to an operation of propositional calculus where a statement is (supposed to be) either 'true' or 'false'. Moreover, the similarity between neurons and logic could also be thought of in terms of electric terms, i.e. as signals that pass or fail to pass through a circuit (i.e., through an artificial system). This result was accordance with Turing's work on a *theoretical* machine that could carry out *any possible conceivable calculation* using a binary code and a set of simple serial instructions (the present-day 'program'). In building upon Turing's conceptions, Von Neumann pursued the notion of devising a program to instruct the Turing machine to reproduce itself. From here came the powerful idea of *stored program* which, ultimately, led to the idea (still holding today) of a computer that could be controlled through a program that itself was stored within its internal 'memory'; in these conditions, the machine (any machine whatsoever, indeed, due to Turing's demonstration) need no longer to be reprogrammed whenever some new task has to be performed.

The convergence of these works was crucial for the setting up of a metaphor which, until quite recently, has pratically ruled the relationships between machines and the brain's activity - the *metaphor of the brain as a serial computer*. As a matter of fact, if (according to Turing and Von Neumann) a computer was able to realize any process that could be unambiguously described and if (following Mc Culloch's conceptions), one at least redoubtable machine - the human brain - could be thought of as operating via the principles of logic, then the brain itself could thus, be also regarded as a powerful computer.

The last, but no less important, step towards the (supposedely complete) identification between computers and minds was afforded by Putnam in the 60's: since different programs, on the same or on different computers, could carry out structurally identical problems then the logical operations themselves (the 'software') could also be described quite appart from the particular 'hardware' on which they happened to be implemented.

The major consequences of this result are still felt today in the ubiquitous and yet unsolved relationships between brains and bodies (the computational 'hardware') on the one hand, and 'minds', 'patterns of thinking', 'problem solving', etc. (the 'mental states' or 'software') on the other, as though the latter could be described as being entirely separatly from the particular constitution of the human nervous system. Furthermore, (according to the lines of the foregoing argument) since human beings, no less then computers, harbored programmes then the same symbolic language can, too, be invoked to describe programmes in *both* entities. Briefly, the *identity* between the artificial 'mind' (of which the serial computer was, and still is in some sense, the foremost representative) and the human mind was thus completed; the famous 'Turing test' was the utmost example of this perspective.

I shall not pursue this briefly, albeit necessary, historical survey of the first decades of cybernetics and AI. I shall only stress that, once such an identification was brought to light, the metaphor 'brain like a computer' was substituted by a new one - the metaphor *'mind like a computer'* - a *pure* 'mind', indeed, altogether governed by *logical rules* and *logical operations* and, because of this, also *timeless* (albeit *logically* dynamic).

From a purely artificial perspective there is, of course, nothing wrong with this conception. *The problem is that these results were also supposed to be extensive not only to all human beings* (hence, its *massifying* character) but also to *human psychology*. In other words, not only the human mind becomes, thenceforth, chiefly regarded as a *device* able to manipulate abstract symbols (according to such logical rules) but also pratically all the features and processes that *adult* observers/designers find in their own minds have been (supposed to be) entailed by such an overall 'top-down' conception; hence, what I named the 'dogma of the adult observer'.

For the ideological movement that has pratically ruled the development of science in the last two centuries, this result was (and still is, for most people) interpreted as being the ultimate confirmation of a way of thinking, which, starting from the realm of physical sciences, endeavoured the unification of the whole world (not only physical but also psychological and social) under the same frame of mind, the so-called 'mechanist' paradigm. But, however fruitfulness this vision has been for such sciences, for a psychology where humans cannot be regarded as simple 'it' (Pask, 1975) but real *sentient beings*, the consequences of the foregoing identification have been regrettably baneful. Because already approached in other works I shall not go into minute details about them. But for the objectives of the essay their overall survey is, undoubtedly, needed. On the one hand the emphasis laid on logics, on 'reasoning' has led to a rather artificial separation between what is supposed to be 'cognitive' (or, commonly 'rational') and 'affective' (or, commonly 'irrational') as though the mind's functioning could be arbitrarily straddled and everything related to affections, wishes, wills, curiosity, deontic beliefs, etc., can be pratically supressed or overlooked in the name of a false 'objectivy' (another 'dogma' which, for centuries, has ruled the most deeply entrenched scientific tradition). On the other hand, also the polimorphous concept of 'thinking' (from Ryle's perspective) was mainly identified with 'deductive reasoning'; thus, excluding too fantasies, day-dreams, the iddle-flitting from one one half-form to another, imagination, remembering, deontic or epistemic beliefs, etc., none of which (with exception, perhaps, of the epistemic beliefs) may be reduced either to 'pure deduction from previously known (or implemented) premises' or even to logic (in the classical sense, at least). On the contrary, and the following listing is by no means exhaustive, what fantasies, day-dreams and the like overemphasize is (amongst other things) the existence of a *stream of autonomous responses to internal changes* (unlikely to be reduced to deduction or to logical rules) whereas remembering, styles of behaviour and deontic beliefs carry the stamp of the living experiences of which *each one of our personal histories* is made (which is another argument against the massifying character of the mechanist frame).

Even 'reasoning' (one of the strongest AI's bastions) is highly ambiguous in meaning. Usually, the term is connoted with the activity of going over a 'neat' argument which has already been worked out and accepted on previous occasions. But, as Ryle and Thomson showed, this is only part of the story. For, if we consider the case of a thinker indulged in some sort of reflective thinking and to whom a (retrospective) description of his/her activities is asked, the answer is certainly far from being 'neat'. From a possible series of calculations, of formation of hypoteses (which, as Mc Culloch demonstrated, is neither a deductive nor inductive activity but, rather, a mix of abduction/distinguishing/inventing), of *conversation with himself/herself* in which he/she may play the roles of 'witness' and 'counsel' posing and answering questions in the light of facts and theories already known and remembered, to goal-seeking situations like consulting books in order to complete some inference (where, again previous learning and previous experience are determinant), etc. - all of these possibilities may be regarded as part and parcel of reasoning but none of them, taken separately, draws a cogent picture of what a thinker does as he/she is 'reasoning'.

These general assertions about 'reasoning lay another emphasis on the poverty of the mechanist conception of the mind as 'device able to manipulate abstract symbols' whenever compared with real, human processes. This new criticism is based on such terms or expressions as 'experience', 'learning' or 'conversation' and stress out various questions among which the following ones are especially poignant. Firstly, the importance which *semantic and pragmatic meanings* have for human life - 'meanings' which have been sidelined or shelved from the purely syntactic approach on which the 'language' of classical computers/programs is entirely based. Secondly, the necessary *reference to an 'external world'* not only 'physical' but also 'social' (where the foregoing 'conversation' is crucially entrenched) without which the setting up of such meanings is not possible. But such a reference besets, in turn:

a) an *appropriate body/brain interface* - a mandatory requirement which not only contradicts the foregoing thesis about the traditional distinction software/ hardware but also re-enhances the crucial role played by the latter (a role which, again, imposes severe restrictions to the dogma of the disembodied mind). This aspect will be developed later, but as Gunderson asserts, cited in [19], "The simulation of some capacities might depend not so much on how a robot is programmed, as on how the program is roboted. No ammount of programming will make my PC print out in red ink; for that we need 'hardware' - a red printer ribbon");

b) what I named the *dogma of the 'in' and the 'out* - meaning with this the underlying processes that in human beings (and, probably, in other forms of life) allow a *distinguo* between what is 'in' and what is 'out' to them. The crucial importance of this distinction (which, in so far as I know, was never the objective of intensive research) stems from the fact that, without it no 'internal' and 'external' worlds exist; consequently, no *theory* about them (philosophical, physical, social, psychological, cybernetic, computational and so forth) could ever have been erected. For, a 'theory' is ultimately nothing but a metaphor, an "as though" (mental) construction based on implication/metaimplication rules whose cogency cannot be checked if never confronted with the 'external reality' (whatever the meaning of 'external reality' may be in this context).

I will not go into more details about this criticism which, I believe, provides duly reasons for supporting my contention about the poverty of the mechanist approach not only to the full modelling of the human mind but also to its (possible) emulation. For (from the countless problems not yet overcome), together with the motivational and affective worlds which, for the reasons previously emphasized have been pratically excluded or overlooked, also 'consciousness', 'self'-knowledge, 'self'-interpretation, 'metaphorizing', the 'frame problem', the *distinguo* in/out, etc., briefly, the faculties with which human *qua sentient beings* have traditionally been endowed - are/continue to be out of court in computational terms.

3. THE EVOLUTIONIST PARADIGM

To solve (some of) these problems I propose to start anew in the light of a fresh and, shall I say, frontally opposed perspective whose major framing lines may be delineated as follows. Instead of beginning with a 'top-down' approach as the defenders of the mechanist paradigm do - where an adult and disembodied 'mind' endowed a priori with logical and timeless rules is supposed to describe and explain all of its entailed processes (with the consequences previously emphasized) - I propose, rather, to follow an overall *holistic, bottom-up approach* (where cognition *as well as* maturation, development, motor control, motivation and affection are inseparably bound together) starting from a biological organization endowed with some innate features which, next, by means of successive 'internal' and 'external' interactions/improvements reaches a stage where some of the features of the adult human mind can be found. In terms of the preceding considerations this means, more precisely, that instead of paying attention to the 'statical', 'disembodied', 'mental', 'logical', 'timeless', 'objective' (and from a computational viewpoint, 'serial') perspectives that the mechanist paradigm unfolds, I shall lay emphasis rather on the *somatic*, *evolutionary*, *idiosyncratic*, *affective*, *interactive* (and, from a computational point of view, *connectionist*) aspects of an artificially replicable organism which initially plays the role of a primitive body/brain system and, next by means of 'internal' and 'external' experiences is acquiring the features of a 'mind' similar to those pertaining to adult human beings.

From the host of biological, physiological and psychological ideas according to which the approach is framed, two overall subject matters (which, in so far as I know have not yet been the objective of the mechanist concernings) are crucially important: firstly, the notions of *self-development* and *willing* while referred to artificial systems (which, in turn, entail some other concepts as those dealing with *purposefulness*, *teleology* and *unpredictableness*) and, secondly, the aforementioned distinction between 'in' and 'out'. In the following considerations I'll approch these matters, beginning with a survey of the similarities and diferences existing between machines (in general) and living systems. Next, I'll review some historical and philosophical notions which the notion of teleology entails. Finally, assuming some physiological and psychological hypotheses I'll present the main reasons on which the theoretical realization of a non-natural, self-developing system with the ability for willing as well as for distinguishing 'in' from 'out' can be based. The pratical realization of such a system, of the responsability of the co-author of the work, will be the content of Section 3.3.

3.1. Self-development, Machines and Living Systems

Inserted into the preceding context (which excludes manifold questions not directly related to the work) the major clue for the introduction of the differences between machines and living* systems proceeds, essentially, from the traditional concept of machine. As a matter of fact, a (classical) machine, 'program' or serial computer is, essentially, a closed system. It can only play the role for which it was designed *after* being finished. *Before* that, say, while is being finished. *Before* that, say, while is being made (or is being written) it does not work as *the* machine its designer intends to realize. So, it can never undergo what may be called 'autonomous' (or 'self') 'development' since this development means to achieve something which the machine has not yet reached *by itself* and a (traditional) machine cannot work *in order to become anything whatsoever outside the bounds of the instructions externally imparted to it.* Thus, the unique type of non-predictable (and undesirable) 'autonomous' change these machines may undergo while being at work is becoming damaged. But this is not 'self-development' since to this damage does not correspond the *acquisition of new functions*, only the loss of the older ones.

In the realm of living organisms things occur in a totally different way. Beyond all the structural and functional differences existing between them and (artificial) machines (whose analysis does not fall under the objectives of the essay) all living organisms are, since their birth, subjected to a process of growth (of mass, shape and size, at least) that minimally besets not only a bodily organization, maturation, permanent interaction between internal and external environments, what Ashby called the possibility of self-repairing, etc., but also *self-development and becoming* as though, at any moment of their existence, all forms of life were pervaded by a *nisus*, 'tendency' or 'effort' *directed towards the realization of something not yet achieved at that moment.*

The most ortodox mechanist defenders claim, of course, that such a development is *entirely* due to some sort of philogenetic 'program', as though all organisms were already endowed at their birth with a complete and definite set of 'stimulus/ response mechanisms' working, each one of them, as some kind of (classical) automation.

* *The use of such terms as those of 'life', 'living' etc., does not imply or presuppose any precise distinction between what is 'living' and 'non-living' since the question it entails (albeit not yet solved) is irrelevant for objectives of the essay. The distinction merely rests upon the fact that in most cases ordinarily met there is agreement as to whether a portion of matter is to be classed in the first or in the second category.*

That *some* of these innate mechanisms exist is, obviously, undeniable. However, that they *entirely* describe the organisms' development (particularly human development) is a conception which may be subjected to severe criticisms, two of which are here especially poignant. The first (put forth in the 60's by Von Foester and Pask) states, briefly, that the (traditional) proposition "the coupled system organism/environment undergoes adaptation" is nothing but a pure tautology. This because the mechanist paradigm unfolds the idea that organism *and* environment can *both* be identified with *two* classical automata. But, in these conditions, neither the environment's internal states can be changed in consequence of any act (like attending something new) that the organism might perform outside the limits of its initial description nor the organism itself may undergo 'adaptation', 'development' or 'evolution' since all possible behavioural interactions are, from the outset, predetermined.

The second criticism lays emphasis on the word 'latent' used by a milder version of the mechanist viewpoint. For, if some of the processes that organisms undergo are said to be in a 'latent' (or 'potencial') state while others are really (or 'actually') occuring then, a) the existence of *parallel processes* and, b) the explicit introduction of *time* (both of which contradict the traditional serial and timeless premises) have necessarily to be assumed. Furthermore: asserting that, the mechanist defenders are also returning to a (descriptive) context where a *reference to the future is necessarily entailed.* As a mater of fact (from the point of view of the organism itself and not of someone who is externally describing its behaviour) no one of its processes may undergo a *definite* (and not a purely random) change unless the final, 'actual', state can influence the whole transformation; this, from its most remote 'potential' or 'latent' beginnings to its real, actual end. But to state the question in these terms is tantamount to use a pure *teleological frame* - which, again, contradicts the causally efficient premises on which the mechanist paradigm has always been based.

As a systematized explanation of the world teleology has been mainly related to the name of Aristotle and to an all-embrancing Image of Nature, a true *Weltanschauung* held for more then 16 centuries in the whole history of the Western Thought until the the XVIIth/XVIIIth centuries. Since those times, however, with the emergence and further development of the Renaissantist movement (on which the roots of the mechanist paradigm lay down) the Greek Image was (almost) totally discredited and Nature was no longer regarded as an alive organism but, rather, as a machine ruled by an external, transcendent God . A detailed analysis of the manifold reasons on which that revolutionary change rest is, obviously, out of the scope of the present essay - particularly in the realm of physical sciences where, indeed, the successs of the mechanist assumptions have, since then, been undeniable. But, as pointed out in the preceding section, one thing is to speak of physics other of biology and psychology, especiall *human psychology.* As a matter of fact when the most advanced researches in genetics show that the genoma contains a plan of the entire organism 'waiting for its full realization', when etologists and experts in the biology of the human species speak of 'innate tendencies' or when psychologists stress the crucial importance of instincts, whishes, desires, intentionality and purposefulness in animal and human life - they are all using a language where the tenets of the Aristotelean philosophy and its entailed teleology are implicitly assumed. This is also one of the reasons why my arguments about 'self-development' are based on the relation between the foregoing Aristotelean conception of *'nisus'* and cybernetics. In the following considerations I'll try to bring to light the standpoints of such a relationship.

3.2. Self-development, Nisus and Cybernetics

To begin with, I'll remark that the problem of self-development is not a novel one. On the contrary, it has been the objective of countless debates (using different terms, of course) from the remote past of the Western Thought - when Plato or Aristotle looked for the ultimate explanation of natural becoming - to the XIXth/XXth centuries when the materialistic currents or the defenders of the so-called 'scientism' were forced to face the deep crisis which the science of that epoch underwent after the introduction of the concept of biological evolution. Behind all its variations the major question lying beneath the whole of such debates was, simply, "What *ultimately* makes things undergo 'change' ?"

For Aristotle (whose reasoning I'll try to condense) this problem was solved by means of the introduction of various concepts amongst which those dealing with 'potentiality', 'actuality', 'grades of Being' (what nowadays may be crudely identified with the hierarchy of biological species) and 'desire' are especially poignant. How they provide answer to that question and simultaneously beset the general notion of 'development' requires that an extremely brief account of the three major standpoints on which the whole Greek Image of Nature rest be previously emphasized.

For the Greek and Medieval thinkers the world of Nature was regarded as a world of self-moving things, of spontaneous movements; thus, an *alive* world altogether pervaded by a 'soul' (or 'vital principle'). Furthermore: since some of these movements were regarded as being 'ordered', eternally cyclic or 'repetitive' (of which the motion of the heavenly bodies was the utmost example) and "order could only be imposed on bodies by the action of minds" then the natural world was also looked upon as being a rational and intelligent organism endowed with a mind. Finally, just like human actions are directed towards (as well as subordinated by) the realization of goals and intentions, then all the natural processes were also conceived as being compelled by (as well as dependent on) the wishes and desires of the world-intelligence (later identified with God). In the Aristotelean system this omnipresent 'desire' was identified with the overall *nisus*, 'effort' or 'tendency' previously referred to. Hence - and this is the crux of Aristotle reasoning - all things of Nature tend to change in a *definite* way because *each one of them and in a certain degree* is affected by a *specific* 'tendency' (nothing but a bare reproduction of the highest, overall *nisus* or desire of Nature towards God) that acts upon each one of their potentialities *in order to* convert them in actualities. In this sense a seed becomes a plant, a pet becomes an adult animal, etc., not only because each one of them is affected by a particular 'effort', *nisus* or 'desire' but also because 'plant', 'adult animal' and so forth are their actualities; and as soon as these actualities are attained then each one of their (self) developing processes cease their influence.

The major consequence of this conception unfolded the idea (held for more than sixteen centuries) that the ultimate nature of the world can be expressed by what may be called 'process', 'growth' or 'development', meaning with this that all of its changes assume successive forms A, B, C, etc., where each one of them is the potentiality of the next. For Aristotle this did not mean 'evolution' since the kinds of change and structure (because nothing but forms) were regarded as constituting an eternal repertory whose characteristic types were *logically not temporally related to*. However, when the idea of biological evolution was put forth in the XVIIIth/XIXth centuries by Lamarck, Darwin, Wallace and others, *these* standpoints of Aristotle's philosophy were rephrased again and men of so high a intelectual standard as Lloyd Morgan, Alexander or Whitehead (among others) have unanimously accepted - and, moreover, *used* - the notion of *nisus* and its entailed teleology in their contemporary conceptions of Nature.

So far I have followed Aristotle; hereupon I' ll return to cybernetics. As a matter of fact when Wiener, Ashby and others gave rise to the (r)evolution we are all presently witnessing - showing that the problem of control in organisms and machines could be altogether interpreted as a mere consequence of a particular type of informative ('feedback') structure - they focused an aspect which, appropriately interpreted and, moreover, not traditionally looked upon contains the possibility of imparting a *nisus* (and, consequently, self-development) to artificial systems. Such an aspect lays emphasis on an expression written by Rosenblueth, Wiener and Bigellow in the paper 18 that preceded Wiener's 'Cybernetics' where they textually assert:

> *"The basis of the concept of purpose is the awareness of 'voluntary activity'. Now the purpose of voluntary activity is not a matter of arbitrary interpretation but a physiological fact. When we perform a voluntary action what we select voluntary is a specific purpose, not a specific movement. Thus, if we decide to take a glass containing water and carry it to our mouth we do not command certain muscles to contract to a certain degree and in a certain sequence; we merely trip the purpose and the reaction follows automatically"*

From my perspective the crucial problem rests, precisely, in these two underlined statements. For, when Rosenblueth, Bigellow and Wiener laid emphasis on the relation between some previously selected purpose and a suitable voluntary action, they were not only stressing out an undeniable fact of our physiological and mental existence as *adult human beings*; they were also touching over a *genetic*, diachronic, question which (amongst manifold ones) may be concisely stated as follows: Being the earliest behaviours of new-born babies essentially reflex, repetitive, involuntary or 'automatic' then what gives rise to the profound changes they undergo throughout their development? In other words, what makes them to overcome this initial stage and to reach the complex, idiosyncratic, intentional, goal-seeking, purposeful and voluntary patterns that adults exhibit?

The answer I venture - crucially important for the entire essay - is based on two major hypotheses. The first emphasizes the *reproductibility* characteristics of the 'mechanism' on which human desiring rests. The second, (which radically changes the functional role traditionally ascribed to our sensors) states that none of these sensors (particularly the external ones) must be regarded as the simple and pure 'receptor' currently admmited in physiology but, rather, as an active and dynamic complex unit which not only receives and sends information to the higher strata of the central nervous system through afferent pathways but also *receives and changes its receptive conditions according to the efferent orders coming from the higher levels of this structure.*

As it will be demonstrated in the conference these cognate assumptions are not only *testifiable* (in the sense that to them corresponds an *effectively realizable* formal/computational/robotic interactive structure) but also allow that the system can make self-distinctions between 'potentiality' and 'actuality', say, has an intrinsic *nisus* or tendency for acquiring new functions (and so, for a permanent self-evolution) as well as for wishing and desiring in a real, human sense (together with everything these concepts beset in terms of affections, frustrations etc.). Finally, the hypotheses also support the mechanism where the former distinctions between what is 'in' and 'out' are based.

Obviously, these hypotheses are inserted into a much more complex structure called BABY II - in reality, an evolutionary neuro-fuzzy system (where homeostatic mechanisms, needs, wishes, wish-fulfilment, pattern identification, relationships between somatic and psychic structures, Conversation Theory and so forth are entailed) - which is being implemented by the co-author of the essay. A brief account of his work is provided in the sequel.

3.3. BABY II: A Survey

BABY II is a complex hierarchical neuro-fuzzy system composed of 3 demarcated structures: the Sensorial/Somatic Structure (SSt), the Associative Knowledge Structure (ASt) and the Linguistic Structure (LSt). The lower level is the SSt. The SSt is a PDP network receiving input signals from several sensors and from a set of somatic generators, implementing a set of biological needs. All these inputs will affect the values of the SSt's units in different ways, allowing the setting up of an overall sensorial state based on the initial definition of simple states of pleasure and displeasure according to the early fulfilment or not of the basic biological needs. This lower level structure is also connected to several effectors. Since the effectors can also actuate upon the sensors a whole set of important considerations is enabled.

The middle level is the ASt. This is an associative structure built upon the development of the structure presented by Rocha[16]. Sometimes, the activation of the SSt's units defining a sensorial state is considered to be important; in these occasions the sensorial state is recorded. To identify this state of activation, an internal symbol is created and stored as a concept on the ASt. The resemblance between these concepts is established by comparing the activation values of the respective sensorial state. The ASt is then able to receive several concepts as inputs, engaging in a series of Spreading Pruning processes starting on these initial concepts, and having as a result the activation of associated concepts in the structure.

The communication between the lower and middle level is possible due to the existence of a PDP network whose units are the concepts on the ASt, the units of the SSt and some additional fuzzy units which measure (according to fuzzy criteria) the importance of the network's units and of their connections, all of them being used to control the communication. This network is called the Psycho-Somatic Network (PSN) and it enables the 2 - way transmission of the activated information, so that a sensorial state on the SSt can trigger Pruning Operations on the ASt, resulting, or not, in the formation of new concepts and, therefore, new connections on the PSN. The result of the pruning operations can, thus, influence the activation of the sensorial units; consequently, creating sensorial states that characterise the pruning result set of concepts rather than only individual concepts or sensor inputs. This permits the formation of new concepts based not only on sensorial inputs but also on the results of pruning processes. Furthermore: since during the process of pruning, sensorial information coming through the SSt can continuously affect the pruning process itself, then this mutual interaction may allow the pruning processes to be different in situations where, despite equal starting concepts, the sensorial/somatic input is different. To the internal symbols can later be ascribed external names through actors' interaction. These names will be stored in the LSt that will form the third level of BABY II. The inclusion of the LSt, enables a more interesting interface with other actors since, instead of establishing a dialogue between actors directly through the associative space, as the standard Conversation Theory does, it not only confers more versatility to the dialogue but also opens the door to the study of hidden aspects of Language, such as those proposed by Henry[8], suggesting a process of relating terms and concepts to associative and sensorial/somatic processes as well as for the setting up of real metaphors[10].

Once the LSt is formed, it will be responsible for pre-processing the inputs from other actors, activating the relevant concepts on the ASt, and also to post-processed concepts activated on the ASt after some pruning process has taken place, and which should be outputted to the actors involved in the dialogue. This communication process is constructed in a manner similar to that which lead to the connections between the SSt and the ASt. That is, another PDP network is formed with associations between names (or other language constituents) in the LSt and the Internal Symbols in the ASt. This is a result which may be interpreted as establishing "corridors" between language and "ideas". This allows, in turn, the construction of other concepts in the ASt, only by its connections to the LSt (in contact with other actors); at some point these concepts might enter in disagreement with the ASt and trigger its RE-structuring, changing the resemblance between concepts as well as its connections to the SSt and LSt. Again, this may be interpreted as being an Intellectual re-structuring of the whole structure, which may lead to a modification of the way by means of which the sensorial and somatic inputs are established and processed. All the processes taken place on the several structures are not isolated. Constantly, information is sent back and forth between the three levels, by virtue of the PDP characteristics. This means that all the processes are always producing intermediate results that will affect other processes in other levels, which, in turn, will affect the initial processes. Very complex and almost unpredictable results are therefore expected as far as the structure of BABY II is acquiring new functions, new terms and new relations throughout its evolution.

The need for establishing an end for some processes that, otherwise, would go on indefinitely, is implemented again by the use of Fuzzy Measures that will judge when the ambiguity or confusion of the intermediate results of the processes taking place, is either resolved to an acceptable level or is an unsolvable task. Relations between these ambiguities or confusions and psycho-somatic processes (like ambivalence, anxiety and so forth) may therefore be also brought to light.

4. OUTLINE

Magic, religion and science have been the tools which, throughout time, mankind has used to flight from the uncertainty that the unpredictableness inherent to the change and evolution of our world (either it deals with human societies or with our 'inner' and 'outer' phenomena) has always begotten in all of us. That the evolutionary aproach we are developing may contribute to a deeper understanging of the mechanisms ruling such an all-embracing world is, the last but not the least, our ultimate hopes not only as cyberneticians but also - and above all - as human *qua sentient beings.*

5. SELECTED REFERENCES

1. J. Anderson and E. Rosenfeld, "Neurocomputing: Foundations of Research", The MIT Press, Cambridge , Massachusetts, London, England (1988)
2. M. Boden, "Computer Models of the Mind", Cambrigge University Press (1988)
3. G. Carpenter and S. Grossberg, "Pattern Recognition by Self-Organizing Neural Networks", The MIT Press, Cambridge , Massachusetts, London, England (1991)
4. S. Freud, "On Metapsychology: the Theory of Psychoanalysis", The Pelican Freud Library (1964)
5. H. Gardner, "The Mind's New Science", Basic Books, New York (1987)
6. R. Gawronski, "Bionics: the Nervous System as a Control System", Elsevier (1971)
7. F. George, "The Brain as a Computer", Pergamon Press (1973)
8. C. Henry, "Non-Literal Aspects of Language and Knowledge Structuring", The Principia Cibernetica Conference , Brussels (1991)
9. G. Klir, "Fuzzy Sets, Uncertainty and Information", Prentice Hall International Editions (1988)
10. H. M. Macedo, "Metaphorical Structure of the Self", XIth European Congress on Cybernetics and Systems Research, Wien (1992)
11. P.R. Medina-Martins, "On the Simulation of Depth Psychological Processes", Xth European Congress on Cybernetics and Systems Research, Wien (1990)
12. P.R.Medina-Martins and I. Faria, "Cognitive Predicate Structuring from Motor Abilities at an Early Level of Language Acquisition", Congress on 'Problems on Support, Survival and Culture', Amsterdam (1990)
13. P.R. Medina-Martins, "Cybernetics and Possibility Theory", Tecnica, July (1990)
14. Y. Pao, "Adaptative Pattern Recognition and Neural Networks", Addison Wesley Publishing Company, Inc. (1989)
15. G. Pask, "The Natural History of Networks", Proceedings of the International Congress on Self-Organizig Systems (1975)
16. L. Rocha, "Fuzzification of Conversation Theory", The Principia Cibernetica Conference, Brussels (1991)
17. A. Rosenblueth and N. Wiener and J. Bigellow, "Behavior, Purpose and Teleolgy", Philosophy of Science, Vol 10 (1943)
18. K. Said and W. Newton-Smith and R.Viale and K. Wilkes, "Modelling the Mind", Clarendon Press (1990)

Chaos, Instabilities
and
Control of Chaos

EFFECTS OF PARAMETRIC PERTURBATION ON THE ONSET OF CHAOS

LEONE FRONZONI
Department of Physics , University of Pisa ,Piazza Torricelli 2
Pisa ,56100, Italy

ABSTRACT

The influence of the modulation parameter on the onset of chaos has been studied in two different dynamical systems .Theoretical and experimental results have shown that in a bistable system parametric modulation can suppress chaos while , in a Josephon -Junction model, chaos was favoured by the parametric perturbation.

1. Introduction

The study of the influence of a parametric perturbation on dynamical systems in the presence of deterministic chaos may have relevant applications in many problems such as the stabilization of particle beams , thermonuclear plasma in tokamak and fluid dynamics. In 1967 O.H. Wehrmann of the Boeing Scientific Research Laboratory observed that it is possible to suppress the turbulence behind a cylinder in moving fluid [1] . By means of a suitable feed-back device the axis of a cylinder was made to oscillate according to the wake fluctuations in the fluid . The use of a feed-back device to suppress the chaotic behaviour indicated that a particular frequency and phase of the parametric perturbation were needed to obtain stability. The theoretical interpretation of this behaviour is not easy because of the presence of many variables in fluid dynamics ; however, one can deal with a very simple model and then proceed to a generalization . We have focused our attention on a bistable system driven by a periodic perturbation . In a recent paper theoretical calculation on a Duffing oscillator [2] showed that parameter modulation can suppress chaotic behaviour . In the following sections the theoretical approach to the Duffing oscillator will be presented according to the analysis by Lima -Pettini [2] . The theoretical results are compared with our experimental ones. The use of a real apparatus to demonstrate the suppression of chaos is a crucial proof in removing doubt as to the possibility of attainig this goal. Moreover a different system will be presented in order to show that , in some cases, parametric perturbation can , indeed, produce chaos , unlike the above-mentioned example.

2.The Duffing-Holms oscillator

2.1. Melnikov's method concerning the Duffing oscillator

Let us consider a nonlinear oscillator with forcing described by this equation

$$\ddot{x} = -\gamma \dot{x} + \alpha x - \beta x^3 + A\cos(\omega t + \phi)\,; \qquad (2.1.1)$$

in the case in which A = 0 , one finds two stationary equilibrium positions

$$x = \pm\sqrt{\frac{\alpha}{\beta}} \quad . \qquad (2.1.2)$$

The solution shows a wide and varied panorama when forcing is applied. In particular one obtains chaos for a wide range of the A and ω parameters.

In a recent theoretical work Lima and Pettini [2] proved that with a suitable condition as to frequency and to the phase of a parametric modulation it is possible to remove the chaos in this model. They write eq. (2.1.1) in the form

$$\ddot{x} + \alpha x - \beta x^3 = -\gamma \dot{x} + A\cos(\omega t + \phi) \; ; \qquad (2.1.3)$$

the Hamiltonian

$$H(p,x) = \frac{1}{2}\left(p^2 - \alpha x^2 + \frac{1}{2}\beta x^4\right) \qquad (2.1.4)$$

corresponds to the left-hand term and the right-hand term is a perturbative one F (dx/ dt , t).

Note that, in the phase space, point (0.0) is a saddle point and for $\alpha = 1$ the homoclinic loop is described by

$$x(t)_0 = \left(\frac{2}{\beta}\right)^{1/2} \operatorname{sech} t \qquad (2.1.5)$$

$$p(t)_0 = -\left(\frac{2}{\beta}\right)^{1/2} \operatorname{sech} t \, \tanh t \quad . \qquad (2.1.6)$$

The parametric perturbation modifies the nonlinear term β according to the following relation

$$\beta = \beta \; [\, 1 + \eta \cos(\Omega t) \,] \qquad (2.1.7)$$

where η is the amplitude of the perturbation and Ω the frequency .
The method for obtaining the prediction of the threshold of chaos is based on the Melnikov technique : it consists in finding the condition for changing the sign of the Melnikov function so defined

$$M(t_0) = \int_{-\infty}^{+\infty} dt \; p_0(t)\, F(\dot{p}_0 , t + t_0) \quad . \qquad (2.1.8)$$

In another way chaos arises when homoclinic intersections appear. It is possible to establish a link between the Lyapunov exponents and the inverse of the time that elapses between two successive homoclinic intersections τ^{-1}. It turns out that this inverse time τ^{-1} goes to zero when frequency Ω approaches the resonant condition

$$\Omega = \frac{n}{m}\ \omega \qquad (2.1.9)$$

where n and m are integer numbers.

This means that for these conditions chaos is suppressed .The authors examine the possibility of obtaining a favouring of chaos in a different way and the shape of the potential might play a crucial role in this connection. In the next section we shall show that these two effects are possible in real systems .

2.2 . The bistable device

The experimental apparatus that we have devised to prove it is possible to suppress deterministic chaos was made up of a still elastic beam clamped at one end . The other end was left free to move in front of two magnets, as shown in fig. 1. The magnets give two equilibrium positions of the beam. The whole apparatus was fixed on a moving frame driven by means of an electromagnetic shaker . In order to obtain a parametric modulation two coils were placed around the magnets and an oscillating current was sent into the coils. An optical device allowed us to measure the position of the beam : a photodiode was put in front a lamp in such a way as to detect the intensity of the light beam and a screen fixed on the steel beam partially obscured the light beam . The current of the photodiode gave an evaluation of the angle that the steel beam made with the vertical axis. A suitable window put in front of the photodiode assured a linear relation between the light intensity detected by the photodiode and the value of this angle.

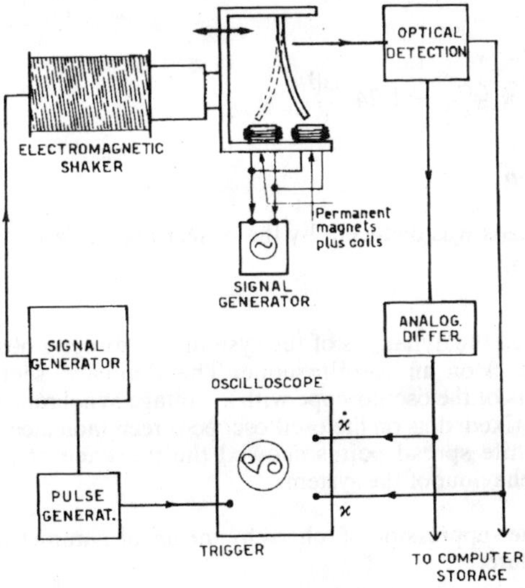

Fig. 1. Outline of the experimental apparatus of the bistable system

2.3. Determination of the Duffing parameters

If the forcing was not too strong one-mode approximation could be carried out, so that the amplitude obeyed the Duffing-Holmes ordinary differential eq.(2.1.1). Measurement of the equilibrium position x_0 of the steel beam and the resonance frequency ν_0 at these values allowed us to obtain the values of the parameters of the eq. (2.1.1) according to :

$$\alpha = \frac{1}{2} (2\pi\ \nu_0)^2 \tag{2.3.1}$$

$$\beta = \frac{1}{2x_0^2} (2\pi\ \nu_0)^2 . \tag{2.3.2}$$

A current I_m flowing through the electromagnets provided the measurement of the parameters corresponding to the modulation terms with frequency modulation ν_m

$$\alpha \rightarrow \alpha\, [\, 1+\ \varepsilon \cos\, (\, 2\, \pi\, \nu_m\, t\,)\,] \tag{2.3.3}$$

$$\beta \rightarrow \beta\, [\, 1+\ \eta \cos\, (\, 2\, \pi\, \nu_m\, t\,)\,] \tag{2.3.4}$$

where

$$\varepsilon\, (\, I_m\,) \quad ; \quad \eta\, (\, I_m\,) \tag{2.3.5}$$

and we found that

$$\frac{\Delta\alpha}{\alpha} = 1.24\ \frac{\Delta\beta}{\beta} \tag{2.3.6}$$

2.4 . Poincare's map

The presence of chaos was deduced by the observation of a Poincare' map obtained by the following technique:

We examined the dynamics of the system by directly observing the phase-space portraits (x, dx/dt) on an oscilloscope. The Poincare' section was obtained by modulating the z-axis of the oscilloscope with a voltage synchronized to the driving signal. The appearance of fixed dots on the oscilloscope screen indicated a periodic behaviour of the steel beam ; while spread points showed the presence of a strange attractor and therefore chaotic behaviour of the system.

To show the suppression of chaos by means of parametric modulation we used the followed procedure:

i) The frequency and the amplitude of the forcing were chosen in such a way as to produce a chaotic behaviour of the steel beam. This was ensured by the appearance of the strange attractor on the oscilloscope screen.

ii) An oscillating current was sent into the electromagnets and the output of the photodiode was connected to the input of a lock-in (EG&G model 5210) .The forcing signal was also sent to the reference input of the lock-in.

Fig. 3 shows the lock-in output as a function of the modulation frequency ν_m. The figure shows how the relative weight of the periodic phase changes in relation to changes in the parametric modulation frequency with respect to the chaotic phases. The two upper curves correspond to different amplitudes of the modulation ($\varepsilon = 0.16$ and 0.1 from top to bottom). The maxima of the peaks correspond to $\nu_m = \nu_f$. Resonance also appeared at $\nu_m = 2\nu_f$.

By means of analysis of the amplitude of the angle beam versus time we observed that :

a) Chaotic vibration is alternated by ordered vibration;

b)The ordered phase is characterized by periodic oscillation around one of the two minima of the potential and the beam is locked at the forcing signal;

c) The average duration of the ordered phase T_{avr} is an inverse function of the modulating and forcing frequencies ν_m and ν_f, respectively :

$$T_{avr} = \frac{1}{|\nu_m - \nu_f|} \quad ; \qquad (2.4.1)$$

d) A special condition of the phase relation between the forcing and the parametric modulation is very important . In particular one obtains completely ordered and periodic behaviour if the modulation is locked in phase with the forcing.

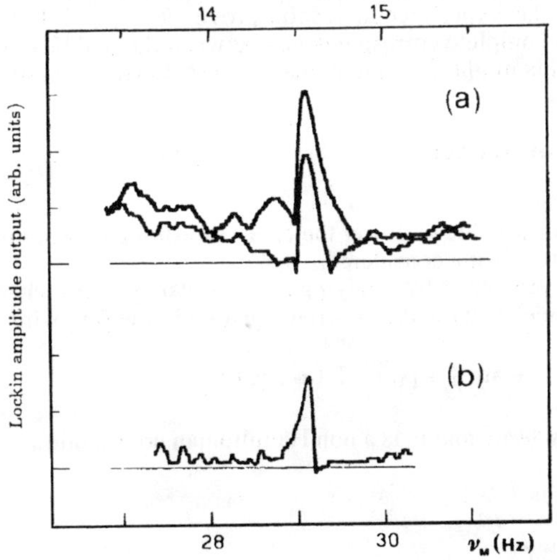

Fig 2. Lock-in output as a function of the forcing frequency

2.5 . The theoretical interpretation

In ref. 2 the authors consider the case in which the modulation is applied to the nonlinear term β . Otherwise, our modulation acts in both parameter α and parameter β and new calculations are need. Using the Melnikov technique we obtain

$$M(t_0) = A(\omega)\sin(\omega t_o) + B(\Omega)\sin(\Omega t_0) + C \qquad (2.5.1)$$

where

$$\omega = 2\pi\nu_m \quad ; \quad \Omega = 2\pi\nu_f \qquad (2.5.2)$$

$$A(\omega) = 2\pi\,\omega A\sqrt{\frac{2}{\beta}}\,\mathrm{sech}\left(\frac{\pi\omega}{2\sqrt{\alpha}}\right) \qquad (2.5.3)$$

$$B(\Omega) = \left[\frac{\pi\varepsilon\alpha^2}{\beta}\left(\frac{\Omega^2}{\alpha} - 1\right) + \frac{\pi\eta}{6\beta}\left(\Omega^4 - 6\Omega^2 + 1\right)\right]\mathrm{cosech}\left(\frac{\pi\Omega}{2\sqrt{\alpha}}\right) \qquad (2.5.4)$$

$$C = \frac{4\gamma\alpha^{3/2}}{3\beta} \,. \qquad (2.5.5)$$

This shows it is possible to prevent the quantity $M(t_0)$ from changing its sign when frequency modulation approaches the frequency of the forcing.

In conclusion the experimental results prove it is possible to suppress chaos although there is not a complete correspondence between the real bistable system and the Duffing relation . This might mean that chaos in real bistable systems is structurally unstable.

3.Josephson-junction model

3.1. Melnikov 's method

In some cases is possible to lower the chaos threshold, depending on the shape of the potential of the system being considered.

We have proved this prediction by studying a very usual system such as a pendulum or a Josephson-junction model [5] with the Newton equation in the following form

$$\ddot{\phi} = -\sin\phi + \rho_0 + \varepsilon F(\phi, \dot{\phi}, t) \qquad (3.1.1)$$

where ρ_0 is a constant term and F is a non Hamiltonian contribution.

The Melnikov function is

$$M(t_0) = \int_{-\infty}^{+\infty} dt\,\dot{\phi}_0(t)\,F[\phi_0(t), \dot{\phi}_0(t), t + t_0] \qquad (3.1.2)$$

where $\phi_0(t)$ indicates the homoclinic orbit when $\varepsilon = 0$.

In the standard Josephson-junction model

$$F[\ \phi_0(t),\dot{\phi}_0(t),t\ ] = -\delta\,\dot{\phi} + \gamma\cos(\omega t) \quad . \tag{3.1.3}$$

When $\rho = 0$ the homoclinic orbit is described by the following relations

$$\phi_0(t) = \pm\ (\ 4 \text{ arctang } e^t - \pi\ ) \tag{3.1.4}$$

$$\dot{\phi}_0(t) = \pm\ 2 \text{ sech } t \quad ; \tag{3.1.5}$$

the Melnikov integral provides a condition for the appearance of chaos

$$R_0(\omega) = \frac{4}{\pi} \text{coh} \frac{\omega\,\pi}{2} \leq \frac{\gamma}{\delta} \tag{3.1.6}$$

where R_0 is the Melnikov ratio.

We observe that the Melnikov method is an approximate technique and the relation in eq.(3.1.6) is in contrast with the experiment where a minima is present in the data, as figure 4 shows .

In the case in which $\rho_0 \neq 0$, the low frequency behaviour of the threshold for chaos drastically changes or, in another way, the effect of ρ_0 is to lower the threshold of chaos.This corresponds to a break in of the symmetry in the system.

When one applies a parametric modulation, eq.(3.1.1) becomes

$$\ddot{\phi} = -[\ 1 + \xi\cos(\Omega t + \theta)\ ]\ \sin\phi + \rho_0 \ - \delta\,\dot{\phi} + \gamma\cos(\omega t) \quad ; \tag{3.1.7}$$

where $\rho_0 = 0$ and the Menlikov distance is

$$M(t_0) = -A\cos(\omega t_0) + B \text{ sen } (\Omega t_0 + \theta) \tag{3.1.8}$$

with

$$A = 8\,\delta \pm 2\,\pi\,\gamma \text{ sech } (\frac{\omega\,\pi}{2}) \tag{3.1.9}$$

$$B = 2\,\pi\,\xi\,\Omega^2 \text{ csch } (\frac{\Omega\,\pi}{2}) \quad ; \tag{3.1.10}$$

if $\xi > 0$ chaos does not appear if

$$\gamma \text{ sech } (\frac{\omega\,\pi}{2}) + \xi\,\Omega^2 \text{ csch } (\frac{\Omega\,\pi}{2}) < \frac{4\,\delta}{\pi} \quad ; \tag{3.1.11}$$

the sign in $M(t_0)$ depends on the sign of ξ or the sign of θ . It is easy to verify that changing the sign in eq. (3.1.9) is the same as changing the sign ξ . This means that modulation plays the same role as angle θ . Therefore modulation introduces asymmetry and on the basis of the remark above, this is the same as to lower the threshold of chaos.

3.2 . Analog simulation

We performed an experiment on the Josephson-junction model using an analog technique based on a phase-locked loop device [5]. Experiments of this kind have been carried out by several authors [6-11]. In our simulation we introduced certain improvements using a "minima component technique" [13].An outline of this device is shown in fig .3. A Miller integrator was connected to a Voltage Oscillator Control (VCO) with multiplier M1 and a low-pass filter in a reaction loop. A second generator was applied at the other input of M1. To obtain the forcing term the output of a generator with frequency ω was sent to the input of the integrator . The modulation term was obtained by means of multiplier M2 introduced in the loop feed-back. A part of the output voltage of the generator of frequency ω was connected to the multiplier in such a way as to obtain a fixed phase of the modulation term. That assured the theoretical condition in which a fixed relation of the phases between forcing and modulation is required.

The sum of the currents in point A at the input of the integrator gives

$$\ddot{\phi} = -\frac{kV_5}{R_5C}\sin\phi - \frac{k\,V_\gamma}{CR_\gamma} - \frac{1}{R_1C}\dot{\phi} \qquad (3.2.1)$$

where k is the VCO modulation coefficient, so that

$$\dot{\phi} = k\ V_1\ . \qquad (3.2.2)$$

Putting switch S in position 2 , the second multiplier was introduced and the modulation term was added at the input of the integrator

$$V_5 = V_0\ [\ 1 + \frac{V_m}{V_0}\cos(\ \omega\, t\ )\ ]\ \sin\phi\ . \qquad (3.2.3)$$

This relation and the previous ones , eq. (3.2.1) ,with a suitable scaling, are equivalent to eq.(3.1.7).

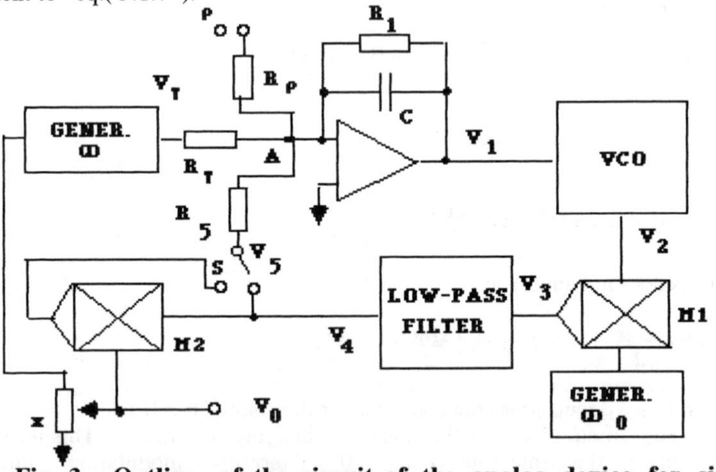

Fig. 3 - Outline of the circuit of the analog device for simulating the Josephson-junction model

3.3 . Experimental procedure

We examined the dynamics by observing the phase-space portraits on the oscilloscope screen . We fixed the amplitude and the frequency of the parametric modulation and the amplitude of the forcing was increased . This was performed for different values of frequency forcing. Because of the presence of hysteresis we recorded only the data for increasing driving voltage. The experimental errors were evaluated by means of a rough standard deviation for a set of 10 data . Because of the presence of both stochastic and deterministic noise, the statistics where not expected to be strictly Gaussian

3.4 . Results of the analog simulation

A comparison of the Melnikov predictions and the experimental results is shown in fig.4. Because of the extreme sensibility of the system in the case in which $\rho_0 = 0$ we compared our data (+) with the previous results obtained by D' Humieres *et al.*[7] (a). In the case in which ρ_0=.1 the experimental results and the Melnikov predictions are in agreement with the values of the drive frequency in a narrow region between 0,6 and 1. However this confirmed that , in this case, the effect of modulation is to lower the threshold of chaos . This is satisfactorily proved by fig. 5 in which the threshold of chaos γ is plotted against the amplitude modulation ξ . The same happened for an increasing ρ_0 and without modulation as indicated in fig. 6 .
If one observes that $\rho_0 \neq 0$ means introducing an asymmetry into the system , we find that this asymmetry causes a lowering of the chaos threshold. So that the asymmetry is also obtained by means of parametric modulation, as indicated in fig. 5 . The disagreement between the experimental data and the prediction data depends on the approximate nature of the Melnikov method .

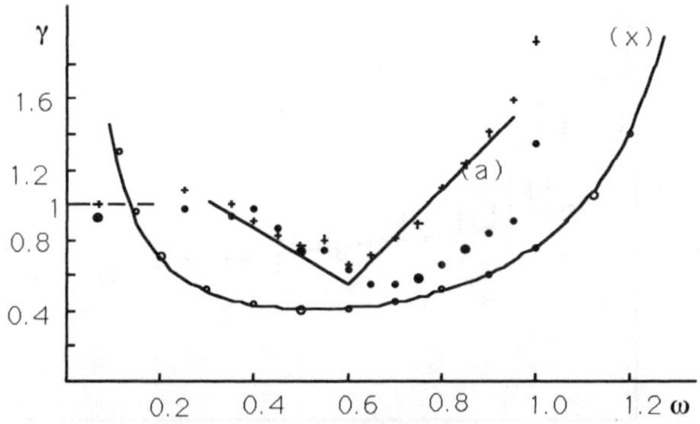

Fig. 4 - Threshold values of amplitude γ for the regular regions as a function of the drive frequency. The small crosses and the solid circles show the experimental values, respectively, when ρ_0 = 0 and ρ_0 = 0.1 .The (x) line means the theoretical values deduced by means of Melnikov's method when ρ_0 = 0.1. Line (a) is the rough boundary line deduced from ref. 7.

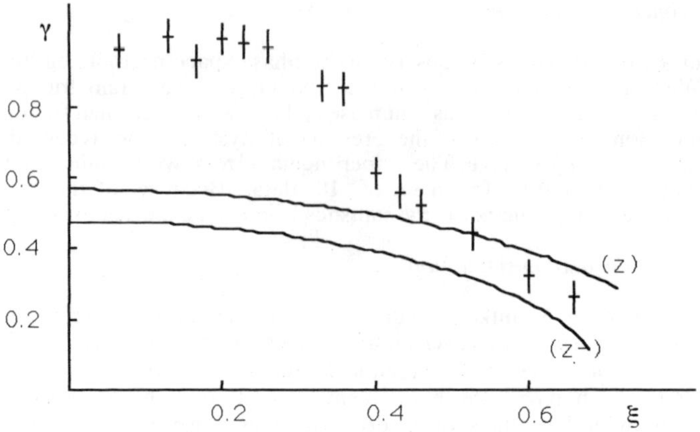

Fig. 5 - Experimental results of the chaos threshold as function of amplitude modulation ξ with $\rho_0=0$ and $\theta=0$.Continuous lines (z) and (z $^-$) refer to the theoretical predictions, respectively, where $\rho_0 = 0$ and $\rho_0 = 0.05$.

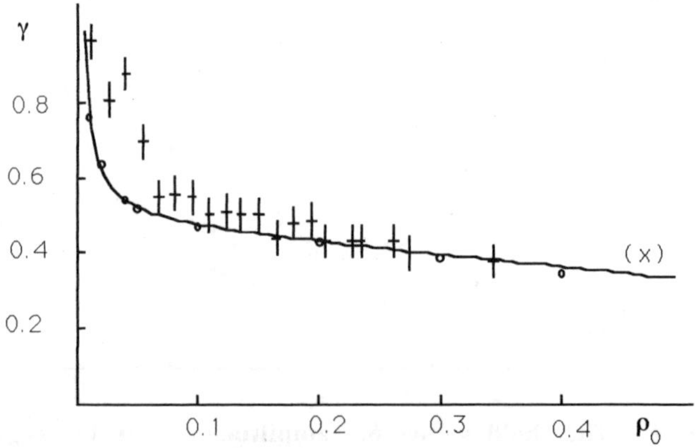

Fig. 6 - Analog results of the threshold of chaos versus dc bias term ρ_0. The curve (x) shows the theoretical predictions.

4. Conclusion.

We have shown that parametric modulation can suppress or induce chaos in two real dynamical systems with forcing. To obtain this it is very important to create special conditions concerning the frequency and the phase between parametric modulation and forcing. In the case of the Josephson -Junction model we have shown that modulation is equivalent to a break in symmetry and therefore to lowering the chaos threshold. In the bistable system we have found that chaos is not structurally stable and that it can be removed by means of a parametric perturbation .

5. References

1. O.H.Wehrmann. Influence of Vibrations on the Flow Field behind a Cylinder.*The phys. of fluids supplement* , S187 (1967)
2. R.Lima and M. Pettini, Suppression of chaos by resonant parametric perturbation. *Phys. Rev.* **A 41,** 2,726 (1990)
3. J.Guckenheimer and P.J. Holmes, *Nonlinear Oscillations, Dynamical System and Bifurcation of vector fields.*(Spring-Velarg, New York, 1983)
4. V.K. Melnikov, Trans. *Moscow Math. Soc.* **12,** 1 (1963).
5. G.Cicogna and L.Fronzoni, Effects of parametric perturbations on the onset of chaos in the josephson-junction model: Theory and analog experiments. *Phys. Rev.* **A 42,** 4,1901 (1990)
6. A. Huberman, j.P. Cruchfieldand N.Packard, *Phys. Lett.* **37**, 750 (1980)
N.F.Peterson and A. Davidson, *Appl. Phys.Letter* **39**,830 (1981)
7. D.D'Humieres, M.R. Beasley, B.A. Hubermn and A.Libchaber, *Phys.Rev* **A 26**, 3483 (1982)
8. M.Cirillo and N.F.Pedersen, *Phys.Letter* **A 90**,150 (1982)
Z.D.Genchev, Z.G. Ivanov and B.n. Todorov, *IEEE Trans. Circuits System* **CAS-30**, 633 (1983)
9. C. Vanneste, C.C Chi and D.C. Cronemeyer, *Phys. Rev.* **B 32**, 4796 (1985).
10. R.l. Kautz and J.C. Macfarlane, *Phys. Rev.* **A 33** , 498 (1986).
11. Bartuccelli, P.L. Christiansen, N.F Pedersen and M.P. Soerensen , *Phys. Rev.* **B 33,** 4686 (1986)
12. F.C. Moon, *Phys.Rev Letter* **53**, 962 (1984)
13. L.Fronzoni, in *Noise in Nonlinear Dynamical System* edited by F.Moss and P.V.E. McClintock (Cambridge University Press, Cambridge, 1989), vol 3,p.222.

CHAOTIC OSCILLATIONS OF A SPHERICAL PENDULUM AS THE EFFECT OF INTERACTION WITH THE EXCITATION DEVICE

Tatyana S. Krasnopolskaya
Institute of Mechanics, Academy of Sciences of the Ukraine, Nesterov str. 3, Kiev, Ukraine 252057

and

Alexander Yu. Shvets
Institute of Automatic, Nagornaya str. 22, Kiev, Ukraine 254107

New features in the parametric oscillations of a spherical pendulum due to the interaction with an excitation device with limited power supply are studied in detail. This interaction results from the energy conservation law and is observed in all physical systems. From the analysis of largest Lyapunov exponent for the composite system pendulum - excitation device,we identify the regions for three steady-state regimes: stationary, periodic and chaotic. Particular attention is given to the properties of chaotic attractors and to the schemes of transition from "order" to chaos. It is shown that the cause of chaos is the interaction with the excitation device.

1. Introduction

A detailed investigation of the main properties of resonance vibrations of mechanical systems with periodic coordinates (cylindrical and spherical shells, rings, circular plates, media in spherical and cylindrical tanks) may be carried out using the model of spherical pendulum, which is the simplest example of an oscillator with two degrees of freedom and equal frequencies. In this paper we present results of an investigation of chaotic steady-state regimes discovered in the averaged equations of vibration of a pendulum parametrically excited at the suspension point. The cause of the appearance of chaotic vibrations is the interaction with an excitation device with limited power supply. It is well known that because of the energy conservation law the vibrating system always interacts with the device that generates the excitation force. In case of a vibrating system with damping (all real system are of this type), the consumption of energy generated by the excitation may change in an essential way the dynamical behaviour of the system. Consequently, the parameters that define the action of the excitation on the vibrating system are changed. For this reason the system dynamics should be described not only by the equation of its free vibrations but also by an additional equation characterizing the changes in the excitation that takes into account the feedback effect. This must be the formulation of problems in the theory of vibrating systems with limited power supply (limited excitation)[1] for which energy losses spent on excitation of the vibrating system "are felt". Onset of chaos by interaction of the pendulum with an excitation device was observed in the averaged equations, which are a good qualitive approximation of the behaviour of the original systems[2]. In the case of ideal (nonlimited) excitation the averaged equations for the parametric oscillations of the pendulum display only regular steady-state regimes[3,4]. The change of behaviour for the averaged equations shows the importance of taking into account the interaction of the vibrating system with the excitation. This is particulary important in the experimental study and identification of dynamical systems. Chaos in the dynamics of many systems is generated only by limited excitation, not by its own properties.

We assume that the excitation device is an electromotor, the shaft of which is connected by a crank-connecting rod mechanism to the pendulum support point. For parametric excitation through vertical vibrations of the pendulum support point

$$v(t) = -a\cos\omega(t), \tag{1.1}$$

$$\varepsilon \equiv \frac{a\omega_0^2}{g} << 1 \tag{1.2}$$

(where ω_0 is the natural frequency of the pendulum) the Lagrangian L has been constructed in Ref. 2. Assuming that $(\dot{\omega}/2)-\omega_0$ is $O(\varepsilon\omega_0)$ one then obtains the evolution equations for the amplitude of the rejection angles and the tuning frequencies $\dot{\omega}/2$ and ω_0 (See Ref. 3).

In Ref. 3 the analysis of the Lyapunov exponents for the composite system "vibrating spherical pendulum - electromotor" is used to determine the regions in parameter space corresponding to three classes of steady-state regimes. The first class corresponds to the fixed points of the evolution equations, the second class to periodical regimes and the third class corresponds to chaotic regimes.

2. The Lagrangian

Assume the excitation device to be an electromotor the shaft of which is connected by a crank-connecting rod mechanism to the spherical pendulum suspension point. In the following we consider the rotation of the electromotor shaft to be described by the angle $\omega(t)$. When the crank turns by the angle $\omega(t)$ the pendulum support point is displaced in the vertical direction by an amount $v(t)=-a\cos\omega(t)$ in Cartesian coordinates 0xyz connected to the support point. The kinetic and potential energies of the composite system (electromotor-pendulum) are:

$$T=\frac{1}{2}H\dot{\omega}^2+\frac{1}{2}m\left[\dot{x}^2+\dot{y}^2+(\dot{z}+\dot{v})^2\right]; \quad V=mg\,(l-z-v). \tag{2.1}$$

H is the moment of inertia of the electromotor shaft[1]; x, y, z - the coordinates of the bob; m the mass of the pendulum; l its length; and g the free acceleration.

We introduce the generalized rejection angles α and γ of the bob. Then

$$x=l\sin\alpha;\ y=l\sin\gamma;\ z=\sqrt{l^2-x^2-y^2}=l\sqrt{1-\sin^2\alpha-\sin^2\gamma}$$

Using Taylor's expansion for small α and γ $\left(\alpha \text{ and } \gamma \text{ are } 0\left(\sqrt{\varepsilon}\right)\right)$ we obtain for the Lagrangian[5,6,7]

$$L=T-V=\frac{1}{2}H\dot{\omega}^2+\frac{1}{2}ml^2\left[\dot{\alpha}^2+\dot{\gamma}^2+2\alpha\gamma\,\dot{\alpha}\dot{\gamma}-2\,(\alpha\dot{\alpha}+\gamma\dot{\gamma})\,\dot{\omega}\frac{a}{l}\sin\omega+\dot{\omega}^2\,\frac{a^2}{l^2}\sin\omega\right]-$$

$$-\,mgl\left(\frac{\alpha^2}{2}-\frac{\alpha^4}{24}+\frac{\gamma^2}{2}-\frac{\gamma^4}{24}+\frac{\alpha^2\gamma^2}{4}+\frac{a}{l}\cos\omega\right) \tag{2.2}$$

From Eq. (2.2) we may construct the Lagrange equations for the angles $\omega(t)$, $\alpha(t)$, and $\gamma(t)$

$$H\ddot{\omega}=H_1\left(\dot{\omega}\right)-H_2\left(\dot{\omega}\right)-mla\left[\ddot{\omega}\frac{a}{l}\sin^2\omega+\dot{\omega}^2\frac{a}{l}\sin\omega\cos\omega+\frac{g}{l}\sin\omega-\left(\dot{\alpha}^2+\dot{\gamma}^2\right)\sin\omega\right.$$

$$\left.-\left(\alpha\ddot{\alpha}+\gamma\ddot{\gamma}\right)\sin\omega\right]\ \ddot{\alpha}+\omega_0^2\left(\alpha-\frac{\alpha^3}{6}+\frac{\alpha\gamma^2}{2}\right)+\delta_0\,\dot{\alpha}+\alpha\left(\dot{\gamma}^2+\gamma\ddot{\gamma}\right)$$

$$-\frac{a}{l}\alpha\left(\dot{\omega}^2\cos\omega+\ddot{\omega}\sin\omega\right)=0$$

$$\ddot{\gamma} + \omega_0^2\left(\gamma - \frac{\gamma^3}{6} + \frac{\alpha^2\gamma}{2}\right) + \delta_0\,\dot{\gamma} + \gamma\left(\dot{\alpha}^2 + \alpha\ddot{\alpha}\right) - \frac{a}{l}\gamma\left(\dot{\omega}^2\cos\omega + \ddot{\omega}\sin\omega\right) = 0 \tag{2.3}$$

Here $H_1(\dot{\omega})$ is the moving moment of the electromotor, $H_2(\omega)$ the moment of the resistive forces of the electromotor[1]; $\omega_0^2 = g/l$; and δ_0 the damping coefficient of the resistive forces in the medium, where the pendulum moves.

The system that is obtained describes the process of interaction of the oscillations of the pendulum with the rotation of the electromotor shaft. It is essentially nonlinear and analytical solutions are not known. For simplicity we introduce the small positive parameter ε of Eq. (1.2). Moreover, we assume that the speed of the electromotor shaft $\dot{\omega}$ in the stationary regime is near the value $2\omega_0$. In this case the detuning of the frequencies $\dot{\omega}/2$ and ω_0 is a small value, proportional to ε,

$$\frac{\dot{\omega}(t)}{2} - \omega_0 = \frac{\varepsilon\omega_0}{2}\nu(\tau) \tag{2.4}$$

The resonant oscillations of the pendulum have the following correlations:

$$\alpha(t) = \sqrt{\varepsilon}\left[p_1(\tau)\cos\frac{\omega(t)}{2} + q_1(\tau)\sin\frac{\omega(t)}{2}\right] \tag{2.5}$$

$$\gamma(t) = \sqrt{\varepsilon}\left[p_2(\tau)\cos\frac{\omega(t)}{2} + q_2(\tau)\sin\frac{\omega(t)}{2}\right]$$

Where τ is a slow time

$$\tau = \frac{\varepsilon}{4}\omega(t)$$

Therefore we go over to the new coordinates $\nu(\tau)$, $p_1(\tau)$, $q_1(\tau)$, $p_2(\tau)$, $q_2(\tau)$.

3. Evolution equation

We use the averaging method which sometimes provides analytical solutions, in particular for stationary regimes. It is a simplification of the original system however, without using the averaging method, it is difficult to find the main features of the interaction process.

Upon introducing Eq. (2.5) into Eq. (2.3) and taking into account the correlations

$$\dot{\alpha} = \frac{\dot{\omega}}{2}\sqrt{\varepsilon}\left[-p_1(\tau)\sin\frac{\omega}{2} + q_1(\tau)\cos\frac{\omega}{2}\right] \text{ and } \quad \dot{\gamma} = \frac{\dot{\omega}}{2}\sqrt{\varepsilon}\left[-p_2(\tau)\sin\frac{\omega}{2} + q_2(\tau)\cos\frac{\omega}{2}\right]$$

Averaging over the fast time $\omega(t)$, we obtain [7]

$$\frac{d\nu}{d\tau} = N_2 - N_1\nu - \mu\left(p_1q_1 + p_2q_2\right)$$

$$\frac{dp_1}{d\tau} = -\delta p_1 - \left(\nu + \frac{E}{8}\right)q_1 - \frac{3}{4}Mp_2 + 2q_1$$

$$\frac{dq_1}{d\tau} = -\delta q_1 + \left(\nu + \frac{E}{8}\right)p_1 - \frac{3}{4}Mq_2 + 2p_1$$

$$\frac{dp_2}{d\tau} = -\delta p_2 - \left(\nu + \frac{E}{8}\right)q_2 + \frac{3}{4}Mp_1 + 2q_2$$

$$\frac{dq_2}{d\tau} = -\delta q_2 + \left(\nu + \frac{E}{8}\right)p_2 + \frac{3}{4}Mq_1 + 2p_2 \qquad (3.1)$$

Here the linear approximation of the static characteristic[1] of the electromotor $[H_1(\dot{\omega}) - H_2(\dot{\omega})]\,[H + (0.5)ma^2]^{-1}$ is used

$$\frac{H_1(\dot{\omega}) - H_2(\dot{\omega})}{H + (0.5)\,ma^2} = \frac{\varepsilon\omega_0}{2}\left(N_0 - N_1\dot{\omega}\right) \qquad (3.2)$$

Then,

$$N_2 = \frac{1}{a}\left(\frac{N_0}{\omega_0} - 2N_1\right) \text{ and } \mu = \frac{2ml}{H + (0.5)\,ma^2}; \quad \delta = \frac{\delta_0}{\omega_0}$$

$$E = p_1^2 + q_1^2 + p_2^2 + q_2^2; \qquad M = p_1q_2 - p_2q_1$$

The first equation in (2.3) is transformed into an equation for $\nu(\tau)$ in (3.1). The interaction of the pendulum vibrations and the rotation of the electromotor shaft is described by the system (3.1) of evolution equations.

4. Asymptotic trajectories

The asymptotic trajectories of the system of equations (3.1) is constructed using numerical methods. In this paper we use a fourth order Runge-Kutta method with the correction procedure of Dormand-Prince. The accuracy of the solutions is of $0(10^{-8})$. During the construction of the asymptotic trajectories care should be taken that the transient behaviour is completed. The construction of Poincaré sections is carried out using Hénon method[8]. For the case of chaotic vibrations the number of the points in the sections is more than 10^4.

One of the main criteria for the existence of chaotic regimes in a dynamical system is the presence of positive Lyapunov exponents characterizing the divergence of trajectories in phase space. The five-dimensional dissipative system (3.1) has five Lyapunov exponents, the sum of which equals the divergence of the system and is a negative value. When at least one of the Lyapunov exponents is greater than zero, chaos is observed in the system. In this case we have global contraction (negative divergence) and local instability (running away of trajectories along one direction). For the construction of the Lyapunov exponents the algorithm of Bennettin et al. was used[9].

The system of equations (3.1) contains the parameters (δ, N_1, N_2, μ) which together with the initial conditions determine the behaviour in the steady-state regime. We assume

$$\delta = 0.5;\ N_2 = 0.5;\ \mu = 1;\ p_1(0) = q_1(0) = 0.1;\ \nu(0) = 0;\ \ p_2(0) = q_2(0) = 1 \quad (4.1)$$

The parameter N_1 was varied with the purpose of determining all possible asymptotic trajectory classes. The magnitude of N_1, called the "steepness parameter of the static characteristic of the electromotor", is equal to the tangent of the angle of incidence of the characteristic and determines the value of the energy losses in the electromotor. In the interval $1.75 \geq N_1 \geq 1.43$ the system has a stable limit cycle of period $T \cong 4.4$. At $N_1 = 1.42$ this cycle looses its stability and a stable double limit cycle of period $T \cong 8.8$ appears. Then at $N_1 = 1.295$ and $N_1 = 1.28$ new bifurcations occur and quadruple and octuple limit cycles appear with periods $T \cong 17.7$ and $T \cong 35.3$. The transformation through period doubling of the trajectory projections is illustrated in Figs.1a - d.

The value $N_1 = 1.27$ is a critical value. At this value of the parameter N_1 a qualitative transformation occurs in the system and a chaotic attractor appears. In Fig.2 we show the projections of the asymptotic trajectories on the planes $p_2-\nu$ (a), $q_2-\nu$ (b), p_2-q_2 (c) as well as the projection on the plane p_2-q_2 (d) of the Poincaré section by the plane $\nu = -11$ at $N_1=1.25$. The largest Lyapunov exponent is 0.27 in this case.

For a further decrease of N_1 a structural reconstruction of the type "chaos-chaos" is observed. Due to inner bifurcations a chaotic attractor of one type changes into a chaotic attractor of another type. At $N_1=1.16$ there is a hard transition from the chaotic attractor shown in Fig.2a-d to a chaotic attractor of another type. In Fig.3 we show the projections

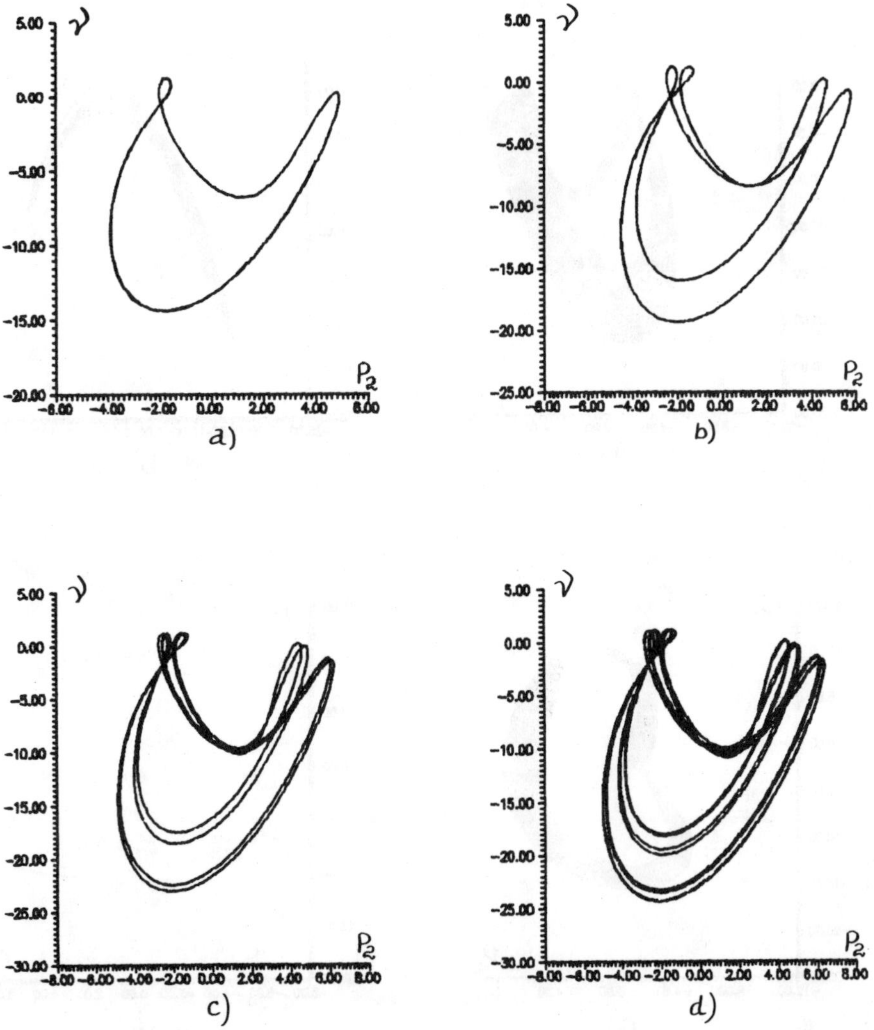

Fig.1 Projection of the trajectories in the plane ν-p_2, corresponding to period-doubling bifurcations of the limit cycles.

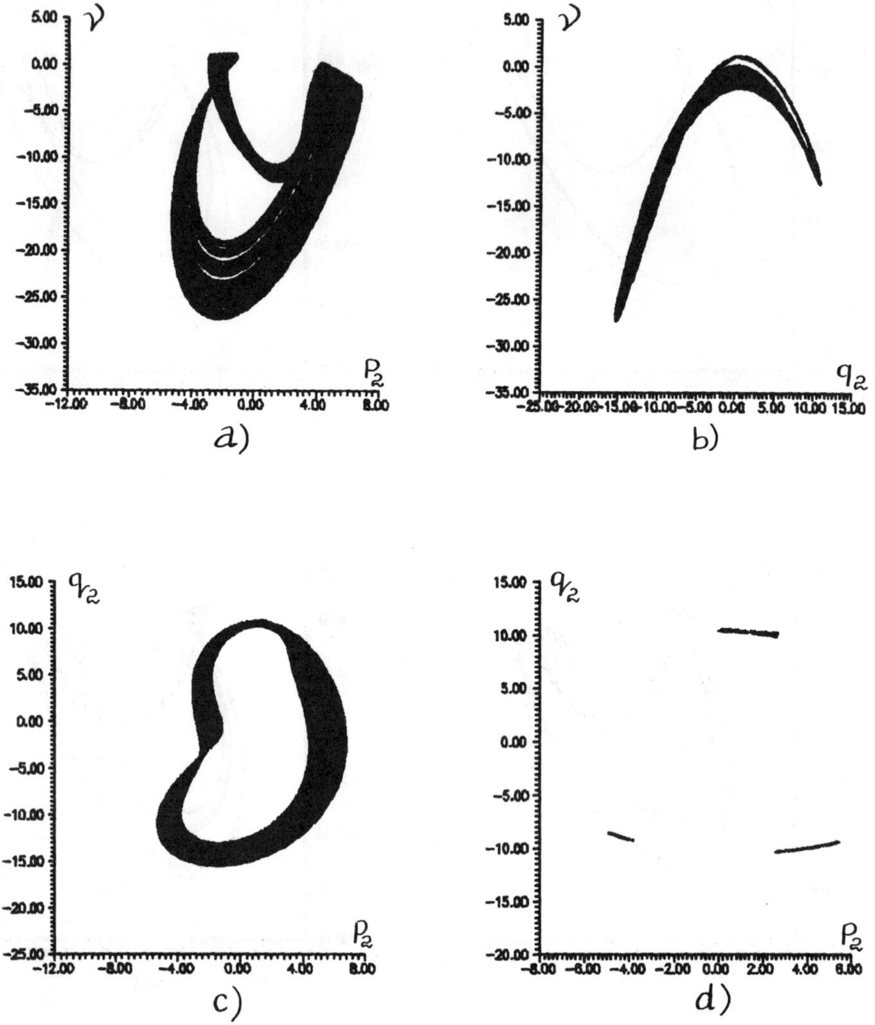

Fig.2 Projection of the trajectories in the plane ν-p_2 (a), q_2-ν (b), p_2-q_2 (c) and p_2-q_2 (d) projection of the Poincaré section by the plane ν=-11 at N_1 = 1.25

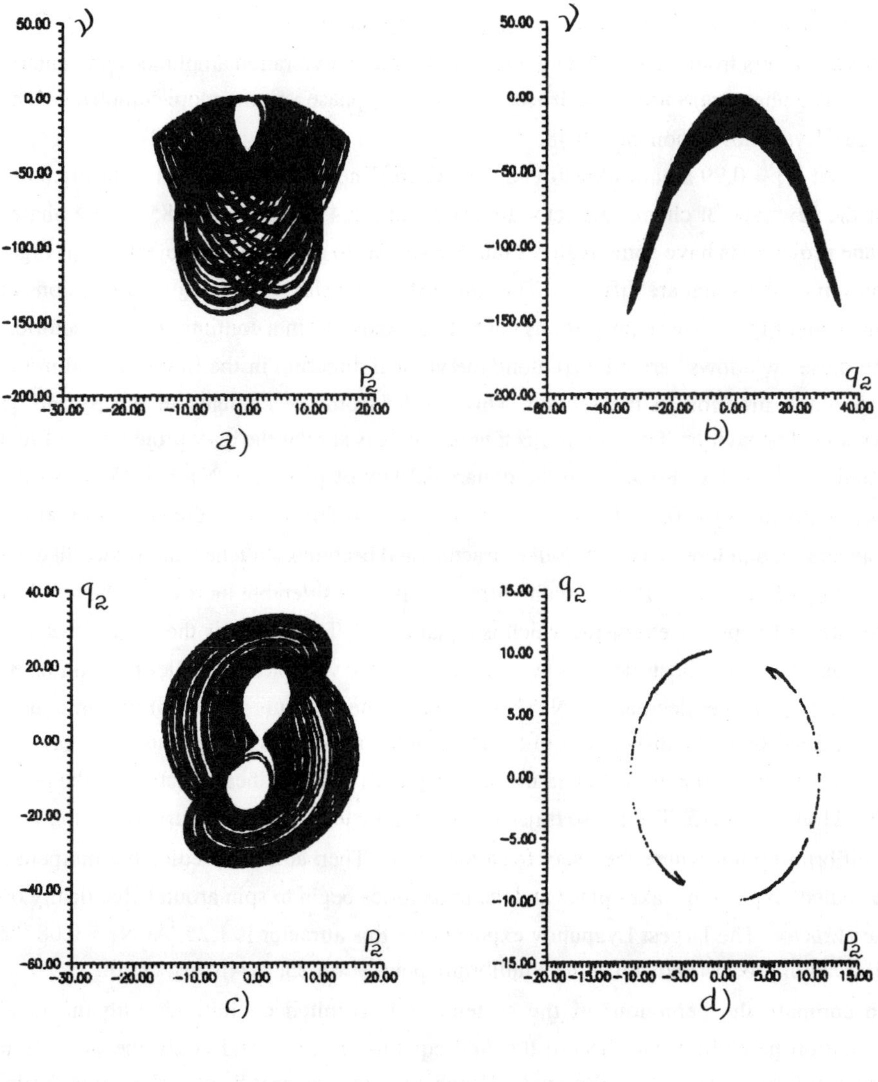

Fig.3 Projection of the trajectories in the plane ν-p_2 (a), q_2-ν (b), p_2-q_2 (c) and p_2-q_2 (d) projection of the Poincaré section by the plane ν=-11 at N_1 = 1.05

of the asymptotic trajectories on the planes p_2-ν (a), q_2-ν (b), p_2-q_2 (c) and the projection on the plane p_2-q_2 (d) of the Poincaré section by the plane ν= -11 at N_1 = 1.05. This attractor differs from the preeceding one by having larger vibration amplitudes particularly for ν. The phase trajectories are denser and cover the phase volume more completely. The largest Lyapunov exponent is 0.48.

At N_1 = 0.99 a hard transition again occurs. The same projections as in Fig.3 a-d for the new type of chaotic attractor are shown in Fig.4 a-d, at N_1 = 0.85. These phase-plane projections have some features that are similar to those of the projections in Fig.3 and some others that are different. The similarity is the shape of the p_2-q_2 projections in Fig.3c and Fig.4c. These projections have "two windows" in a continuum of trajectories. But these "windows" are aligned along the vertical direction in the first case and in the horizontal direction in the second. This results from the changes in the p_2 and q_2 dynamical behaviour. The p_2-ν projection in Fig.3a is similar the q_2-ν projection in Fig.4 b and Fig.3b is like Fig.4 a. So the dynamical law of p_2 (q_2) at N_1 = 1.05 is like the q_2(p_2) law at N_1 = 0.85. Chaos in the last case also differs from the other in that the Poincaré section looses its ribbon-like structure and becomes stretched and folded like the kneading of the dough. But the main difference is a considerable increase in the value of the largest Lyapunov exponent which is equal to 7.7. This refflects the higher speed of divergence of the trajectories, but the amplitudes of the variables is smaller than in Fig. 3.

For a further decrease of N_1 chaos acquires the structure typical of intermittency. In Fig.5 we show the projections of the asymptotic trajectories in the planes p_2-ν(a), q_2-ν(b), p_2-q_2(c) and the projection in the plane p_2-q_2(d) of the Poincaré section by the plane ν = -11 at N_1 = 0.15. The phase trajectories are attracted to a neighbourhood of the zero equilibrium point where they stay for a long time. Then at an unpredictable moment a "turbulent explosion" takes place and the trajectories begin to spin around chaotically on the attractor. The largest Lyapunov exponent for this attractor is 1.25. At N_1 = 0.08 the bifurcation process ends up at the equilibrium point ν = 6.25; p_1=q_1=p_2=q_2=0.

To compare the behaviour of the system. (3.1) (limited excitation) with the ideal excitation pendulum, we discard the first equation in (3.1) and study the other four equations for constant ν. We studied such a system at δ=0.5 with the same initial conditions as in Eq. (4.1). We varied the value of the parameter ν in the interval $-150 \leq \nu \leq 2$, which previously covered chaotic regimes. In all cases a stable equilibrium position is observed. At $\nu \leq -3$ this is ν=const, p_1=q_1=p_2=q_2=0. In this case, as in Ref. 3, we have no periodic or chaotic regimes in the dynamics of the pendulum. This shows the

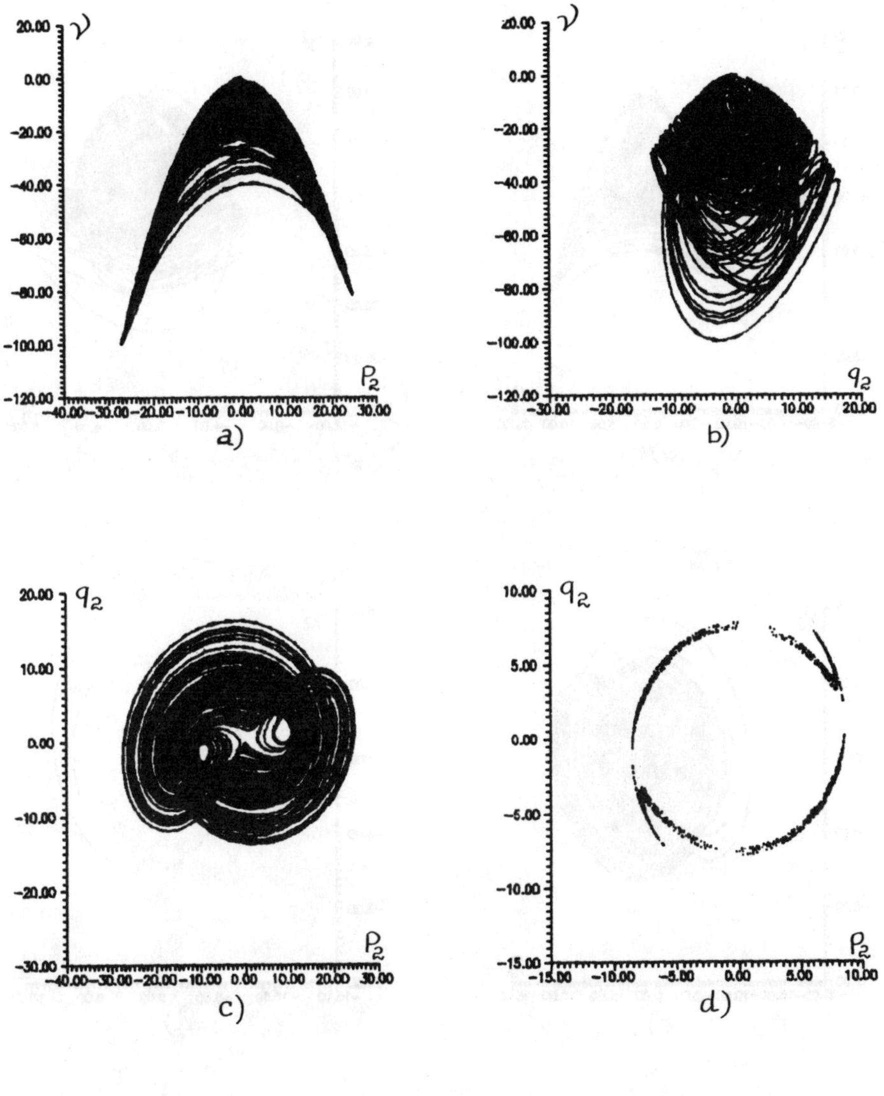

Fig.4 Projection of the trajectories in the plane ν-p_2 (a), q_2-ν (b), p_2-q_2 (c) and p_2-q_2 (d) projection of the Poincaré section by the plane ν=-11 at $N_1 = 0.85$

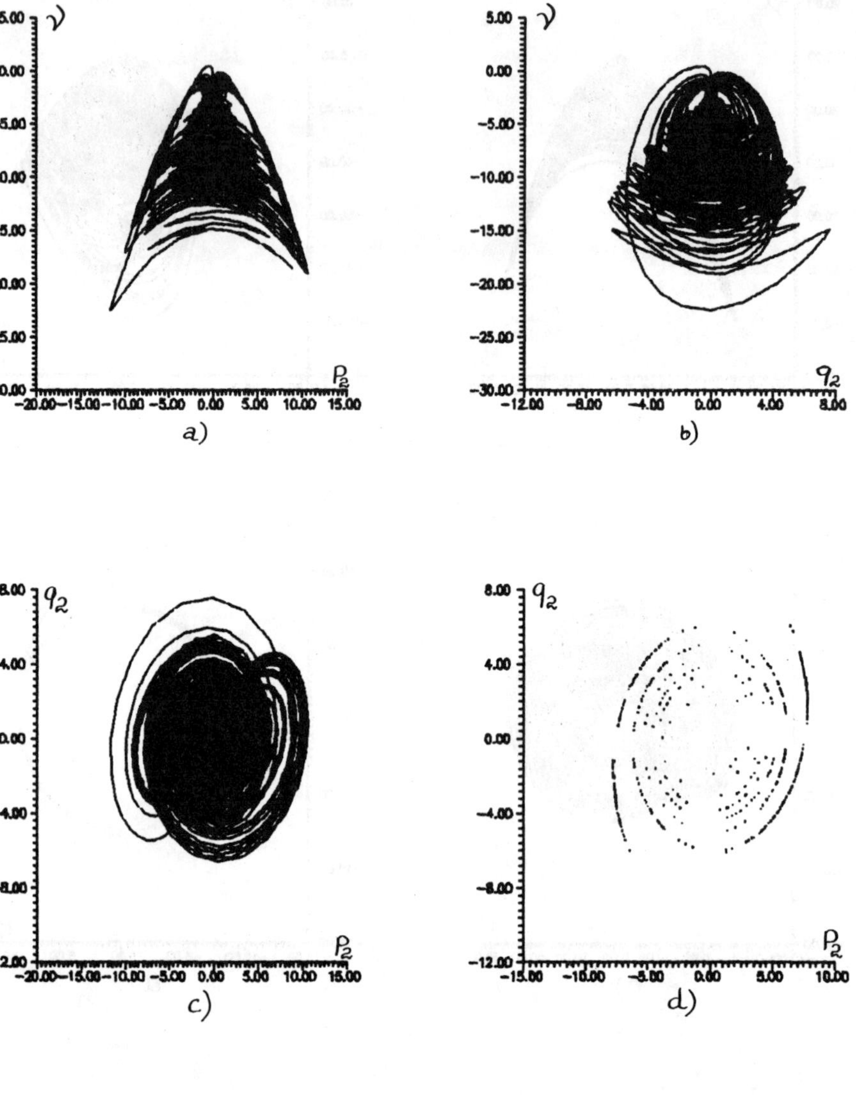

Fig.5 Projection of the trajectories in the plane ν-p_2 (a), q_2-ν (b), p_2-q_2 (c) and p_2-q_2 (d) projection of the Poincaré section by the plane ν=-11 at $N_1 = 0.15$

importance of correctly taking into account the interaction of the pendulum with the excitation.

A specific feature of the steady-state parametric oscillation of a spherical pendulum both ideal and with limited excitation is the similar character of the amplitudes in the x direction $p_1(\tau)$, $q_1(\tau)$ and in the y direction $p_2(\tau)$, $q_2(\tau)$. As is obtained in the numerical steady-state regimes the correlation $p_1/p_2 = q_1/q_2 = \text{const}$ is realized. Because we have no angular moment in the vertical displacement of the pendulum suspension point, we may consider that the angular moment of pendulum oscillations M is zero[4]. So it suffices to study projections on the plane p_2-q_2.

5. Conclusions

From the study of the nonlinear interactions in the system "spherical pendulum-electromotor with limited power-supply" we obtain the following main results:

It is established that the chaos originate only as a result of the interaction of the pendulum oscillation with the electric drive. Chaos cannot be obtained from the averaged equations for parametric oscillations of a pendulum with ideal (unlimited) excitation.

We establish the existence of several types of chaotic attractors. The transition from regular to chaotic motion occurs by different scenarios: cascade of period doubling bifurcations, intermittency and hard transitions.

It is shown that an essential feature of the parametric oscillations lies in the changes of behaviour of the dynamical laws for p_2 and q_2 (or p_1 and q_1). It depends on the initial conditions and the history of the process.

6. References

1. V. O Kononenko, Vibrating System with a Limited Power-supply, (London, 1969) p. 256.
2. A. K. Bajaj and J. Johnson, Int. J. Non-Linear Mech. 25 (1990) 211.
3. J. W. Miles, Z. Angew. Math. Phys. 36 (1985) 337.
4. J. W. Miles, J. Fluid Mech. 146 (1984) 285.
5. J. W. Miles, Quart. Appl. Math. 20 (1962) 21.
6. J. W. Miles, Physica 11D (1984) 309.
7. T. S. Krasnopolskaya and A. Yu. Shvets, Soviet Appl. Mech. 26 (1990) 500.
8. M. Hénon, Physica 5D (1982) 412.
9. G. Bennettin, L. Galgani and I. M. Strelcyn, Phys. Rev. A41 (1976) 2338.

MOTION OF A SATELLITE IN AN ELLIPTIC ORBIT

Paolo Santini
Remo Benedetti
Paolo Gasbarri

Dipartimento Aerospaziale
Universita' di Roma 'La Sapienza'
Via Eudossiana 18 - 00184 Rome, Italy

ABSTRACT

The general equations of a body orbiting around the Earth are written, where terms of two different orders of magnitude appear. The equations are filtered in such a way that the two orders of a magnitude are separated.Special attention is devoted to the equatorial orbit, in order to illustrate phenomena associated with orbit ellipticity. The apparent amplification due to the orbital acceleration is shown, and the reasons for it are discussed. For elastic bodies at a very low stiffness (a case that is not likely to occur in practice) instability associated with elastic modes may appear.

1. Introduction

The equilibrium of a spacecraft orbiting around the Earth is governed by the interaction of the following forces:

(i) inertia forces;
(ii) Earth gravitational forces;
(iii) perturbing forces (air drag; solar pressure; celestial bodies);
(iv) elastic forces, present, of course, in flexible spacecrafts only.

As far as Earth gravitational forces are concerned, one must distinguish between 'large' or 'mean' gravity, which is the one corresponding to gravity acceleration acting at a specific point C of the spacecraft (e.g., the center of mass) and the little variations associated with gravity gradient. The latter has very little effect on the orbit of the center of mass (which is chiefly ruled by the former) but it is practically the only responsible for attitude motion, via gravitational torque and modal forces which, in turn,depend on the distance of C from the Earth center. For a circular orbit, the attitude motion is very similar to that of a pendulum of constant length at high amplitudes, whereas, for elliptic orbits, the system is nonautonomous, since the length of the pendulum is a known function of time. Furthermore, in elliptic orbits, the angular acceleration on the orbit (which is missing in circular conditions) acts as an input to the spacecraft, so that the motion can be very different from the simple circular case.

Stability of attitude and of modal amplitudes time history is also very important, since pointing accuracy is directly related to it. Gravity gradient plays a key role also in this context, as is well known, depending on the attitude of the spacecraft and on its flexibility. It behaves like a nonlinear elastic system coupled to the main one,and it can be either stabilizing or destabilizing (as it happens for other phenomena, like aeroelasticity and buckling), depending on the position and on the attitude of the spacecraft. In other words, around a given position, the original stiffness matrix must be combined with a gravity matrix, so that a new matrix is produced which is not necessarily a definite positive one. This is the reason why instability conditions may arise.

2. A Basic Formula

In considering the motion of a spacecraft around the Earth three reference systems are needed: (i) the inertial Earth centered system $T_0(X_1, X_2, X_3)$, (ii) the local intrinsic geographic system $T_1(\rho,\Lambda,\omega)$; the reference body coordinates system $T_2(x_1, x_2, x_3)$. We denote by E_{jk} the eulerian matrix allowing to pass from the unit vectors of T_j to the unit vectors of T_k. The basic property is:

$$\dot{\mathrm{E}}_{jk} = \{ \Omega_{jk} \} \, \mathrm{E}_{jk} \tag{1}$$

where dot denotes differentiation w.r. to time t.

Here we denote by the symbol $\{ V \}$ as applied to a vector $V(v_1,v_2,v_3)$ the matrix

$$\begin{vmatrix} 0 & v_3 & -v_2 \\ -v_3 & 0 & v_1 \\ v_2 & -v_1 & 0 \end{vmatrix} \tag{2}$$

In (1) Ω is the vector $(r_{jk,1}, r_{jk,2}, r_{jk,3})$, where $r_{jk,s}$ is the s-th component of the angular velocity of T_j with respect to T_k as viewed on the axes of T_j.

Let now x, X denote the position vectors of a generic point P in T_2 and in T_0 respectively. We have:

$$X = X_c + E_{20}\, x \tag{3}$$

where X_c is the position vector of the spacecraft center of mass. Hence we may obtain the acceleration of P as viewed in T_0 as:

$$\ddot{X} = \ddot{X}_c + E_{20}(\ddot{x} + 2\, R_{20}\dot{x} + [\dot{R}_{20} + R_{20}^2]\, x) \tag{4}$$

where we have put $R_{20} = \{\, \Omega_{20}\}$. The acceleration, as viewed in T_0, reads:

$$a = E_{02}\, \ddot{X}_c + (\, \ddot{x} + 2\, R_{20}\, x + [\dot{R}_{20} + R_{20}^2]\, x) \tag{5}$$

It should be noticed that quantities like X are of the order of $L \equiv R_0$, the Earth's equatorial radius, whereas x 's are of the order of l, a typical dimension of the satellite. We will retain, of course, $\lambda = L \,/\, l << 1$.

3. Forces Acting on an Element of the Spacecraft

Forces acting on the spacecraft, as already said in Introduction, may be reduced to

(i) gravitational forces $G(x)$ associated to the Earth's potential $U = U(\gamma)$, where γ are geographic coordinates, related to the components of X via the well known relationships:

$$\begin{cases} X_1 = \rho \cos \Lambda \cos \omega \\ X_2 = \rho \cos \Lambda \sin \omega \\ X_3 = \rho \sin \Lambda \end{cases} \tag{6}$$

(The explicit expressions for E_{21} are readily obtained, **[1]**: here ρ stands for radius vector, Λ for latitude, ω for longitude)

(ii) forces arising from perturbations other than higher harmonics of G (air drag; solar pressure; celestial bodies **[2]**)

(iii) other external (e.g.,control) forces.

In this paper we shall consider (ii) as included in (iii), and we shall also restrict (i), as we will see later. For $G(x)$ we write, neglecting quantities of order λ^2 :

$$G(x) = G_0 + G_1 x \tag{7}$$

where $G_0 = G(C)$. It should be noticed that forces appearing in (7) are viewed in the body axes, and that G_1 is independent of x. The expression of G_1 is given in Appendix I.

Now we want to write the expression of the "lost force" $d\mathscr{L}$ in the element of the body, of volume $d\mathbf{V}$ and of mass $d\mu = \mu^* d\mathbf{V}$, located at $\mathrm{P}(x)$ (μ^*denotes the mass per unit volume). For it we must also consider the reaction of the surrounding elements, given by $S(w)d\mathbf{V}$ where S is the structural operator transforming aeroelastic displacements w into internal forces. So we have:

$$d\mathscr{L} = \mathscr{S}(w)\, d\mathbf{V} + d\mu\, [\underset{\#}{G_0} + G_1\, x - \ddot{x} - 2\, \mathrm{R}_{20}\, \dot{x} - (\ddot{\mathrm{R}}_{20} + \mathrm{R}_{20}^2)\, x - \underset{\#}{\mathrm{E}_{02}}\, X_c + \underset{\#}{\mathrm{F}}] \tag{8}$$

where F is the force (iii) per unit of mass of the body.

We note that terms labeled # are of the order of L and they are therefore overwhelming with respect to the other terms.

Translational equation of equilibrium is obtained by stipulating that

$$\int_V d\mathscr{L} = 0; \; V = \text{body volume} \tag{9}$$

We neglect variations of the position of C within the body frame on account of elastic deformation, and we obtain:

$$\begin{aligned} X_c &= \mathrm{E}_{20}\, (G_0 + \bar{\mathrm{F}}) \\ \bar{\mathrm{F}} &= \int_V \mathrm{f}\, d\mu \Big/ \int_V d\mu \end{aligned} \tag{10}$$

Thus Eq.(8) becomes now:

$$d\mathscr{L} / d\mathrm{V} = S(w) + \mu^{*} [G_1 \; x - \ddot{x} - 2 \, \mathrm{R}_{20}\dot{x} - (\dot{\mathrm{R}}_{20} + \mathrm{R}_{20}^{2}) \, x - f]; \tag{11}$$

$$f = \mathrm{F} - \overline{\mathrm{F}}$$

where *L*-quantities have been filtered and all the remaining terms are of the same order of *l*. We further specialize x as the sum of ξ, the coordinates in the undeformed body, and of w, the elastic displacement, described through the normal modes of the free-free structure:

$$x = \xi + \sum_{1}^{\infty}{}_{n} A_n(t) \; U_n(\xi) \tag{12}$$

For a linear operator:

$$S(U_n) + \mu^{*} \sigma_n^2 \; U_n = 0$$

where the σ_n's are the natural frequencies of the spacecraft, so we have now

$$\begin{aligned} d\mathscr{L}/d\mu = \; & e + [G_1 \; \xi - (\dot{\mathrm{R}}_{20} + \mathrm{R}_{20}^{2}) \; \xi] \; + \\ & - \sum_{1}^{\infty}{}_{n} (\ddot{\mathrm{A}}_n + \omega_n^2 \, \mathrm{A}_n) \, \mathrm{U}_n - 2 \sum_{1}^{\infty}{}_{n} \dot{\mathrm{A}}_n \, \mathrm{R}_{20} \, \mathrm{U}_n \\ & + \sum_{1}^{\infty}{}_{n} \mathrm{A}_n \, [G_1 - (\dot{\mathrm{R}}_{20} + \mathrm{R}_{20}^{2})] \, \mathrm{U}_n \end{aligned} \tag{13}$$

In Eq.(13) we can clearly distinguish gravitational forces associated with rigid and elastic motion respectively, inertia terms, forces associated with elastic deformations.

4- Rotational Equilibrium

Rotational equilibrium is stipulated via the condition:

$$\int_{\mathrm{V}} \{ \; \xi \; \} \; d\mathscr{L} = 0 \tag{14}$$

Remembering Eqs. (12)(13) we obtain

$$\int_V \{\xi + \sum_1^{\infty} A_n U_n\} e \, d\mu + \int \{\xi\} [\underset{=}{G_1} \underset{=}{\xi} - (\dot{R}_{20} + R_{20}^2) \xi] \, d\mu +$$

$$\int_V \{ \sum_1^{\infty} A_n U_n \} [G_1 \xi - (\dot{R}_{20} + R_{20}^2) \xi] \, d\mu +$$

$$\int_V \{ \xi \} [\underset{\$}{G_1} \underset{\$}{\xi} - (\dot{R}_{20} + R_{20}^2) \xi] \sum_1^{\infty} A_n U_n(t) \, d\mu + \tag{15}$$

$$-2 \sum_1^{\infty} \dot{A}_n \int_V \{ \xi \} R_{20} U_n \, d\mu = 0$$

In Eq.(15) terms labeled = are those corresponding to the gravitational torque that would act if the body were rigid, and those labeled $$ are the gravitational torque associated with the elastic deformations. Other terms are self-explaining [1].

Note also that terms of the order of $A_n * A_m$ have not been considered, on account of the assumption of small elastic displacements, and the basic properties of free-free modes have been taken into account.

5 - Modal Amplitudes Equations

They are obtained through the conditions:

$$\int_V U_{nT} \, d\mathscr{L} = 0 \qquad ; \, n = 1,2,\ldots\infty$$

and this is the final result:

$$\int_V U_{nT} e \, d\mu + \sum_1^{\infty} \int_V U_{nT} [G_1 - (\dot{R}_{20} + R_{20}^2)] [\xi + \sum_1^{\infty} A_m U_m(t)] \, d\mu \tag{16}$$

$$-(\ddot{A}_n + \sigma_n^2 A_n) - 2 \sum_1^{\infty} \dot{A}_n \int_V U_{nT} R_{20} U_n \, d\mu = 0$$

Subscript T denotes the operation of transposing a matrix.

6 - The Equatorial Orbit

The foregoing equations, written as they are, are of little utility from a practical (i.e. computational) standpoint. On the other hand, explicit expressions in terms of the parameters of motion (coordinates of the center of mass; Euler's angles; modal amplitudes) are very complicated, although perfectly possible and quite elementary [1]. The lengthy exercise of giving the full operational equations will not be repeated here, since the aim of this paper is to illustrate the basic phenomena described in the Introduction. Therefore we shall confine ourselves to study equatorial orbits, for which the gravity potential is written now (Appendix I):

$$U = U(\rho,\omega) = -(\upsilon/\rho)\ [\ 1 - \sum_{1}^{\infty} r^{-n}\ Q_n(\omega)] \tag{17}$$

The expression of the functions Q_n appearing in Eq.(17) are given in Appendix I, and r is defined in subs.(18). The symbol υ of course denotes the gravitational constant $= g_0\ R_0^{\ 2}$, where g_0 is the gravity acceleration at the radius R_0.

(a) We define the following nondimensional quantities:

$$\tau = t\sqrt{\frac{g_0}{R_0}}\ ; \qquad r = \rho\ /\ R_0\ ; \quad (\ )' = d\ (\ )/\ d\tau \tag{18}$$

and we introduce the nondimensional angular momentum:

$$K = r^2\ \omega' \tag{19}$$

which is constant in keplerian conditions. So Eq.(9) yields now the fourth-order ordinary differential system

$$r'' - r\omega'^2 = -\ (1 + \gamma_{r0})\ /\ r^2 + e_r$$

$$K' = -\ \gamma_{\omega 0}\ /\ r^2 + e_\omega \tag{20}$$

$$\omega' = K\ /\ r^2$$

In Eqs.(20) the term 1 within brackets corresponds to the newtonian potential, whereas γ_{r0} and $\gamma_{\omega 0}$ provide the contribution of higher harmonics of the Earth's gravitational potential. Their lengthy expression is not reported here, for sake of brevity. Finally, e_r and e_ω denote the components of the (iii) forces on r- and ω- direction respectively.

(b) We consider the motion of a bar of length d (so in the foregoing equations r and ω are the geographic coordinates of its midpoint). As far as its attitude motion is concerned, we have only one parameter of motion, the angle θ of the undeformed position of the bar with the radial direction at C. The relevant (scalar) equation reads now :

$$I\ (\theta'' + \omega'') = C_T \ / \ r^3 \tag{21}$$

where I is the moment of inertia of the bar w.r. to its center of mass.

The expression of the gravitational torque coefficient C_T in terms of the parameters of motion is given, for the case under concern, in Appendix II.

7 - The Rigid Body

We will substantiate the investigation by showing some typical results. We shall first consider a rigid body, consisting of a bar of length d. For center of mass motion we have Eqs.(20) to be completed with the appropriate initial conditions. We have to prescribe initial height, whereas longitude is inessential in the present symmetric case; we assume injection at perigee, and apogee height is defined by means of a factor χ multiplying the local orbital velocity, so that $\chi = 1$ corresponds to a circular orbit, and $\chi = \sqrt{2}$ to a parabolic, i.e. escape, orbit.

As already said, if we neglect λ, attitude motion does not influence orbital motion. The converse is not true, however.

As a further simplification, to be removed later, we start considering newtonian potential, i.e. deleting higher harmonics. For the relevant motion, and in absence of other forces, we obtain the very simple equation:

$$\theta'' + \omega'' + 3 \sin\theta \cos\theta \ / \ r(\tau)^3 = 0 \tag{22}$$

For the circular case r = const., ω'= const, and it is found immediately, and well known, that the bar will perform undamped oscillations which, in the linear case, have a period of *(orbital period)*/ $\sqrt{3}$. Inspection of Eq.(22) reveals that there is an infinite number of equilibrium positions,those corresponding to even multiples of $\pi/2$ (stable) and those corresponding to odd multiples of $\pi/2$ (unstable).

Thus, for sufficiently small initial angles, the phase-plane plot (θ, θ') is approximately an ellipse of semiaxes $\left(\theta_0, \theta_0\sqrt{3}\right)$. Fig.1 provides the time history for $\chi = 1.02$ (corresponding to an eccentricity of $e = .04$) and Fig.2 describes for the same case the phase-plane plot. We see a strong amplification due to the orbital forcing terms as described above. The situation is even more evident, of course, for $\chi = 1.05$ shown in Figs.(3,4). It should be noticed that the above illustrations refer to initial conditions $\theta_0 = .01$, $\theta_0' = 0$, $r_0 = 1$. The nonsymmetry of the latter diagram is due to the fact that orbital period and frequency of rigid motion of the bar are in an irrational ratio. The nonsymmetry increases with the ellipticity of the orbit, Figs (5,6).

We introduce now higher harmonics, up to $n = 2$ only (App.I). The first coefficient, J_2, is of the order of .001, so one would expect an effect of the same order of magnitude. The differences are bigger, as can be seen from Figs.(7,8).The reason for this apparently strange behavior is in the fact that introduction of higher harmonics changes the orbital geometry, so that a circular orbit becomes, for the case under concern, an ellipse of small eccentricity which, as seen above, produces an amplification of the amplitude of θ. For an elliptic orbit the effect of J_2 is much less discernible. The other effect, associated with the nonnewtonian gravitational torque has a much smaller effect, as one can see from the comparison of Figs 1 and 9, which refer to the case $\chi = 1.02$.

All the foregoing results were relevant to initial values near to the stable equilibrium position, $\theta = 0$. With $\theta_0 = \pi/2 + .01$, in the circular case, we obtain the periodic solution described in Fig. 11, where no higher harmonics have been introduced. With $\chi = 1.02$ the diagram moves to that of Fig.12, where one can see successive oscillations of the bar about successive unstable equilibrium positions of amplitude π.

With $\chi = 1.05$ we have the awkward plots represented in Figs.(13,14), where we see that the succession of passages in the vicinity of, and through, the successive unstable equilibrium positions is not very clear. Is that chaos? May be, but it is certainly a big casino. If not, this is for sure the characteristics of complexity of the problem.

8 - The Elastic Body

In the above described conditions, the influence of modes on the orbital and attitude motion is very small. If we consider careful inspection of Eq.(16) would reveal that its essential version is

$$\ddot{A}_n + (\sigma_n^2 - [\dot{\omega} + \dot{\theta}]^2)\, A_n = Q_n \tag{23}$$

where Q_n is the modal force associated with perturbing forces only, since modes 1 and ξ, characteristics of the $G(\xi)$ distribution are orthogonal to free-free modes. In practical applications, $\dot{\omega} + \dot{\theta} << \sigma_n$, thus orbital motion has but a very little effect on modal amplitudes time history. If the stiffness of the bar is very small, however, it may happen that the coefficient of A_n in Eq.(23) becomes negative, at least in some portions of the orbits (or everywhere, for circular orbits). Just to give a conceptual view of the phenomenon, (which, as said, has almost no practical importance) we show, in Fig.15, the phase plane-plot for the first mode of the free-free beam in a circular orbit of radius $r = 1$ for several values of the ratio $\sigma_1 / \dot{\omega}$ which is constant along the orbit, whereas of course θ is not. An analysis conducted on an unstable position for θ would reveal occurrence of instability for a greater value of σ_1.

9 - CONCLUSIONS

The analysis has led to some important conclusion. Firstly, it has been shown that terms of different order of magnitudes in length can be separated, so that attitude motion has practically no influence on c.g. motion. The converse is not true, and orbital geometry has a tremendous impact on the amplitude of attitude motion. When one starts from an unstable position, attitude motion in an elliptic orbit becomes almost chaotic. Also modal time histories may become unstable as the stiffness of the body becomes too low, but, for that to happen, stiffness should be low beyond every practical limit, so that the last conclusion has just a theoretical value.

APPENDIX I

(For symbols, see Main Text)

1. The general expression for the Earth's gravitational potential reads

$$U = U(\rho,\Lambda,\omega)) = -(g_0\, R_0 / r)\, [1 - \sum_{2}^{\infty} J_n\, r^{-n}\, P_n \cos \Lambda +$$
$$- \sum_{2}^{\infty}{}_n \sum_{1}^{n}{}_m J_{nm}\, r^{-n}\, P_n^{(m)} \cos \Lambda \cos m(\omega - \omega_{mn}) \qquad \text{(A.I.1)}$$

The constants J_n, J_{mn}, ω_{mn} are known for the first values of m,n **[3]**. In the present case, $\Lambda = 0$ and we can write:

$$U = U(\rho,\omega) = -(g_0\, R_0 / r)\, [1 - \sum_{2}^{\infty}{}_n r^{-n}\, Q_n(\omega)] \qquad \text{(A.I.2)}$$

where

$$Q_n(\omega) = J_n\, P_n(1) + \sum_{1}^{n}{}_m J_{nm}\, P_n^{(m)}(1) \cos m\, (\omega - \omega_{nm}) \qquad \text{(A.I.3)}$$

2. Hence the radial and tangential components of the gravity are written:

$$G_r = g_0\, [-1 / r^2 + D_r(r,\omega)]$$
$$G_\omega = g_0\, [D_\omega(r,\omega)] \qquad \text{(A.I.4)}$$

where

$$D_r = \sum_{2}^{\infty}{}_n (n+1)\, r^{-(n+2)}\, Q_n(\omega)$$
$$D_\omega = - \sum_{2}^{8\infty}{}_n r^{-(n+2)}\, dQ_n / d\omega$$

3. Now consider Fig. A.1 and let P be a point of coordinates (x_1,x_2). Denote by $(\Delta\rho\ ,\ \rho\ \Delta\omega)$ the difference in geographic coordinates from *C to* P. We have of course:

$$\begin{vmatrix} \Delta\rho \\ \Delta\omega \end{vmatrix} = \mathrm{B} \begin{vmatrix} x_1 \\ x_2 \end{vmatrix} \qquad \text{(A.I.5)}$$

where the matrix B is the rotation matrix of the angle θ.

Again from Fig.A.1, if we define the two unit vectors in radial and in tangential directions respectively I_1, I_2, since :

$$dI_1/d\omega = I_2 \quad ; \quad dI_2/d\omega = -I_1 \tag{A.I.6}$$

we find for the change in gravitational force on passing from C to P the following expression:

$$\begin{vmatrix} \Delta G_r \\ \\ \Delta G_o \end{vmatrix} = C \begin{vmatrix} \Delta \rho \\ \\ \rho \ \Delta\omega \end{vmatrix} \tag{A.I.7}$$

The elements of the matrix C are as follows:

$$\begin{aligned} c_{11} &= b_{11} + 2 / r^3 \\ c_{12} &= b_{12} \\ c_{21} &= b_{21} \\ c_{22} &= b_{22} - 1 / r^3 \end{aligned} \tag{A.I.8}$$

where

$$\begin{aligned} b_{11} &= \partial D_r / \partial r \\ b_{12} &= \partial D_r / r\partial\omega - D_\omega / r \\ b_{21} &= \partial D_\omega / \partial r \\ b_{22} &= \partial D_\omega / r\partial\omega + D_r / r \end{aligned}$$

By combining Eqs.(A.I.5) (A.I.7) (A.I.8) and projecting on the body axis we have finally the matrix G_1, whose elements are written in the form $g_{jk} = (g_o/R_0)\ h_{jk}$, where:

$$\begin{aligned} h_{11} &= k_{11} + (1 / r^3)\ (2\ c^2 - s^2) \\ h_{12} &= k_{12} + (3 / r^3)\ c\ s \\ h_{21} &= h_{12} \\ h_{22} &= k_{22} + (1 / r^3)(\ 2\ c^2 - s^2) \end{aligned} \tag{A.I.9}$$

where:

$$
\begin{aligned}
k_{11} &= b_{11}c^2 + b_{22}s^2 + (b_{12} + b_{21})\, c\, s \\
k_{12} &= b_{12}c^2 - b_{21}s^2 + (b_{22} - b_{11})\, c\, s \\
k_{21} &= k_{12} \\
k_{22} &= b_{22}c^2 + b_{12}s^2 - (b_{12} + b_{21})\, c\, s
\end{aligned}
\qquad \text{(A.I.10)}
$$

$(c = \cos\theta;\ s = \sin\theta)$

In the foregoing equations the newtonian terms are clearly distinguished from the nonnewtonian ones. For a complete expression of G_1 in the three dimensional case, see **[1]**.

APPENDIX II

The gravitational torque on the bar for the case under concern is written:

$$T = \int \{x\}\ G\ d\mu \qquad \text{(A.II.1)}$$

With Eqs.(7) it is seen that the the G_0 term gives no contribution. Thus, combining Eqs.(12) and (A.I.9), and denoting by η the abscissa along the beam, we have

$$T = (g_0 / R_0) \int [\eta \quad -w] \begin{bmatrix} h_{11} & h_{12} \\ h_{21} & h_{22} \end{bmatrix} \begin{bmatrix} w \\ \eta \end{bmatrix} d\mu$$

Finally

$$T = C_T / r^3 \qquad \text{(A.II.2)}$$

where

$$C_T = (g_0 I / R_0)\ (k_{12}\ r^3 + 3) \qquad \text{(A.II.3)}$$

The term with k_{12} gives the entity of the gravitational torque associated with the higher harmonics, and its importance as compared with the classic coefficient 3 of the newtonian counterpart. For a first evaluation let us consider the term corresponding to $n = 2$ in the first summation in Eq.(A.I.1), for which $J_2 = .001$. It is easy to obtain,for the term within brackets of Eq.(A.II.3), the expression $(3 - 15\ J_2/r^3)$; so the relative importance of the relevant harmonic, which is .001 in the potential is multiplied by a factor of 5 in the torque. This is due to the fact that higher harmonics, although smaller, have a gravity gradient increasing more rapidly than the lower ones.

REFERENCES

[1] Santini P., "*Stability of Flexible Spacecrafts*", Acta Astronautica, 3,pp 685-713 1976.

[2] Cornelisse J.W., Shöyer H.F.R., Wakker K.F.,"*Rocket Propulsion and Spacecraft Dynamics*", Pitman, London 1979.

[3] Kaplan, M.H., "*Modern Spacecraft Dynamics and Control*", J.Wiley & Sons, 1976.

[4] Wiesel, W.E, "*Spaceflight Dynamics*", Mc Graw-Hill, 1989.

[5] Herrick, S., "*Astrodynamics*", Vol I, Van Nostrand, London, 1971.

[6] Battin, R.H.,"*Astronautical Guidance*", New York, 1964.

[7] AA.VV.,"*Dynamics of Satellites (1969)*", Springer-Verlag 1970.

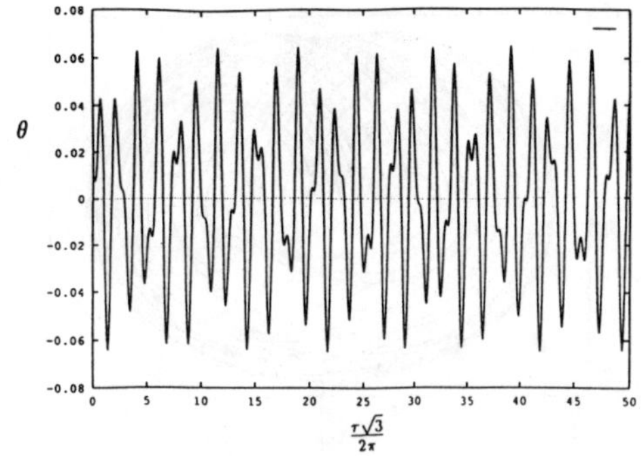

Fig. 1 Bar rotation time-history. (No higher harmonics); $\theta_0 = 0.01$; $\theta_0' = 0$; $\chi = 1.02$; $\tau_0 = 1.0$

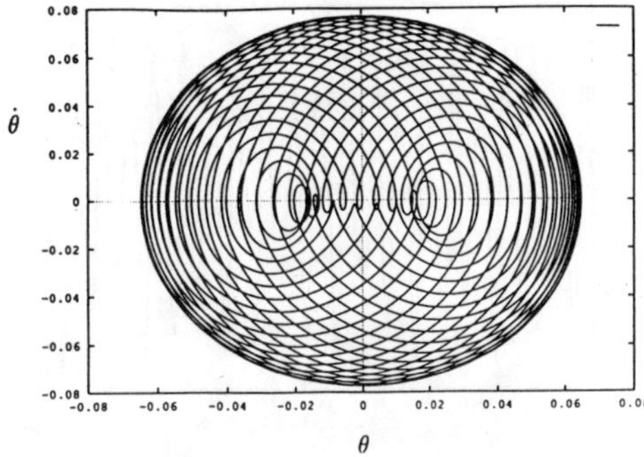

Fig. 2 Bar rotation phase-plane. Same conditions as in Fig. 1.

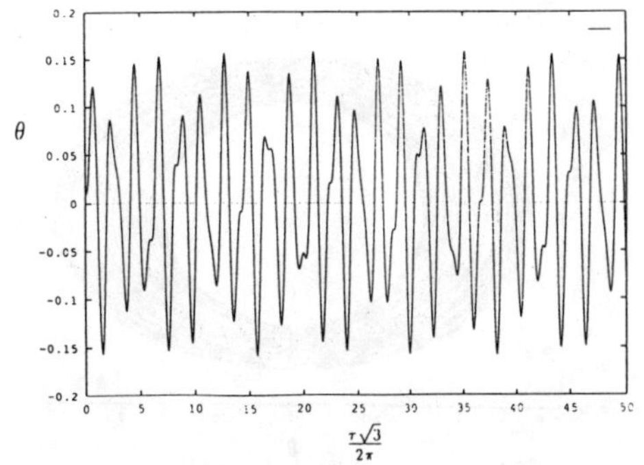

Fig. 3 Bar rotation time-history. (No higher harmonics); $\theta_0 = 0.01$; $\theta_0' = 0$; $\chi = 1.05$; $\tau_0 = 1.0$

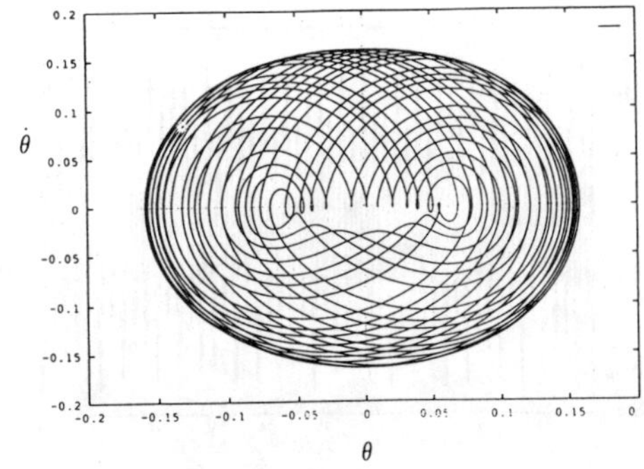

Fig. 4 Bar rotation phase-plane. Same conditions as in Fig. 3.

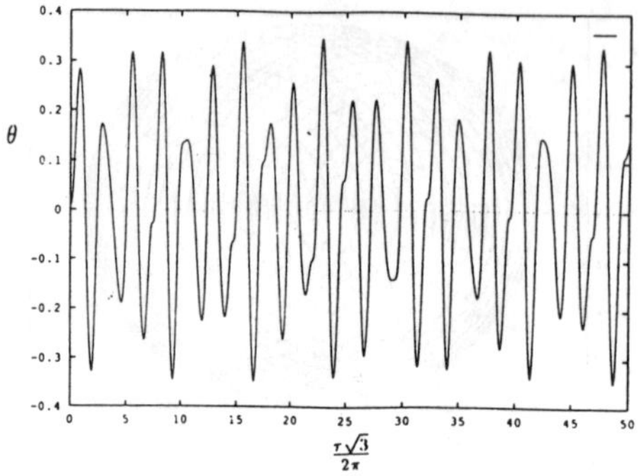

Fig. 5 Bar rotation time-history. (No higher harmonics); $\theta_0 = 0.01$; $\theta_0' = 0$; $\chi = 1.1$; $\tau_0 = 1.0$

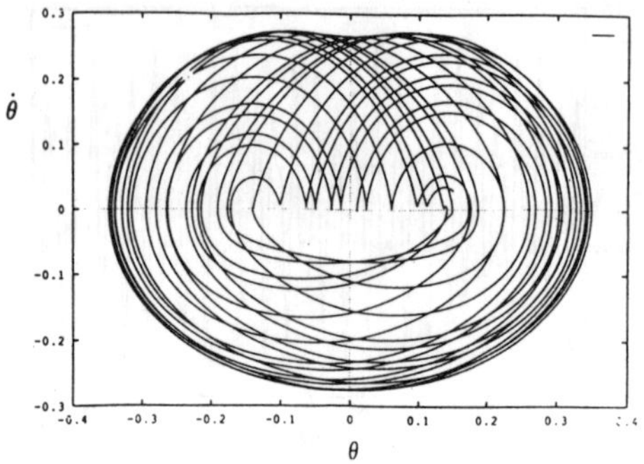

Fig. 6 Bar rotation phase-plane. Same conditions as in Fig. 5.

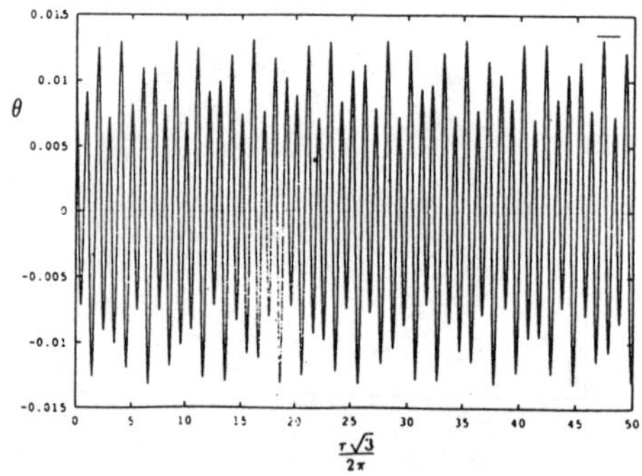

Fig. 7 Bar rotation time-history. (Including J_2 effect); $\theta_0 = 0.01$; $\theta_0' = 0$; $\chi = 1.0$; $\tau_0 = 1.0$

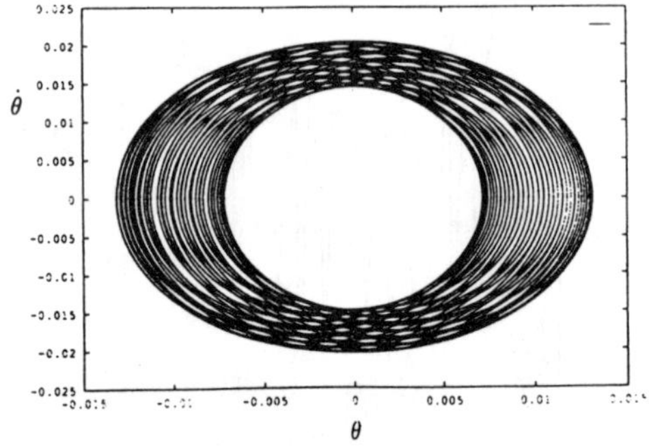

Fig. 8 Bar rotation phase-plane. Same conditions as in Fig. 7.

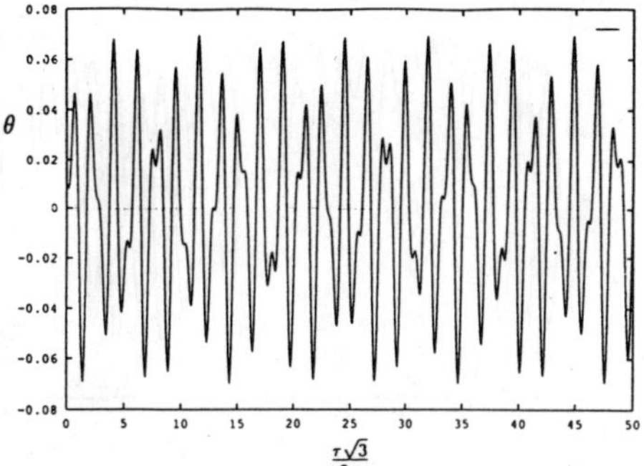

Fig. 9 Bar rotation time-history. (Including J_2 effect); $\theta_0 = 0.01$; $\theta_0' = 0$; $\chi = 1.02$; $\tau_0 = 1.0$

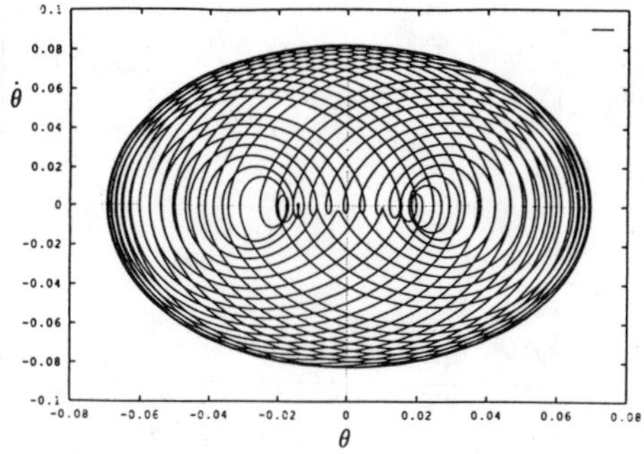

Fig.10 Bar rotation phase-plane. Same conditions as in Fig. 9.

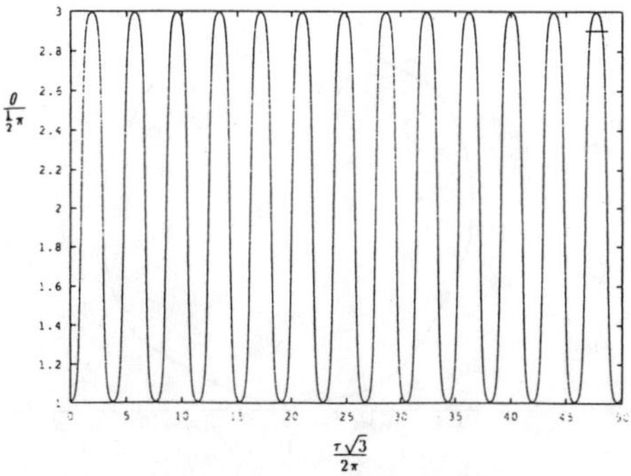

Fig.11 Bar rotation time-history near unstable position (No higher harmonics); $\theta_0 = 0.01+\pi/2$; $\theta_0' = 0$; $\chi = 1.0$; $\tau_0 = 1.0$

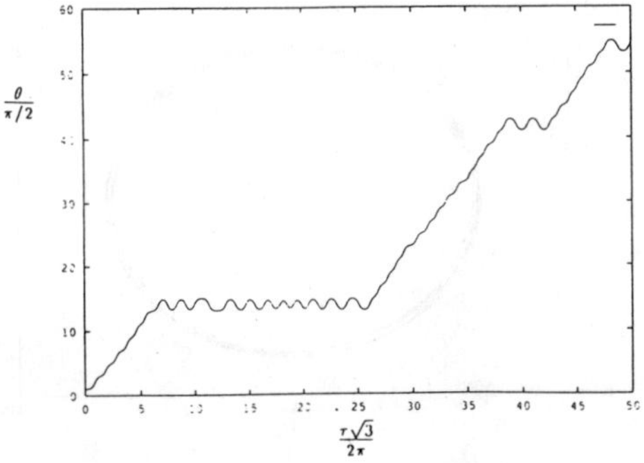

Fig.12 Bar rotation time-history near unstable position (No higher harmonics); $\theta_0 = 0.01+\pi/2$; $\theta_0' = 0$; $\chi = 1.02$; $\tau_0 = 1.0$

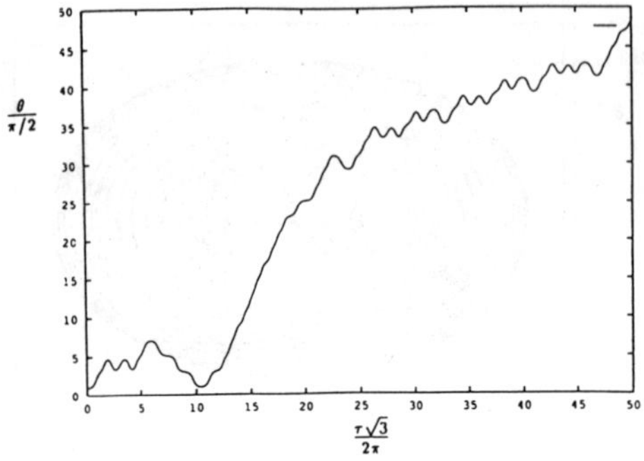

Fig.13 Bar rotation time-history near unstable position (No higher harmonics); $\theta_0 = 0.01+\pi/2$; $\theta_0' = 0$; $\chi = 1.05$; $\tau_0 = 1.0$

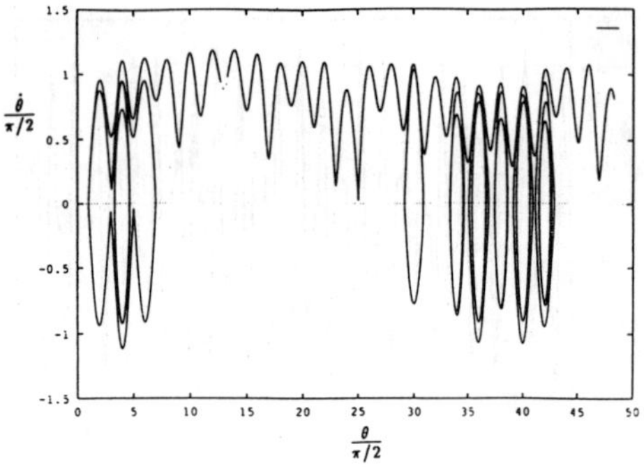

Fig.14 Bar rotation phase-plane. Same conditions as in Fig.13.

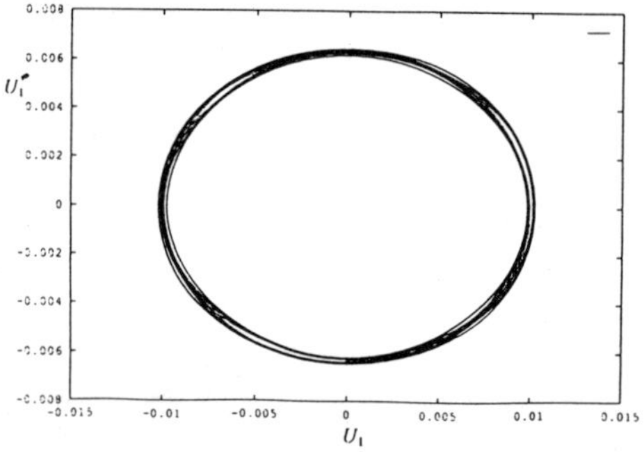

Fig.15a First Mode phase-plane. $\sigma_1 = 0.25\omega'$; $U_{10} = 0.01$; $\theta_0 = 0.01$; $\theta_0' = 0$; $\chi = 1.05$; $\tau_0 = 1.0$

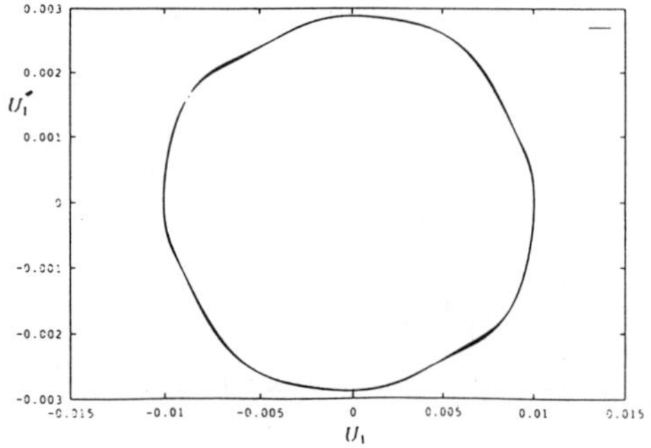

Fig.15b First Mode phase-plane.Same condition as in Fig.15a; with $\sigma_1 = 0.21\omega'$

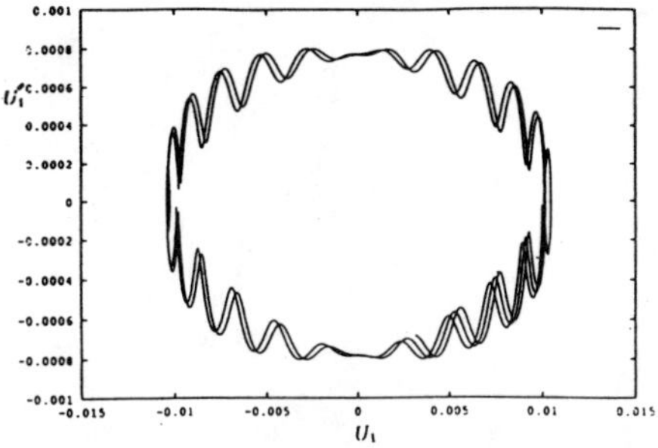

Fig.15c First Mode phase-plane.Same condition as in Fig.15a; with $\sigma_1 = 0.211\omega'$

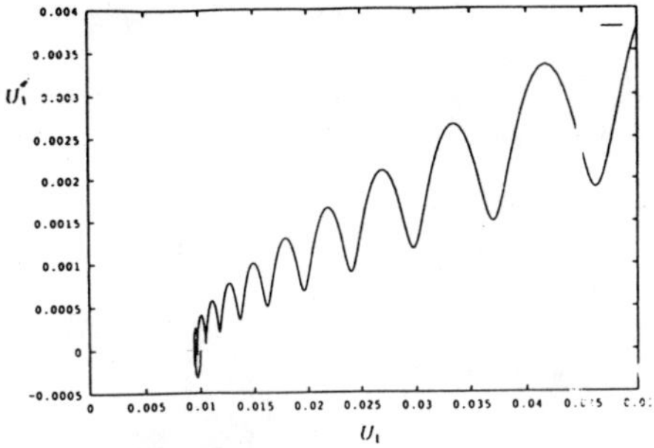

Fig.15d First Mode phase-plane.Same condition as in Fig.15a; with $\sigma_1 = 0.212\omega'$

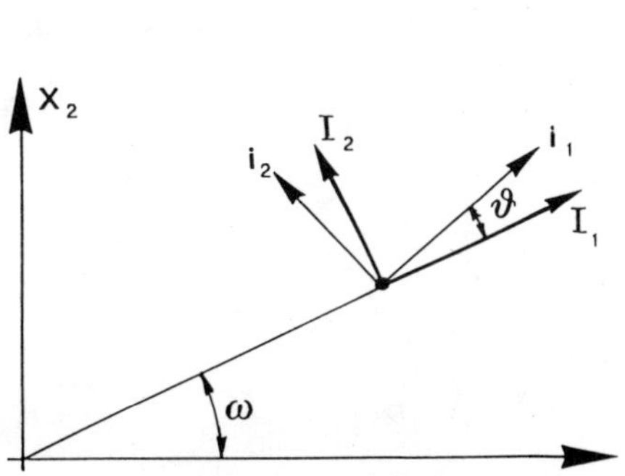

FIG. A.1

Oscillator Chains

and

Propagation in Non-linear Media

EQUIPARTITION TRANSITION AND LYAPUNOV EXPONENTS IN CLOSED HAMILTONIAN SYSTEMS

Roberto Livi
Dipartimento di Fisica, Universitá di Firenze
and Istituto Nazionale di Fisica Nucleare, Sezione di Firenze
L.go E.Fermi, 2 50125 FIRENZE, ITALY

Stefano Ruffo
Dipartimento di Energetica, Universitá di Firenze
and Istituto Nazionale di Fisica Nucleare, Sezione di Firenze
L.go E.Fermi, 2 50125 FIRENZE, ITALY

Abstract

We show the presence of an equipartition transition in a chain of nonlinearly coupled oscillators (Fermi-Pasta-Ulam problem) starting from a far from equilibrium initial condition. More specifically, this effect indicates that the mechanism of relaxation to thermal equilibrium is characterized by very different time scales according to the energy of the system. We introduce a time dependent order parameter, which allows to discover two distinct transitions, which we call instability transition and equipartition transition, respectively. Only the second one is expected to be relevant in the thermodynamic limit. Lyapunov exponent analysis confirms the same conclusions.

1. Introduction

In this paper we consider the relaxation to equilibrium in a closed Hamiltonian system, where 'closed' means absence of external forcing. Historically Fermi, Pasta and Ulam (FPU) [1] already in 1954, performed a numerical experiment on a chain of nonlinearly coupled oscillators and observed a very slow, or, possibly, the absence of relaxation to the equipartition state, characterized by a uniform distribution of energy among the Fourier modes. The experiment was performed exciting initially only one mode (a plane wave) and the nonlinearity was very small, justifying the reference to Fourier modes. Later, it was realized [2],[3],[4] that a threshold in the strength of the initial excitation (i.e. the energy of the excited mode) was present, above which the system relaxed to equipartition quite rapidly. Little is still known on the time evolution of the states below threshold and on their characterization from the statistical ensemble point of view. Let us quote the work of Galgani and collaborators [5], who have related the slow relaxation present in these models to Nekhoroshev's theorem of quasi-integrable Hamiltonian systems, which predicts exponentially long time scales (Arnold diffusion). All these results refer to the finite size case, i.e. the case where the number of oscillators is kept constant to a finite N in the FPU model. Few results are available at $N \to \infty$ (thermodynamic limit), so that the relevance of the existence of these slowly relaxing non-equilibrium states to the foundations of classical statistical mechanics remains still to be established. Quite recently Kolmogorov-Arnold-Moser theorem has been extended to $N \to \infty$ systems for particular initial states and/or for special interactions [6].

In Section 2 we introduce the FPU β model and define an order parameter for the equipartition transition. Section 3 is devoted to the analysis of the dependence of this order parameter on the energy density. Let us stress that, at variance with equilibrium phase transitions, our order parameter depends also on time. In Section 4 we report the Lyapunov exponent spectrum analysis for a 8 particle system. A sharp increase of the maximal Lyapunov exponent detects the transition suggested by the previous analysis. Section 5 contains some conclusions.

2. The Model

Let us consider the following Hamiltonian:

$$H = \sum_{n=1}^{N} \frac{p_n^2}{2} + V\{q_{n+1} - q_n\} \tag{1}$$

where

$$V\{x\} = \frac{x^2}{2} + \frac{\beta}{4} x^4 \tag{2}$$

with periodic boundary conditions $q_{N+1} = q_1$. Let us perform the following linear transformation for even N :

$$q_n = \sum_{j=-\frac{N}{2}}^{\frac{N}{2}} e^{ik_j n} \tilde{q}_j$$

The reality condition for q_n requires $\tilde{q}_j^* = \tilde{q}_{-j}$. The equations of motion in mode space read:

$$\frac{d^2}{dt^2}\tilde{q}_j = -\Omega_j^2\tilde{q}_j + \beta\sum_{l,m} T_{jlm}\tilde{q}_j\tilde{q}_l\tilde{q}_m \tag{3}$$

where, due to p.b.c.:

$$\Omega_j = 2sin(\frac{k_j}{2}) = 2sin(\frac{\pi j}{N}) \tag{4}$$

Some of the elements of the tensor T_{jlm} are zero because of selection rules which depend on the form of potential (2).

The linear energy in mode j is:

$$E_j = \frac{|\tilde{p}_j|^2}{2} + \frac{\Omega_j^2}{2}|\tilde{q}_j|^2 \quad , \quad \tilde{p}_j = \frac{d}{dt}\tilde{q}_j \tag{5}$$

If $\beta = 0$ then $\dot{E}_j = 0$; in fact E_j is the action variable of the harmonic chain. The corresponding angle variable $\Phi_j(t)$ is the phase of the solution of (3) with $\beta = 0$ and it varies linearly with time:

$$\Phi_j(t) = \Omega_j t + \phi_j \tag{6}$$

If $\beta = 0$, the Hamiltonian in mode space depends only on the action variables and reads:

$$\tilde{H}_{\beta=0} = \sum_{j=-\frac{N}{2}}^{\frac{N}{2}} E_j \quad with \quad E_j = E_{-j} \tag{7}$$

If $\beta \neq 0$, then the E_j's are no more constants of motion and the system is no more integrable. In what follows we will consider initial conditions that correspond to the excitation of packets of modes, each with the same energy, i.e. we initially excite the set of modes $\{-\bar{j}, -\bar{j}+1, \cdots, \bar{j}-1, \bar{j}\}$, giving each of them the same energy $E/(2\bar{j}+1)$. The total energy of the system is therefore E. The energy may be given to each mode in the packet in the form of kinetic and potential energy. We are interested to study the N dependence of the relaxation process. The question is how to vary the initial condition as the size N of the system is increased. A physically reasonable requirement is that the center of the packet and its width remain of unchanged frequencies. That is we require that $j \propto N$ and $\Delta j \propto N$, such that Ω_j and $\Delta\Omega_j = \Omega_{j+\frac{\Delta j}{2}} - \Omega_{j-\frac{\Delta j}{2}}$ are fixed as the size of the system increases. Let us define the following weights:

$$p_k = \frac{E_k}{\sum_k E_k} \tag{8}$$

where E_k is defined in (5). It is clear that $0 \le p_k < 1$. Let us consider the Shannon entropy in terms of these weights:

$$S(t) = -\sum_{k=1}^{\frac{N}{2}} p_k ln(p_k) \qquad (9)$$

$S(t)$ is bounded from above and from below:

$$0 \le S(t) \le ln(\frac{N}{2}) \qquad (10)$$

It reaches its maximum value if the system attains equipartition of energy, while it goes to zero if energy is concentrated in a single mode. The increase in S detects thermalization among the modes. We are now ready to introduce a time dependent order parameter, that detects the equipartition transition:

$$\eta(t) = \frac{S_{MAX} - S(t)}{S_{MAX} - S(0)} \qquad (11)$$

It is clear that:

$$0 \le \eta(t) \le \frac{1}{1 - log_{N/2}\Delta j} \qquad (12)$$

In the limit $N \to \infty, \eta$ is unbounded from above. A value of η near zero means that the system is reaching equipartition. If η remains near one then $S(t) \sim S(0)$ and the system is far from equipartition.

3. The Equipartition Transition

In this section we study the N dependence of the equipartition transition of model (1). In Ref. [7] we reported the first analysis of the equipartition transition in terms of this order parameter, showing the presence of an energy density threshold $\epsilon_c = E_c/N$ below which equipartition is not reached on physically reasonable time scales. We suggested therefore the importance of the existence of "metastable" states far from equilibrium in the thermodynamic limit. More recently [8] it has been shown, using the same methods, that a more refined finite time analysis shows a much richer scenario. Following this study we have performed, in collaboration with Holger Kantz, a further analysis in order to enphasize the relevant features of the equipartition transition. At present our knowledge of the finite time dependence of the order parameter η on energy and energy density is resumed in Figs. 1 and 2,

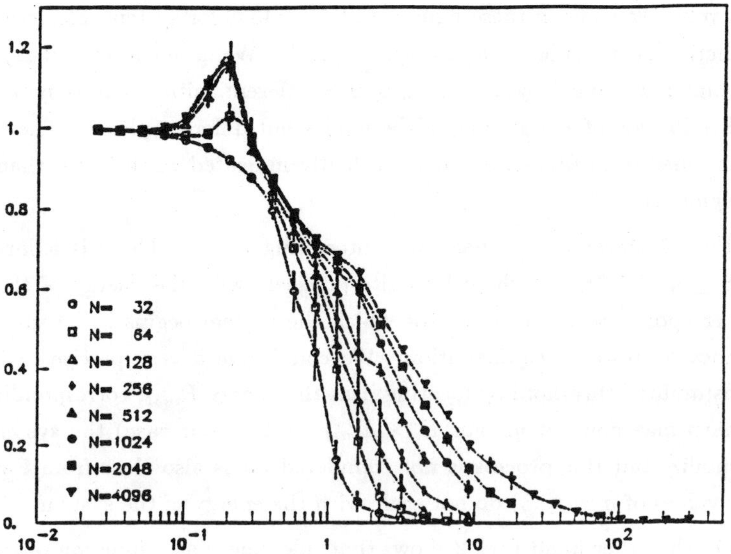

Fig.1 - η *at* $t = 14336$ *for the FPU model as a function of* βE

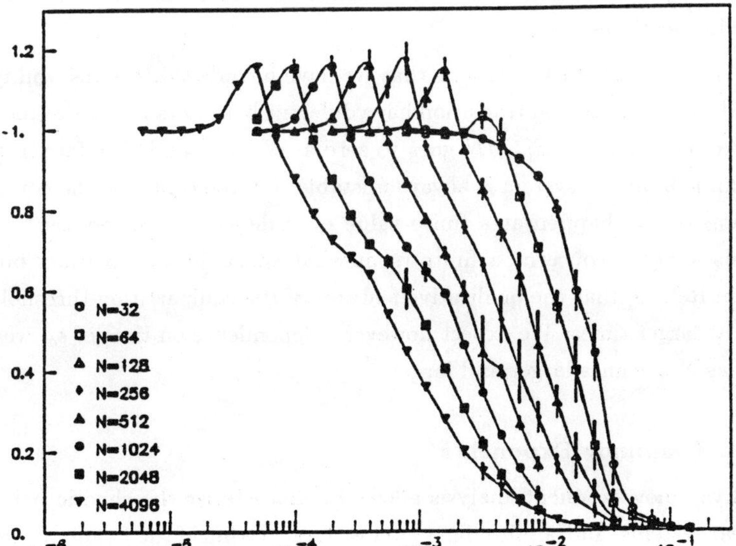

Fig.2 - η *at* $t = 14336$ *for the FPU model as a function of* $\beta E/N$

respectively. We show in these figures η at $t = 14336$ for systems of increasing size as a function of βE (Fig. 1) and $\beta E/N$ (Fig. 2). We have chosen $j = \Delta j = N/16$. Each point is obtained by averaging over 8 different initial conditions having the same distribution of energy among the modes but different phases. The statistical error obtained from this ensemble is explicitly indicated when larger than the size of the symbols.

Fig. 1 shows a very neat and interesting result. There is a large region of η ($.8 \leq \eta \leq 1.2$) , which scales quite precisely with the energy of the system. This corresponds to the energies for which the system begins to be unstable and starts moving towards equipartition. We could give a very precise definition of this "instability" threshold by locating it at the energy E_{inst}, corresponding to the very sharp maximum of η. For $E > E_{inst}$ ($\simeq 2$ in our case) the system begins to thermalize but this process is not completed. It is also clear from Fig. 1 that smaller values of η ($\eta \leq .8$) do not scale with the energy of the system.

On the other hand Fig. 2 shows that, plotting η as a function of the energy density ϵ, one begins to observe a timid scaling for small values of η ($\eta \leq .1$). One could estimate an energy density threshold $\epsilon_c \simeq 1$ for the equipartition transition. Further analysis is needed in order to confirm this result for systems of larger size and for longer times.

The two distinct transitions that we have introduced: the instability transition and the equipartition transition have different behaviors in the thermodynamic limit. Actually, $\epsilon_{inst} = E_{inst}/N$ goes to zero as $N \to \infty$ and therefore in the thermodynamic limit the system is always unstable. On the contrary, the equipartition transition should happen at a finite value ϵ_c, independent of the size of the system.This analysis concerns a finite (somewhat short) evolution time, but it was shown in Ref. [9] that the qualitative features of the equipartition threshold do not change at larger times. We expect however a dependence on time of ϵ_c, weaker and weaker as N becomes larger and larger.

4. Lyapunov Exponents

Lyapunov exponent analysis allows to characterize the chaotic behaviour of nonlinear systems. In Hamiltonian systems the spectrum of the Lyapunov exponents is symmetric and the sum of the positive exponents gives the Kolmogorov-Sinai (K-S) entropy [10]. We have performed the calculation of the full spectrum for model

(1), with $N = 8$. The initial condition is randomly chosen of the form

$$q_i(0) = \alpha r_i \quad ; \quad p_i(0) = \alpha s_i \tag{13}$$

where r_i, s_i are random numbers uniformly distributed in the interval [-1,1] and α is the energy scale parameter.

In Fig.3 we show the maximal Lyapunov exponent as a function of E, where different values of E have been obtained by scaling α in the initial condition. The curve increases sharply up to $E \simeq 10$ and then it continues to increase smoothly. Below $E = 10$ we report also values obtained with different random initial conditions (diamonds), which do not lye on the same curve.

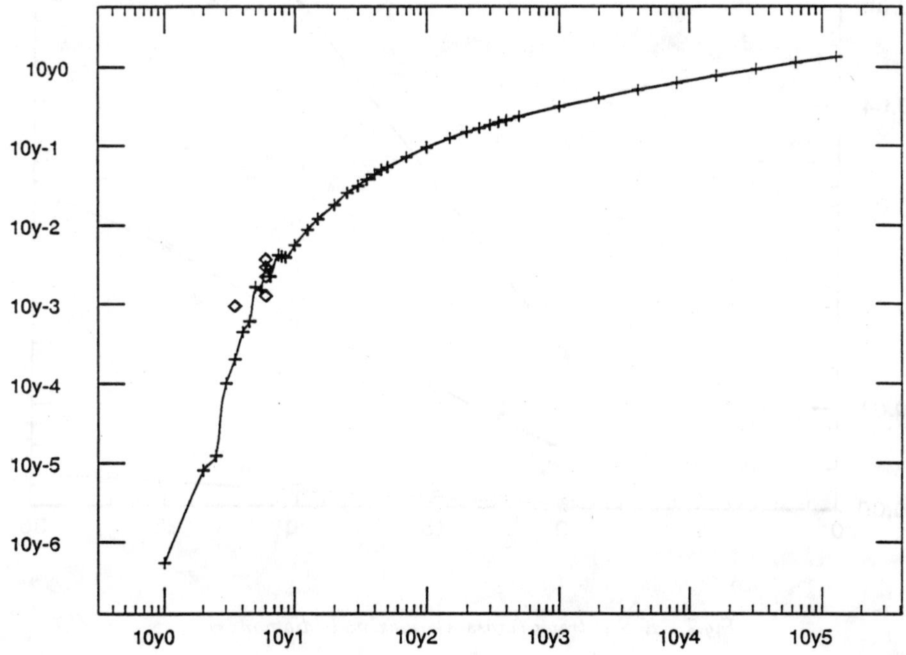

Fig.3 - The maximal Lyapunov exponent versus E for N = 8

These results show that below $E = 10$ two scenarios are possible. Either different ergodic components are present corresponding to different values of the

Lyapunov exponents or, due to Arnol'd diffusion, only one ergodic component is present, but the time needed to reach the value of the exponent common to any initial condition lying in the chaotic sea is too large. Fig.4 shows the K-S entropy (crosses), the maximal Lyapunov exponent (diamonds) and the minimal non zero Lyapunov exponent (squares) as a function of E on a linear scale. It is interesting to observe that all the curves extrapolate to zero at approximately the same value of $E \sim 6$.

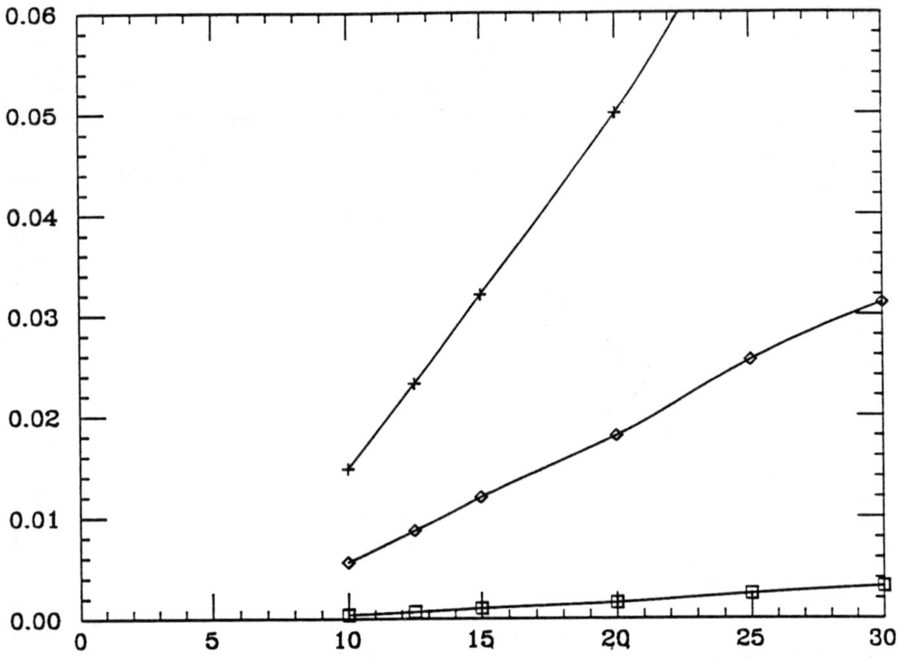

Fig.4 - K-S entropy (crosses), maximal (diamonds) and minimal (squares) Lyapunov exponent vs. E

Of course, we know that even below this value the Lyapunov exponent is positive; however, a threshold-like behaviour is present also in the dependence of the Lyapunov spectrum on the energy. We have also measured the order parameter

η for the system with $N = 8$, considering an initial condition where the modes $j = 2, 3$ are excited. The result is shown in Fig.5, where the statistical error has been obtained as in Fig.s 1 and 2. One observes that η departs from the value 1 for $E \simeq 6$. Therefore, the threshold revealed by the Lyapunov analysis coincides with what we have defined as the "instability" transition.

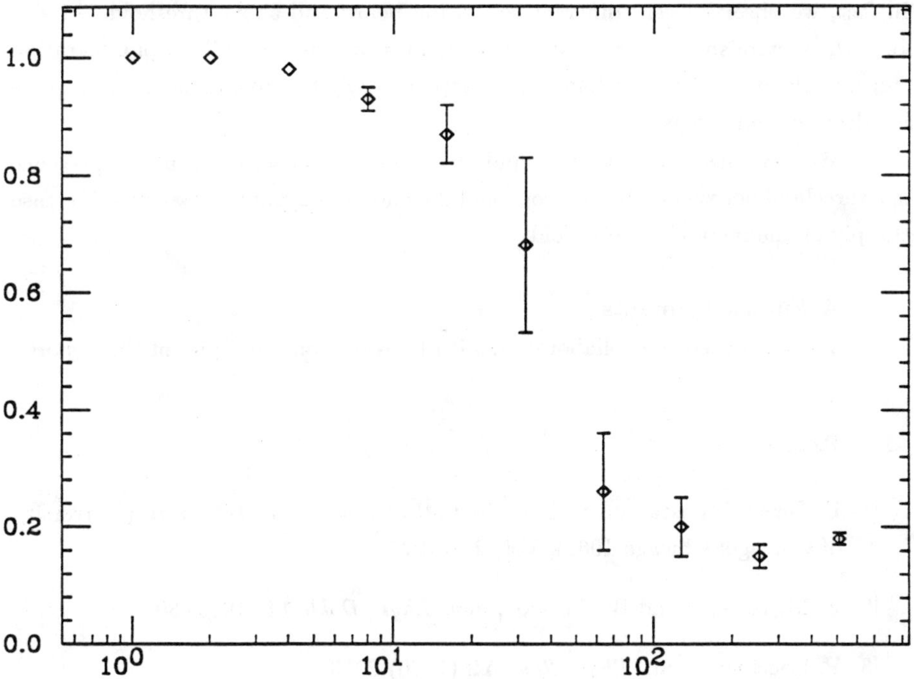

Fig.5 - η versus E for $N = 8$

We have repeated this analysis with $N = 16$, confirming these results. We think that this property should extend to larger values of N. The phase-space mechanism which determines the instability transition leads to a sharp increase of the Lyapunov spectrum. This result is in agreement with a similar one obtained recently for the discrete Sine-Gordon equation [11].

5. Conclusions

The finite time analysis reported in this paper has allowed to discover two different thresholds in the transition to equipartition. An analysis at larger times should confirm this result. An interesting question is however what is the time scale for the relaxation to equipartition, since it could be so large below the equipartition threshold to let the system remain out of equilibrium for physically observable times. In fact, we observed very different relaxation below and above threshold. At $\epsilon \simeq 10^{-2}$ the system shows no relaxation (η remains near one) for millions of integration steps, while at $\epsilon \simeq 1$ the system relaxes exponentially fast to equipartition in some hundreds of time steps.

We have also shown that Lyapunov exponent analysis confirms the presence of a threshold behaviour. K-S entropy and the maximal Lyapunov exponent increase sharply at the instability threshold.

Acknowledgements

We acknowledge a collaboration with H. Kantz on the topics of this report.

References

[1] E. Fermi,J. Pasta and S. Ulam in *Collected papers of E. Fermi*, (University of Chicago, Chicago 1965), Vol. 2, p. 978.

[2] F. M. Izraelev and B. Chirikov, *Sov. Phys. Dokl.* **11** (1966) 30.

[3] P. Bocchieri et al., *Phys. Rev.* **A2** (1970) 2013 .

[4] B. Callegari et al., *Nuovo Cimento* **54** (1979) 463 .

[5] G. Benettin, L. Galgani and A. Giorgilli, *Nature* **311** (1984) 444 and refs. therein.

[6] J. Frölich, T. Spencer and E. C. Wayne, *J. Stat. Phys.* **42** (1986) 247 ; J. Bellissard and M. Vittot, *Ergodic Theory and Dynamical Systems*, (1987).

[7] R. Livi et al., *Phys. Rev.* **A28** (1983) 3544.

[8] H. Kantz , *Physica* **D 39** (1989) 322.

[9] R. Livi et al., *Phys. Rev.* **31A** (1985) 1049 .

[10] J.P. Eckmann, D. Ruelle, *Rev. Mod. Phys.* **57** (1985) 617.

[11] C.G. Goedde, A.J. Lichtenberg and M.A. Lieberman, UCB/ERL M91/76 *preprint*, University of California, Berkeley , Sept. (1991).

DARK OPTICAL SOLITONS

YURI S. KIVSHAR
Institute for Low Temperature Physics and Engineering
310164 Kharkov (Ukraine)
Institut für Teoretische Physik I, Heinrich-Heine-Universität Düsseldorf
D-4000 Düsseldorf 1 (Germany)

ABSTRACT
The paper reviews theoretical and experimental results on propagation of dark-pulse solitons in nonlinear self-defocusing optical media. Using an analytical approach based on the nonlinear Scrödinger equation, I describe properties of dark solitons, features of their generation by different input pulses, and the dynamics of dark solitons in the presence of perturbations. I also briefly review recent experimental results which show a possibility of dark-soliton propagation in optical fibers (temporal dark solitons), laser beams passed through a sodium vapor and nonlinear slab waveguides (spatial dark solitons).

1. Introduction

Although solitons arise in many areas of physics, the single-mode optical fibers have been found as the most convenient quasi-one-dimensional objects to investigate properties of solitons. As was shown by Hasegawa and Tappert [1], in the region of the anomalous (negative) group-velocity dispersion (GVD) of optical fibers it is possible to have a propagation of the so-called *bright solitons.* Since then, soliton propagation of bright pulses has been verified in a number of elegant experiments in the anomalous GVD region of the fiber spectrum, i.e. at $\lambda_0 \leq 1.3\mu m$ in the case of silica, λ_0 being a carrier wavelength (see the pioneer work by Mollenauer *et al.*[2] and the books by Agrawal [3] and Hasegawa [4]). Recently, bright soliton propagation over a 10.000 km distance was achieved using Er-doped fiber amplifiers to compensate fiber loss [5].

For the normal (positive) GVD there are no bright solitons, instead the pulses undergo enhanced broadening and chirping. But in the normal GVD region it is possible to have propagation of the so-called *dark solitons*, consisting of rapid dips in the intensity of a constant-wave (cw) background, the cw being stable in this case. The possibility to propagate dark solitons in optical fibers was also theoretically predicted by Hasegawa and Tappert [6], but these solitons remained a mathematical curiosity until approximately three years ago, when techniques for controlling both the phase and amplitude of subpicosecond pulses were applied in order to generate odd-symmetry dark pulses that propagated unmistakably as solitons [7]. The earlier experiments by Emplit *et al.*[8] recognized the necessary condition of odd symmetry to obtain a fundamental dark soliton but they did not prove clearly that the dark pulse they observed propagates as a soliton; additionally, the background pulse with a phase-step being created was not long enough to compare the experimental results with theoretical predictions valid for non-zero asymptotics of the background pulse. However, it was, in fact, the first experimental observation of a dark-soliton pulse in fibers. Later Krökel *et al.*[9] showed experimentally that an even dark pulse evolves into a pair of low-contrast (small-amplitude) dark solitons, the result has been predicted numerically in 1985 by Blow and Doran [10]. Dark *spatial* (in particular, three-dimensional) solitons were also recently discovered experimentally [11–13].

Some properties of dark-pulse solitons were discussed in a number of theoretical papers [6,10,14–35], however, in experimental studies [7–9,36–40], unlike the theory, dark solitons were observed on a broad bright pulse with a rapid intensity dip generated by a driving pulse [9], or utilizing a specially shaped antisymmetric input pulse of finite extent [7,8]. But as was demonstrated numerically [17] and analytically [21], a finite duration of the background pulse is not a principal limitation for the dark-pulse propagation, and the dark pulses created on a finite extent background have the same main properties as dark solitons on a cw background, for the latter case we may apply the inverse scattering transform [41] as a powerful tool to study properties of solitons in exactly integrable models.

Dark solitons have a number of interesting properties. For example, unlike bright solitons [42–44], dark solitons may be created without a threshold value in the input-pulse power [14,20]. As a result, the dynamics of such soliton pulses differs from the corresponding dynamics of bright optical solitons.

The present paper reviews various aspects of dark-soliton propagation in self-defocusing nonlinear optical media having in mind comparison of analytical, numerical, and experimental results. Such a consideration allows us to demonstrate that the dark-pulse solitons are responsible for nonlinear dynamics in the normal dispersion region. The problems considered in this review are mostly devoted to dark-pulse propagation in optical fibers, but, however, we have included also a discussion of some recent experimental results on propagation of dark spatial solitons.

2. Bright and dark solitons

To discuss the dark-soliton phenomena in optical fibers, we need to start from the basic dynamical equation for the complex envelope amplitude of the pulse which may be derived taking into account the nonlinearity arising from the Kerr effect. This derivation is well-known and it it *exactly the same* as in the case of bright solitons (see, e.g., Refs.[3,4,45].

The derivation is based on the well-known Maxwell's equation for a dielectric medium in which it is assumed that the electric displacement vector **D** can be split up into two parts, the linear and nonlinear ones. The nonlinearity arises from the Kerr effect alone, and the nonlinear part $\mathbf{D}_{nl}$ may be presented in the form, $\mathbf{D}_{nl} = \epsilon_2 |\mathbf{E}|^2 \mathbf{E}$, where $\mathbf{E}$ is the electric field, and ϵ_2 is the nonlinear Kerr coefficient. As a result, it is possible to demonstrate that propagation of short optical pulses in single-mode optical fibers is described by the well-known nonlinear Schrödinger (NLS) equation [1–4]. In an appropriate system of normalized coordinates, this equation is

$$iu_z - \sigma u_{tt} + 2|u|^2 u = 0, \tag{1}$$

where u is the (complex) amplitude envelope of the pulse, z is the distance along the fiber, and the time variable t is a retarded time measured in a frame of reference moving along the fiber at the group velocity. The normalizing units are well known and they may be found, e.g., in Ref.[4].

Solutions of the NLS equation devide into two different classes depending on the sign of σ. In the case $\sigma = -1$ (negative GVD) the appropriate boundary conditions in Eq.(1) is $|u| \to 0$ at $t \to \pm\infty$. For these conditions Zakharov and Shabat [41] showed that the Eq.(1) is exactly integrable one and it possesses bright soliton solutions. The bright soliton,

$$u(z,t) = a \frac{\exp[iVt/2 - i(V^2/4 - a^2)z]}{\cosh[a(t - Vz)]} \tag{2}$$

which at $V = 0$ has the well-known sech-type shape

$$u(z,t) = a\,\mathrm{sech}(at)\exp(ia^2 z). \tag{3}$$

Here a is the soliton amplitude, and V is its velocity (in the t-space).

At $\sigma = +1$ (positive GVD) the solution $|u| = u_0$ is stable, and, as a result, Eq.(1) has soliton solutions in the form of localized nonlinear "dark" pulses on the cw background. The NLS equation with the boundary conditions $|u| \to u_0$ is exactly integrable [41] and its one-soliton solution (in this case it is a dark soliton) may be presented in the form,

$$u(z,t) = u_0 \frac{(\lambda + i\nu)^2 + \exp Z}{1 + \exp Z} \exp(2iu_0^2 z), \tag{4}$$

$$Z = 2\nu u_0 (t + 2\lambda u_0 z), \quad \nu^2 = 1 - \lambda^2 \tag{5}$$

which corresponds to the boundary conditions

$$\begin{cases} u \to u_0 \exp(2iu_0^2 z), & \text{at } t \to +\infty, \\ u \to u_0 \exp(2iu_0^2 z + i\phi), \tan(\phi/2) = \nu/\lambda^2, & \text{at } t \to -\infty. \end{cases} \tag{6}$$

The solution (4) has, unlike the bright soliton (2), the only parameter ν which characterizes the soliton intensity. The value $2\lambda u_0$ has the sense of the soliton velocity in the t-space, i.e. the relative velocity between the dark pulse and the cw background on pictures $u(t)$ at fixed z. The intensity $|u|^2$ has the special shape,

$$|u|^2 = u_0^2 \left(1 - \frac{\nu^2}{\cosh^2(Z/2)}\right), \tag{7}$$

describing a low intensity dip in the cw background produced by the soliton.

It is convenient to distinguish two limit cases of the dark solitons (4). The *fundamental dark soliton* at $\lambda = 0$

$$u(z,t) = u_0 \tanh(u_0 t) \exp(2iu_o^2 z), \tag{8}$$

is the anti-symmetric function of time with the π phase shift and zero intensity at its center. Another limit case, when $\nu \ll 1$, corresponds to a small-amplitude dark soliton, when the solution (4) may be presented in the form,

$$u(z,t) \approx \left[u_0 - \frac{1}{2}u_0\nu^2 \text{sech}^2(Z/2)\right] \exp[2iu_0^2 z + i\Phi(z,t)], \tag{9}$$

where

$$\Phi(z,t) = -2\nu/(1+\exp Z), \quad \text{and} \quad Z = 2\nu u_0(t \mp 2u_0 z \pm u_0\nu^2 z)$$

and its velocity is close to the velocity of linear excitations of the background.

3. Mathematical tools

3.1. *Inverse scattering transform*

As was discovered by Zakharov and Shabat [41,46], the NLS equaton (1) may be analysed by means of the so-called inverse scattering transform (IST) method either for $\sigma = -1$ or for $\sigma = +1$. The mathematical tools related to the IST method allow us to solve the Cauchy problem for the NLS equation if the input pulse is given. According to the IST method for Eq.(1) at $\sigma = +1$ (see Ref. [41]), to find which type of input pulse $u(0,t)$ generates solitons and to determine their parameters, one has to solve the eigenvalue Zakharov-Shabat problem for the auxiliary two-component eigenfuncton (Ψ_1, Ψ_2),

$$\begin{cases} \frac{\partial \Psi_1}{\partial t} = i\mu\Psi_1 - iu(0,t)\Psi_2, \\ \frac{\partial \Psi_2}{\partial t} = -i\mu\Psi_2 - iu^*(0,t)\Psi_2, \end{cases} \tag{10}$$

with the boundary conditions

$$\begin{cases} u(0,t) \to u_0, & \text{for } t \to +\infty, \\ u(0,t) \to u_0 e^{i\theta}, & \text{for } t \to -\infty, \end{cases} \tag{11}$$

where the asterisk denotes the complex conjugation, u_0 is the background pulse amplitude, and θ is a constant phase. As was shown by Zakharov and Shabat[4]1, the spectral problem (10),(11) has a descrete spectrum lying in the region of real values $|\mu| < u_0$, and the descrete spectrum is conserved in z, i.e. when the coordinate evoluton is included. Each real discrete eigenvalue $\mu_n \equiv \lambda_n u_0$ corresponds to a dark soliton (4) with the amplitude $w_n = (u_0^2 - \mu_n^2)^{1/2}$ moving with the velocity $2\lambda_n$ [see Eq.(4)]. Thus, the asymptotic evolution of the input pulse is mostly described by eigenvalues of the discrete spectrum of the scattering problem (10), so that any input pulse will be transformed into a certain set of dark solitons for large enough time. From the viewpoint of experimental verifications, it is necessary to know results of the IST method for input pulses, which can be easily created. For example, if the pulse shape is

$$u(0,t) = u_0 \tanh(at), \tag{12}$$

then the eigenproblem (10) may be solved exactly [18] and the eigenvalues of the discrete spectrum are given by $\mu_1 = 0$, $\mu_{2n} = -\mu_{2n+1} = \sqrt{u_0^2 - w_n^2}$, where $w_n = u_0\nu_n = u_0(1 - na/u_0)$, $n = 1, 2, ..., N_0$, and N_0 is the largest integer number satisfying the condition $N_0 < u_0/a$. The total number of eigenvalues, i.e. the total number of the solitons created, is $N = 2N_0 + 1$.

3.2. Small-amplitude limit

As was mentioned in Section 2, it is convenient to distinguish analytically two limiting cases of the dark solitons (4). The fundamental dark soliton (8) is the anti-symmetric function of time with the π phase shift and zero intensity at its center. Another limit case, when $\nu^2 \ll 1$, corresponds to a small-amplitude dark soliton, when the solution (4) may be presented in the form (9), and its velocity is close to the velocity of linear waves propagating on the background pulse. For analytical results it is important that the dynamics of such small-amplitude pulses may be described by the KdV equation (see, e.g., [23,26]), because its properties are well known.

To discuss the dark-soliton dynamics in the small-amplitude limit, we look for a solution of Eq.(1) at $\sigma = +1$ in the form,

$$u(z,t) = [u_0 + a(z,t)] \exp[2iu_0^2 z + i\phi(z,t)]. \tag{13}$$

Substituting Eq.(13) into Eq.(1), we may obtain two nonlinear equations assuming that the amplitude a is small enough,

$$(a_z - u_0\phi_{tt}) - (2a_t\phi_t + a\phi_{tt}) = 0, \tag{14}$$

$$u_0(\phi_z - 4u_0 a) + a_{tt} + a\phi_z - (u_0 + a)\phi_t^2 - 6u_0 a = 0, \tag{15}$$

the indices "t" and "z" stand for partial derivatives.

As follows from Eqs.(14), (15), the linear excitations of the cw background may be described by the wave equation supporting wave propagation in two directions with the velocities $\pm|C|$ (in the t-space), where $C^2 = 4u_0^2$. If we will use new variables which allow us to devide the propagation directions and to consider the nonlinear dynamics for both of them separately,

$$\tau = \varepsilon(t - Cz), \quad y = \varepsilon^3 z, \tag{16}$$

ε being an arbitrary small parameter connected with the soliton amplitude (see below), and we will look for solutions for the functions $a(z,t)$ and $\phi(z,t)$ in the form of the asymptotic series in the same small parameter ε ,

$$a = \varepsilon^2 a_0 + \varepsilon^4 a_1 + \ldots, \quad \phi = \varepsilon\phi_0 + \varepsilon^3\phi_1 + \ldots. \tag{17}$$

then we may obtain the set of equations produced by coefficients at the different powers of ε (see details, e.g., in Ref.[23]). In the lowest approximation those equations yield the simple equation for the phase,

$$\frac{\partial\phi_0}{\partial\tau} = -\frac{Ca_0}{u_0}. \tag{18}$$

and the KdV equation for the pulse amplitude,

$$2C\frac{\partial a_0}{\partial y} + 24u_0a_0\frac{\partial a_0}{\partial\tau} - \frac{\partial^3 a_0}{\partial\tau^3} = 0. \tag{19}$$

Equation (19) has the soliton solution in the form

$$a_0(\tau,y) = -(\frac{\kappa^2}{2u_0})\mathrm{sech}^2\left[\kappa(\tau + \frac{2\kappa^2 y}{C})\right], \tag{20}$$

κ being the KdV-soliton amplitude. Direct comparison of the solutions (17), (18), (16), and (20) with the small-amplitude dark soliton (9) leads to the simple relation between the dark-soliton parameter ν and the formal perturbative parameter ε, $\nu u_0 = \varepsilon\kappa$, which demonstrates that the dynamics of small-amplitude dark solitons (9) is described by the KdV equation (19). This property of dark solitons is useful to analyse the influence of different perturbations of their dynamics using the known analytical results for the solitons of the KdV equation (see, e.g., Ref.[49] as a review of the soliton perturbation theory).

4. Generation of dark solitons

To describe creation of dark solitons by different input pulses, let us start from the simplest example when the input pulse is an arbitrary phase step on the cw background, i.e.

$$u(0,t) = \begin{cases} u_0e^{i\theta_1}, & t < 0, \\ u_0e^{i\theta_2}, & t > 0. \end{cases} \tag{21}$$

Such an example allows us to show the main properties of dark solitons.

The eigenfunctions and eigenvalues of the spectral problem (10) with the input pulse (21) have been found in Refs.[21,24]. The spectral parameter μ of the discrete spectrum lies in the region $|\mu| < u_0$, and it may be written in the form,

$$\mu = -u_0 \cos\psi, \quad \psi = \frac{1}{2}(\theta_2 - \theta_1). \tag{22}$$

The eigenvalue (22) corresponds to a dark soliton with the intensity $w = u_0|\sin\psi|$ and the velocity $2u_0 \cos\psi$. In the case $\theta = \theta + \pi$ the step (21) will produce the fundamental, so-called "black", soliton (8), i.e. the pulse with zero intensity at its center, plus some radiation. In another case the soliton created will be gray, i.e. it will have the lower-contrast intensity, see Eq.(7). In the paper [35] a lot of numerical results on creation of dark solitons, in particular, by phase steps, are presented.

To generate a number of dark solitons in the same way, one needs to choose a cw background with a variable function. In the case when the background is excited by series of N phase steps, in the asymptotic region it is possible to produce N dark solitons provided [24]

$$\Delta\tau_j > (2u_0)^{-1}|\cot(\theta_{j+1}/2) + \cot(\theta_j/2)|, \tag{23}$$

where $\Delta\tau_j$ is the time duration between the $j+1$-th and j-th steps, and θ_j is the j-th phase jump. Some examples of such a generation were numerically investigated in Ref. [35]. In particular, two steps of different signs always produce two dark solitons of opposite velocities [21].

The most interesting property of dark solitons is their generation without a threshold condition to the pulse intensity (see Refs. [14,20]). As is well known, the process of the generation of bright solitons is threshold one [4,42,44]. Namely, the bright solitons will be created by a localized input pulse provided $\int_{-\infty}^{+\infty} |u(0,t)|dt > \pi/2$. Unlike bright solitons, dark solitons may be created by an *arbitrary* initial small dip $u_1(t)$ on a cw background, for example, when the input pulse has the following general form,

$$u(0,t) = u_0 e^{i\theta} + u_1(t), \quad u_1 \to 0 \ \text{ at } \ t \to \pm\infty. \tag{24}$$

Investigation of the IST results obtained for the case (24) was performed in Refs. [14,20]. The main result is following. For an arbitrary $u_1(t)$ (which falls off fast enough at $t \to \pm\infty$), and for negative

$$\Delta \equiv \mathrm{Re}\left\{e^{-i\theta}\int_{-\infty}^{+\infty} u_1(t)dt\right\} < 0, \tag{25}$$

there always exist two eigenvalues of the discrete spectrum of the spectral problem (10), $\mu_{1,2} = \pm\mu_0 \equiv \pm u_0(1 - \frac{1}{2}\Delta^2)$, which correspond to a pair of dark solitons with equal amplitudes $u_0\Delta$ and the opposite velocities $\pm 2\mu_0$. As a consequence of this result we may conclude that an even input pulse always produce at least a pair of dark solitons.

To generate dark solitons by an anti-symmetric input pulse one needs to produce the pulse shape in the form

$$u(0,t) = u_0 \tanh(t). \tag{26}$$

The spectral problem (10) with the input pulse (26) was solved in Refs. [18,30]. As was shown, the pulse (26) with $u_0 = N - b$, where $N \geq 1$ is an integer number and $0 \leq b < 1$ is arbitrary, the central tangent-like pulse will always produce the fundamental dark soliton (8) with $u_0 = N - b$ and additional pairs of gray solitons with parameters $\nu_n = 1 - n/(N-b)$, $n = 1, 2, \ldots, N-1$, propagating with the opposite velocities from the central black (primary) soliton. The number of gray soliton pairs is defined by the intensity of the background pulse (26). If the intensity of such a pulse is not large enough, the input pulse produces only one fundamental soliton, but if the intensity is increased, it is possible to obswerve generation of secondary dark solitons (these secondary dark solitons are always generated in pairs as follows from the theory).

Generation of dark-pulse solitons by a random input pulse or by pulse which has a random component was discussed in Refs. [20,21]. In fact, such a case correspond to experimentally observed noise-induced generation reported by Grudinin *et al.*[36]. In particular, as was shown by Gredeskul *et al.*[21], the input background pulse with the random component $v(t)$ in the form of white noise applied to the interval $0 < t < T$ may generate dark solitons, and the number of their pairs N may be estimated as [21] $N \sim u_0 T (D/u_0)^{1/3}$ for $D \ll u_0$, and $N \sim u_0 T$, for $D \geq u_0$, $2D$ being the binary correlator of the random component $v(t)$: $< v(t) >= 0$, $< v(0)v(t) >= 2D\delta(t)$. Thus, for strong fluctuations the number of the soliton pairs N does not depend on D.

The effect of initial fluctuations (noise) on the dynamics and propagation of a single dark soliton was studied analytically by Konotop and Vekslerchyk [30]. They used the perturbation theory to obtain statistical characteristics of soliton parameters. Earlier the similar problem was considered numerically by Zhao and Bourkoff [18]. As was shown, the small-amplitude noise level has a larger effect on the pulse shape of the bright soliton than of the dark soliton, and in the dark-pulse case, the noise affect primarily on the constant background. However, an effect of noise on a dark soliton may demonstrate a more drastical effect: noise may produce additional (secondary) dark solitons in the system [30], and this effect is similar to the previously discussed one of dark-soliton creation by a random pulse.

5. Dark solitons under perturbations

In this section we will describe effect of some physically important perturbations on the dark-soliton dynamics.

In real optical fibers solitons propagate in the presence of loss which may be described by the additional term $-i\gamma u$ in the r.h.s. of the NLS equation (1),

$$iu_z - \sigma u_{tt} + 2|u|^2 u = -i\gamma u. \tag{27}$$

As is well known (see, e.g., Ref.[4]), in the presence of loss the intensity of a bright soliton decreases according to an exponential law. Indeed, let us consider the loss-induced evolution of the fundamental bright soliton (3). Multiplying Eq.(27) by u and integrating over t, we may derive the equation for the change of the total pulse intensity $I = \int_{-\infty}^{+\infty} |u|^2 dt$ in the form,

$$\frac{dI}{dz} = -2\gamma I. \tag{28}$$

Equation (28) is exact, but it may be applied to the soliton pulse (3) assuming its *adiabatic* evolution along the fiber. Thus, we may obtain the decay law for the soliton amplitude, $a(z) = a(0)\exp(-2\gamma z)$ (see, e.g., Ref. [4]). For width of a bright soliton defined as $\tau_b = a^{-1}$ [see Eq.(3)], it means that

$$\tau_b(z) = \tau_b(0)e^{2\gamma z}. \tag{29}$$

To obtain the similar result for a dark soliton we will use another approach. First of all, let us consider a time-independent solution of the NLS equation (27) at $\sigma = +1$. It may be easily proved that this solution has the form, $u(z,t) = u_0(z)\exp[2i\int^z u_0^2(z')dz']$, where $u_0(z) = u_0(0)e^{-\gamma z}$. Let us look for a dark-pulse soliton of Eq.(27) at $\sigma = +1$ in the form

$$u(z,t) = u_0(z)A(z,t)\exp[2i\int^z u_0^2(z')dz']. \tag{30}$$

Substituting Eq.(30) into Eq.(27) at $\sigma = +1$ we may derive the equation for the function $A(z,t)$,

$$iA_z - A_{tt} + 2u_0^2(z)A(|A|^2 - 1) = 0. \tag{31}$$

To analyse a dark-pulse evolution, let us take the function $A(z,t)$ in the form of a fundamental dark soliton [cf. Eq.(8)] with unchanged boundary conditions, $A(z,t) = \tanh[L(z)t]$, which is not an exact solution of Eq.(31) but it may be considered as an approximate one, assuming the parameter $L(z)$ is slowly changing. Applying a simplest version of the variational approach (e.g., based on the effective Lagrangian calculated with the help of Eq.(31), see, e.g., Ref. [48]) we may obtain the result, $L(z) = u_0(z) = u_0(0)e^{-\gamma z}$, i.e. for the soliton width, defined as $\tau_d = L^{-1}$, we have

$$\tau_d(z) = \tau_d(0)e^{\gamma z}. \tag{32}$$

Comparing (29) and (32), we conclude that dark solitons spread *more slowly* than bright ones (exactly, at half the rate). This fact was observed by Zhao and Bourkoff in direct numerical simulations (see Fig.2 in Ref.[18]), and it has been proved by Giannini and Joseph [27] using a rigorous perturbation expansions up to the second order in the perturbation parameter γ. The physical reason of such a difference in bright and dark solitons is evolution of the background supporting the dark-pulse propagation: It fades *slower* than the amplitude decay of a bright-soliton pulse.

To describe effect of other perturbations on the dark-pulse dynamics, it is convenient to use the small-amplitude limit discussed above. For example, such an approach describes with a good accuracy the experimentally observed temporal self-shift of dark solitons produced by the Raman self-pumping. The Raman contribution to the nonlinear refractive index may be taken into account by the term,

$$n_2|u|^2 \to n_2 \left[(1-\tilde{\alpha})|u|^2 + \tilde{\alpha} \int_{-\infty}^{+\infty} dt' f(t-t')|u(t')|^2 \right], \tag{33}$$

where $\tilde{\alpha}$ is the fraction of the total (low frequency) nonlinearity with a delayed response, and $f(t)$ is the Raman response function (see, e.g., Ref.[47]). The Raman response function of fused silica is extremely short, so that it may be considered as a local function, expanding the function $|u(t-s)|^2$ from Eq.(33) (here $s = t - t'$) in the Taylor series about t to obtain the perturbed NLS equation of the form,

$$iu_z - u_{tt} + 2|u|^2 u = \alpha u(|u|^2)_t, \tag{34}$$

α being in proportion to the Raman gain parameter $\tilde{\alpha}$.

Using the small-amplitude approach presented above, we may calculate the contribution of the term $\alpha u(|u|^2)_t$ to the amplitude equation (19). The resulting first-order equation for the amplitude a_0 has the following form,

$$2C\frac{\partial a_0}{\partial y} + 24 u_0 a_0 \frac{\partial a_0}{\partial \tau} - \frac{\partial^3 a_0}{\partial \tau^3} = 2\left(\frac{\alpha}{\varepsilon}\right) u_0^2 \frac{\partial^2 a_0}{\partial \tau^2} \tag{35}$$

Equation (35) is the Korteweg-de Vries-Burgers (KdVB) equation arising in different branches of physics (see, e.g., Ref. 49 and references therein). The dynamics of the KdV soliton in the presence of the small perturbation from (35) may be investigated by the perturbation theory for solitons [49,50]. Using this approach, it is possible to demonstrate that evolution of the dark-soliton amplitude is described by the equation [23,26],

$$\nu^2 = \frac{\nu_0^2}{1 + \mathrm{sgn}C(z/z_0)}, \quad z_0^{-1} = \frac{8}{15}\alpha u_0^3, \tag{36}$$

where $\text{sgn}C = +1$ for $C > 0$ and $\text{sgn}C = -1$ for $C < 0$.

The result (36) means that the evolution of the dark-soliton parameters is connected with a direction of the soliton motion. In such a situation, two dark solitons with opposite velocities but equal amplitudes, which are produced by a localized input pulse on a cw (or finite-extent) background, will change their velocities and amplitudes in different ways. The dark soliton propagating to the right ($C > 0$) decreases its amplitude and increases its velocity (in the t-space). But the dark soliton propagating to the left ($C < 0$) increases its amplitude and decreases its velocity. As a result, the initially symmetric pulse will demonstrate an asymmetric shapes as was observed in real experiments [37].

The KdVB equation (35) has, instead of decaying soliton pulses, another type steady-state moving solutions in the form of shock waves with oscillating or non-oscillating shapes (see, e.g., Ref. 51 and references therein). Looking for a solution of Eq.(35) in the form $a_0(\tau - Wy)$ and using the condition that $a_0(-\infty)$ is zero (i.e. $|u| \to u_0$), one may obtain the equation ($\varepsilon = 1$) $-2CWv + 12v^2 - v'' = 2\alpha u_0^2 v'$, where $v \equiv u_0 a_0$, and the prime stands for the differentiation in respect to the new variable $\xi = \tau - Wy$. This equation describes the motion of a unit-mass particle in the effective potential $CWv^2 - 4v^3$ in the presence of the friction force $2\alpha u_0^2 v'$. Straightforward analysis demonstrates that the solution with proper asymptotic behaviour is the above-mentioned oscillating shock wave [23], [31]. The amplitude of the shock wave is proportional to its velocity W, and the shape depends on a relation between the wave parameters. In particular, the shock wave is monotonic provided $\alpha^2 > 4|W|/u_0^3$.

In communication systems, it is desirable to work near the zero of the second-order dispersion [4] because there the power required for creating bright solitons is significantly lower. Although exact analytical solutions describing the soliton propagation near the zero-group-dispersion (ZGD) point are not available, the numerical and perturbative studies were used to analyze the main features of the soliton propagation near the point (see, e.g., Refs. [53–55]). In the case of dark solitons, analytical results may be obtained in the small-amplitude limit (see Refs. [28,34]).

To describe the influence of the ZGD point on the dark solitons, we start from the modified NLS equation,

$$iu_z - \alpha u_{tt} + 2|u|^2 u = i\beta u_{ttt}. \tag{37}$$

where the parameters α and β describe the second- and third-order dispersion effects, respectively, $\alpha = k''/2T|k'|$, $\beta = k'''/6T^2|k'|$, [cf. Eqs.(1)-(2)], and k, as above, is the propagation wave number, so that, $k^{(n)} \equiv \partial^n k/\partial\omega^n$. At $\beta = 0$ Eq.(37) is the same as Eq.(1) with a minor rescaling. Considering Eq.(37) in the framework of the small-amplitude approach described above, we may obtain the KdV equation

[28],

$$2C\frac{\partial a_0}{\partial y} + 24\alpha u_0 \left(1 + \frac{\beta C}{2\alpha^2}\right) a_0 \frac{\partial a_0}{\partial \tau} - (\alpha^2 + 2\beta C)\frac{\partial^3 a_0}{\partial \tau^3} = 0, \tag{38}$$

where y and τ are the same as above, but this time $C^2 \equiv 4u_0^2\alpha$. The result depends on the sign of the velocity C. In particular, when $C < 0$, in the region $1 < \alpha^{3/2}/\beta u_0 < 4$, a dark soliton changes the sign of its amplitude to be, in fact, *an anti-dark soliton* [28,34]. This anti-dark soliton exists only for a certain propagation direction. It means that there is a possibility to observe interactions between *different* types of optical solitons, in fact, between dark and anti-dark solitons, the former solitons are really dark because they propagate in the opposite direction. Collision between dark and anti-dark solitons of equal intensities has been investigated in Ref.[34]. Such a collision generates only small emission which may be explained by non-integrability of the NLS equation including the third-order dispersion (see, e.g., Ref.[35], where the non-integrability was discussed for the anomalous dispersion regime).

6. Interaction of dark solitons

As is well known, bright solitons are two-parametric pulses and their interaction in a single-mode optical fiber depends on their relative phase, so it may be either attractive or repulsive (see, e.g., Ref. [56]). It means that bright solitons may form a bound state of a few solitons when their relative velocities are small[42].

Interaction of two bright solitons belonging to different optical modes of a two-(or many-) mode optical fiber is connected with the type of a coupling between modes, and it is usually attractive (see, e.g., some examples in Refs. [4,57]). As a result, bright solitons in the two-mode (or birefringent) fiber may also form a coupled state in the form of a vector soliton (see, e.g., Ref. [58] and references therein).

The interaction of dark solitons in the framework of the exactly integrable NLS equation was firstly discussed by Blow and Doran [10]. They published analytical solutions which describe multiple dark solitons on a finite background, noting that, in contrast to bright solitons, multiple dark solitons do not form bound states and their interaction is simpler. The problem was studied numerically by Zhao and Bourkoff [19], who considered a pair of fundamental dark solitons given as $u(0,t) = \tanh(t+\Delta) - \tanh(t-\Delta)$, where 2Δ is the intial separation between the dark pulses. Based on numerical results, they deduced am empirical formula for the soliton separation r as a function of the initial separation Δ and the coordinate z,

$$r = \Delta + \frac{1}{2}\ln\{\frac{1}{2}[1 + \exp(4ze^{-2\Delta})]\}. \tag{39}$$

From the result (39) there are some useful properties of the interaction force between dark solitons, e.g., the force is obvious repulsive and the separation increases monotonically instead of periodically for the bright solitons.

In the recent paper [29], Thurston and Weiner reported numerical simulations that lay the groundwork for possible experiments aimed at the observation of dark soliton collisions. Such collisions are possible when, unlike the case discussed previously, dark solitons are grey. The experiment discussed in [29] would test whether an actual collision is adequately described by the NLS equation. Thurston and Weiner [29] included also the effects of a finite-width background pulse and of the Raman contribution to the nonlinear index. This detailed analysis demonstrated that the experimental observation of dark soliton collisions should be possible, and they indicated suitable experimental parameters.

New problems arise when two pulses with different frequencies interact due to cross-phase modulation (XPM) or a nonlinear intermode coupling. The effect of XPM may introduce important novel features in nonlinear pulse interactions and propagation, and it has attracted considerable attention during the last years for different applications (see, e.g., references in Ref. [59]). The XPM effects give rise to strong interactions of dark and bright solitons belonging to different optical modes [57,59–65]. As was first demonstrated in Ref. [57], the interaction between dark and bright solitons are repulsive. However, an intersting effect arises when two pulses with wavelengths situated below and above the zero-dispersion point propagate together in a fiber while interacting through XPM. In this so-called 'reverse case', it is assumed that the bright optical soliton is in the normal GVD region and a dark optical soliton is in the anomalouis GVD refgion [60–62,64,65]. Unlike the standard situation, the interaction of the dark and bright solitons in this reverse case is attractive, and they may form a bound state, i.e. a symbiotic soliton pair, which is stable with respect to mismatches in group velocities and initial positions of the pulses [64]. However, in this reverse case the background pulse supporting the symbiotic soliton pair is unstable due to modulational instability in the anomalous dispersion region of the fiber spectrum. In Ref.[65] the influence of a symbiotic soliton pair on modulational instability was analysed numerically, and it was shown that the symbiotic solitons may strongly stimulate development of the modulational instability in fibers.

In the standard case, when a bright pulse is in the anomalous GVD region and a dark pulse is in the normal GVD region, a solitary wave exists due to mutual support of the soliton pulses through XPM. However, a mismatch in group velocities or initial positions of the separate bright and dark pulses can destroy the solitary waves because such coupled states are unstable [57]. Such an effect shows a generation of dark solitons by the input bright pulse due to the XPM coupling between modes [59,65]. It is interesting to note that the similar interaction of two bright pulses, one of them being not a soliton pulse, is attractive, but after some distances the bright soliton generates a dark dip on the broad non-soliton pulse, and the created dark soliton repulses the bright one. During such an effect dark solitons may be produced so that the method was proposed for generation of dark solitons.

Therefore, dark solitons, in contrast to bright ones, do not form bound states in a single-mode fiber and their interaction is simpler. In many-mode fibers the interaction between solitons is mostly determined by an inter-mode coupling, so that the difference between dark and bright solitons is not so evident in this case.

7. Experimental verifications of dark-pulse propagation in optical fibers

Although both bright and dark solitons in optical fibers were predicted in 1973 by Hasegawa and Tappert [1,6], and bright solitons were experimentally observed by Mollenauer *et al.*in 1980 [2], dark solitons remained a mathematical curiosity until recently. This is because of the difficulty of generating a short dark pulse on a cw or long-pulse background to initiate soliton evolution and to control its phase. The first experimental studies of the dark pulses in optical fibers were carried out by Emplit *et al.*[8], who used the phase step to create a condition of odd symmetry and observed the dark pulse which had properties of a fundamental dark soliton. The experimental results were reported in Ref. [8] for significantly long dark pulse (~ 5 psec). In this case the fiber attenuation cannot be neglected because the normalizing fiber length (~ 220 m) was greater than the fiber attenuation length of ~ 140 m. Thus, those experiments were unable to provide clear evidence and rigorous proof for the phenomena of the dark-soliton propagation described by the NLS equation for the normal GVD region.

Later Krökel *et al.*[9] showed experimentally that an even dark pulse evolves into a pair of low-constant (small-amplitude) dark pulses that propagate unmistakably as solitons: the decay of the even dark pulse into two symmetric dark solitons as observed numerically in 1985 by Blow and Doran [10]. Krökel *et al.*[9] used a YAG laser with an output of 100 psec pulse in which 0.3 psec holes were produced by a modulator into a 10 m-single mode optical fiber. The output signals were measured using the autocorrelation technique for various input levels and the results were compared with the numerical solution of the NLS equation (1) at $\sigma = +1$. The driving pulse generated a pair of dark solitons which were distinguished directly.

The propagation of a pair of dark solitons with opposite velocities and equal amplitudes is a consequence of the fact that the input pulse does not have the phase jump which is needed for a dark soliton, as seen from Eq.(4). If the input pulse has the same phases at $t \to \pm\infty$, a pair of dark solitons appers, so that the one dark soliton is reversal in its phase to the other one. To create a single dark soliton, one needs to prepare a background with a phase jump (see Section 4). Weiner *et al.*[7] prepared specially shaped, antisymmetric input pulses which closely correspond to the form of the fundamental dark soliton propagation generated by the 185-fsec dark pulse at the wavelength 0.65 μm in a 1.4 m-length of single-mode optical fiber, and they observed qood quality pictures demonstrating propagation of a fundamental dark soliton. For higher input powers they also observed generation of a pair of secondary dark solitons with opposite velocities and equal amplitudes in accordance with the results of the IST method for the input pulse (26). The data provide the first evidence for fundamental dark-soliton propagation in optical fibers.

Generation of dark solitons on a background with a noise component was reported by Grudinin *et al.*[36]. Such an experimental observation may be refered to the case considered in Section 4 where it was shown that any dip (deterministic or random) an a cw or long-pulse background will evolve into dark solitons when GVD is normal.

Weiner *et al.*[37] reported also the discovery of temporal and spectral self-shifts of dark solitons propagating in single-mode fibers. The effect is well-known for bright optical solitons [4]. In the case of dark solitons these shifts become increasingly pronounced as the intensity and the fiber length are increased. These experimental data are in a good agreement with numerical simulations in the framework of the modified NLS equation that includes the Raman contribution to the nonlinear refractive index. Analytical description of the effect was proposed in Refs. [23,26] (see also Section 5).

8. Spatial dark solitons

Soliton solutions of the two-dimensional NLS equation with self-focusing nonlinearity have been known since the paper of Zakharov and Shabat [46]. These solitons (the so-called self-trapped channels) are characterized by zero boundary conditions as the transverse coordinate (i.e. beam size) goes to infinity. The spatial solitons are similar to the temporal bright solitons in fibers - in both cases the related phenomena are described by the same normalized NLS equation. This is also true for the self-defocusing nonlinear optical medium. In particular, experimental studies of spatial bright solitons demonstrate the main features which may be described by the NLS equation [66,67]. However, experimental investigations of spatial dark solitons have been hampered by the apparent need for extremely large beem powers in order to form the power constant background on which the dark solitons may propagate. On the other hand, three-dimensional (two transverse and one longitudinal) systems were not expected to produce stable solitons. For example, it is well-known that bright excitations are not stable in three-dimensional Kerr media [68]. However, as was demonstrated recently by Swartzlander *et al.*[11,12], dark solitons can exist as spatial structures (e.g., stripes, crosses, and grids) in the transverse cross sections of a cw optical beam propagating in a material with a self-defocusing nonlinearity. Although no (2+1)-dimension analytical solutions for dark solitons in the NLS equation is known to date, the experimental and numerical data [12] with various two-dimensional amplitude and phase masks provided strong evidence that the phenomenon is indeed due to spatial dark solitons. Furthermore, the study of quasi-(1+1)-dimension propagation of such solitons have shown excellent agreement with the analytical results for the (1+1)-dimension dark solitons obtained by Zakharov and Shabat [41].

In the original experiment [12] (see also Ref.10 in Ref. [12]) a cw frequency-stabilized dye laser beam of power $P_{in} \sim 100$ mW was passed through a wire mesh and then imaged with a lens into a $L = 18$-mm-long cell containing $\sim 10^{12}$ atoms/cm^3 of sodium vapor, which is known to be a strongly nonlinear material with the negative nonlinear coefficient n_2 (in experiments reported in Ref.[12] it was $\simeq -3\text{x}10^{-7} cm^2/W$). Far-field intensity profiles were recorded at the distance of > 1 m. Further cw experiments [11,12] were done using the so-called thermal nonlinearity exhibited by many slightly absorbing liquids under action of low-power radiation. Assuming that the self-defocusing medium is characterized by the intensity-dependent refractive index $n(|\mathcal{E}|^2) = n_0 + n_2|\mathcal{E}|^2$, one may write the electric field $\mathcal{E}$ in the form

$$\mathcal{E}(x, y, z; t) = E(x, y, z)e^{ikz}e^{i\omega t} + \text{c.c.},$$

to derive the (2+1)-dimensional NLS equation for the slowly varying envelope E of the laser beam propagating in the nonlinear medium,

$$2ik\frac{\partial E}{\partial z} + \frac{\partial^2 E}{\partial x^2} + \frac{\partial^2 E}{\partial y^2} + k^2(\frac{n_2}{n_0})|E|^2 E = 0, \tag{40}$$

z being the propagation coordinate. The numerical simulations for the far-field region show the same features as the experimental and numerical results is remarkable. Analysing far-fiels patterns the authors naturally concluded that the dark solitons, i.e. (2+1)-dimension dark stripes, behave amazingly similar to the analytical (1+1)-dimension dark solitons. The authors of Ref. [11,12] examined the near-field patterns at the output face of the nonlinear material using various wire meshes and other boundary conditions, in particular, phase jump transparent masks. To verify that the observed phenomenon was indeed attributable to spatial dark solitons, Swartzlander *et al.*have investigated two important cases: the first case, when the (1+1)-dimension dark solitons are created by a wire and their transverse structure may be described by the exactly integrable NLS equation, and the second case, when (2+1)-dimension dark solitons are created by a phase mask, i.e. crossed phase steps. In particular, in the latter case the input cross does not split into two (or more) ones, and it may be regarded as a cross of two fundamental dark solitons (with a phase step in each) i.e. "the fundamental dark-soliton cross" [12]. A strong manifestation of the soliton nature of the observed phenomenon was achieved by comparising experimentally measured parameters (e.g., the transverse velocity of the soliton) with the theory [11,12]. Thus, the (2+1)-dimension structures observed in the papers [11,12] behave as if they consisted of two almost independent and and noninteracting (1+1)-dimension soliton structures.

Recently, the fundamental dark soliton was also observed by Allan *et al.*[13] in a bulk semiconductors ZnSe in the region of picosecond time scales. The material ZnSe was chosen bacause it is known to have an instantaneous defocusing component of the nonlinear refractive index at $\lambda = 532$ nm, owing to the dispersive change associated with two-photon absorbtion. The soliton-profile pulses were launched in bulk semiconductors using an initial condition with odd symmetry. In order to form the initial condition, a glass platelet was positioned in one half of an incident 30-psec pulse, creating a π-phase shift in the spatial profile. To confirm the observation of dark spatial solitons experimentally, Allan *et al.*[13] measured a number of propagation parameters for the irradiance minima, namely, the phase profile of the electric field near the irradiance minimum, propagation direction, conservation of the product of the square of the width and the peak dark irradiance (the so-called soliton constant), and the conservation of the total number of dark structures. The authors mentioned that in their experiments the fundamental dark soliton, and the number of such irradiance minima was conserved.

Finally, we would like to mention an effect which, probably, may be also understood with the help of the dark soliton phenomenology. In 1985 Jerominek *et al.*[70,71] observed an interesting nonlinear phenomenon in photorefractive slab waveguides made by titanium metal diffusion into $LiNbO_3$ crystals, which they called the "optical branching effect". They observed that when the power of an injected optical beam reaches a certain threshold value, the TE_0 mode propagating in a $Ti : LiNbO_3$ slab waveguide splits into three (or more) beams across the direction of the c^-axis of the crystal. These beams propagate at different angles on each side of the direction of the original (input) beam. When the power is increased further, more beams appear. Assuming that the photorefractive effect can be associated with a Kerr effect, Belanger and Mathieu [72] proposed a qualitative explanation of this effect. Indeed, since in a defocusing medium dark solitons can be generated, it is natural to comjecture that the splitted of the beam is simply associated with the production of dark solitons. In their experiments Jerominek *et al.*[70,71] injected a small Gaussian beam, so that for this reason it is difficult to compare directly their experimental data with the model based on an incident plane wave [72]. However, as we have seen for temporal dark solitons, when dark solitons are produced on a background pulse with a constant phase, one can expect that they will be produced in pairs. It means that to the lowest order, we could anticipate the production of two (symmetric) dark solitons, giving rise to a three-beam structure as was observed experimentally as a result of an increase in the intensity [70,71]. Belanger and Mathieu [72] estimated the width of the hole between the central beam and the first beam and they concluded that the soliton model of the optical branching effect is qualitatively consistent with the experimental result. However, more exact comparison may be done after more precise experimental results are obtained because the photorefractive effect is more complicated and it may be described as a Kerr effect only in the lowest approximation.

9. Conclusions

In conclusion, it has been shown that dark solitons demonstrate a number of interesting properties which may be important for different applications in nonlinear optics. Dark-soliton pulses are less affected by loss or background fluctuations than are bright solitons, but dark solitons are destroyed under the action of the Raman self-scattering. It is also important that dark solitons are stable near the zero-dispersion point, and there is a region in the dispersion parameters where these solitons may change their amplitudes to be transformed into anti-dark solitons. The theory which assumes the infinite boundary conditions to the NLS equation may give good predictions for dark solitons with finite-width background pulses when the background pulses are only 10 times wider than the soliton width. Series of numerical simulations demonstrate such a property, and also dark pulses observed in experiments behave like dark solitons. I have briefly described experimental results which show a possibility of dark-soliton propagation in nonlinear optical fibers (temporal dark solitons), laser beams passed through sodium vapor, and nonlinear slab waveguides (spatial dark solitons).

Thus, all results presented lead to the following conclusions. One can easily create dark optical solitons by different driving pulses and these pulses are rather stable against small perturbations. However, on the other hand, the dark solitons may propagate on distances limited by the self-induced Raman effect, so that applications of dark solitons in long optical communication systems likely are not possible. At the same time, dark solitons may be useful for other applications in nonlinear optics (especially, spatial dark solitons) because they have many interesting properties along with their structure stability. Such solitons are responsible for different nonlinear effects in the normal dispersion region because they may be created without a threshold and naturally arise from any dip on a finite-width pulse.

Acknowledgments

The work was partially supported by the Universidad Complutense de Madrid (Spain) through Programa Sabaticos de Complutense and Alexander von Humboldt-Stiftung (Germany).

References

1. A. Hasegawa and F. Tappert, *Transmission of stationary nonlinear optical pulses in dispersive dielectric fibers. 1. Anomalous dispersion*, Appl. Phys. Lett. **23**, 142 - 144 (1973).

2. L.F. Mollenauer, R.H. Stolen, and G.P. Gordon, *Experimental observation of picosecond pulse narrowing and solitons in optical fibers*, Phys. Rev. Lett. **45**, 1095 - 1098 (1980).

3. G.P.Agrawal, *Nonlinear Fiber Optics* (Springer-Verlag, Berlin, 1989).

4. A. Hasegawa,*Optical Solitons in Fibers* (Springer-Verlag, Berlin, 1989).

5. L.F. Mollenauer, M.J. Neubelt, S.G. Evangelides, J.P. Gordon, J.R. Simpson, L.G. Cohen,*Experimental study of soliton transmission over more than 10.000 km in dispersion-shifted fiber*, Opt. Lett. **15**, 1203 - 1205 (1988).

6. A. Hasegawa and F. Tappert, *Transmission of stationary nonlinear optical pulses in dispersive dielectric fibers. 2. Normal dispersion*,Appl. Phys. Lett. **23**, 171 - 172 (1973).

7. A.M. Weiner, J.P. Heritage, R.J. Hawkins, R.N. Thurston, E.M. Kirschner, D.E. Leaird, and W.J. Tomlinson,*Experimental observation of the fundamental dark soliton in optical fibers,* Phys. Rev. Lett. **61**, 2445 - 2448 (1988).

8. P. Emplit, J.P. Hamaide, F. Reynaud, C. Froehly and A. Barthelemy,*Picosecond steps and dark pulses through nonlinear single mode fibers*,Opt. Commun. **62**, 374 - 379 (1987).

9. D. Krökel, N.J. Halas, G. Guiliani and D. Grischkowsky, *Dark-pulse propagation in optical fibers*,Phys. Rev. Lett. **60**, 29 - 32 (1988).

10. K.J. Blow and N.J. Doran,*Multiple dark soliton solutions of the nonlinear Schrodinger equation*,Phys. Lett. A **107**, 55 - 58 (1985).

11. D.R. Andersen, D.E. Hooton, G.A. Swartzlander, Jr., and A.E. Kaplan,*Direct measurement of the transverse velocity of dark spatial solitons*, Opt. Lett. **15**, 783-785 (1990).

12. G.A. Swartzlander. Jr., D.R. Andersen, J.J. Regan, H. Yin, and A.E. Kaplan,*Spatial dark-soliton stripes and grids in self-defocusing materials*, Phys. Rev. Lett. **66**, 1583-1586 (1991).

13. G.R. Alan, S.K. Skinner, D.R. Andersen, and A.L. Smirl, *Observation of fundamental dark spatial solitons in semiconductors using picosecond pulses*, Opt. Lett. **16**, 156-158 (1991).

14. S.A. Gredeskul and Yu.S. Kivshar, *Generation of dark solitons in optical fibers*,Phys. Rev. Lett. **62**, 977 (1989).

15. R.H.Enns and L.J. Mulder*Bistable holes in nolinear optical fibers*, Opt. Lett. **14**, 509-511 (1989).

16. W. Zhao and E. Bourkoff, *Periodic amplification of dark solitons using stimulated Raman scattering,*Opt. Lett. **14**, 808 - 810 (1989).

17. W.J. Tomlinson, R.J. Hawkins, A.M. Weiner, J.P. Heritage, and R.N.Thurston, *Dark optical solitons with finite-width background pulses,*J. Opt. Soc. Am. B **6**, 329 - 334 (1989).

18. W. Zhao and E. Bourkoff, *Propagation properties of dark solitons,*Opt. Lett. **14**, 703 - 705 (1989).

19. W. Zhao and E. Bourkoff,*Interaction between dark solitons,*Opt. Lett. **14**, 1371 - 1373 (1989).

20. S.A. Gredeskul and Yu.S. Kivshar, *Dark-soliton generation in optical fibers,*Opt. Lett. **14**, 1281 - 1283 (1989).

21. S.A. Gredeskul, Yu.S. Kivshar, and M.V. Yanovskaya, *Dark-pulse solitons in nonlinear optical fiber,*Phys. Rev. A **41**, 3994 - 4008 (1990).

22. S.A.Gredeskul and Yu.S.Kivshar*Generation of dark solitons from a Raman noise*, Pisma Zh. Tekhn. Fiz. (Sov. Techn. Phys. Lett.) **16**, 25-29 (1990).

23. Yu.S. Kivshar, *Dark soliton dynamics and shock waves stimulated by the Raman effect in optical fibers,*Phys. Rev A **42**, 3292 - 3296 (1990).

24. Yu.S. Kivshar and S.A. Gredeskul,*Dark-solitons produced by phase steps in nonlinear optical fibers,*Opt. Commun. **79**, 285 - 290 (1990).

25. Yu.S.Kivshar,*Dark solitons in optical fibers*, In: *Solitons and Applications*, Eds. V.G.Makhankov, V.K. Fedyanin, and O.K.Pashaev (World Scientific, Sigapore, 1990), pp. 298-315.

26. Yu.S. Kivshar, *Perturbation-induced dynamics of small- amplitude dark optical solitons,*Opt. Lett. **15**, 1273- 1275 (1990).

27. J.A. Giannini and R.I Joseph*The propagation of bright and dark solitons in lossy optical fibers*, IEEE J. Quantum Electron. **26**, 2109-2124 (1990).

28. Yu.S. Kivshar*Nonlinear dynamics near the zero-dispersion point in optical fibres*, Phys.Rev.A **43**, 1677-1679 (1991).

29. E.N. Thurston and A.W.Weiner,*Collisions of dark solitons in optical fibres*, J.Opt.Soc.Am. B **8**, 471-477 (1991).

30. V.V. Konotop and V.E.Vekslerchyk,*Randomly modulated dark solitons*, J.Phys.A **24**, 767-785 (1991).

31. Yu.S. Kivshar and V.V. Afanasjev,*Decay of dark solitons under the stimulated Raman scattering*, Opt.Lett. **16**, 285-287 (1991).

32. V.E. Vekslerchyk,*Dark solitons of the generalized nonlinear Shrödinger equation*, Phys Lett.A **153**, 195-198 (1991).

33. Yu.S. Kivshar,*Dark optical solitons near the zero-group-dispersion wavelength*, Opt.Lett. **16**, 892- 894 (1991).

34. Yu.S. Kivshar and V.V. Afanasjev*Dark optical solitons of the reverse-sign amplitude*, Phys.Rev.A **44**, R1446- R1449 (1991).

35. V.V. Afanasjev, Yu.S.Kivshar, and V.N. Serkin,*Generation and dynamics of dark optical solitons*, J. Opt. Soc. Am. B (1991) to be published.

36. A.B. Grudinin, E.M. Dianov, A.M.Prokhorov, and D.V. Khaidarov, Pisma Zh. Techn. Fiz. (Sov. Tekhn. Phys. Lett.) **14**, 1010-1014 (1988).

37. A.M. Weiner, R.N. Thurston, W.J. Tomlinson, J.P. Heritage, D.E. Leaird, E.M. Kirschner, and R.J.Hawkins,*Temporal and spectral self-shift of dark optical solitons*, Opt.Lett. **14**, 868-970 (1989).

38. J.E. Rothenberg*Colliding visible picosecond pulses in optical fibers*, Opt. Lett. **15**, 443-445 (1990).

39. J.E. Rothenberg*Dark soliton trains formed by visible pulse collisions in optical fibers*, Opt. Commun. **82**, 107-111 (1991).

40. J.P.Hamaide, Ph. Emplit, and M.Haelterman,*Dark-soliton jitter in amplified optical transmission systems*, Opt. Lett. **16**, 1578- 1580 (1991).

41. V.E.Zakharov and A.B.Shabat,*Interactions between solitons in a stable medium*, Zh. Eksp. Teor. Fiz. **64**, 1627-1635 (1973) (Sov. Phys. JETP **37**, 923-930 (1973).

42. J.Satsuma and N.Yajima,*Initial value problem of one-dimensional self-modulation of nonlinear waves in dispersive madia*, Progr. Theor. Phys. Suppl. **55**, 284-306 (1974).

43. V.E. Zakharov, S.V. Manakov, S.P. Novikov, and L.P. Pitaevsky,*Theory of solitons. The inverse scattering transform* (Nauka, Moscow, 1980) (English translation by Consultant Bureau, 1984).

44. Yu.S. Kivshar,*On soliton generation in optical fibers*, J.Phys.A **22**, 337-340 (1989).

45. A. Kumar,*Soliton dynamics in a monomode optical fiber*, Phys. Reports **187**, 63-108 (1990).

46. V.E. Zakharov and A.B. Shabat,*Exact theory of two-dimensional self-focusing and one-dimensional self-modulation of waves in nonlinear media*, Sov. Phys. JETP (Zh. Eksp. Teor. Fiz.) **34**, 62-69, 1972.

47. R.H. Stolen, J.P. Gordon, W.J. Tomlinson, and H.A. Haus, *Raman responce function of silica-core fibers*, J. Opt. Soc. Am. B **6**, 1159-1166 (1989).

48. D. Andersen,*Variational approach to nonlinear pulse propagation in optical fibers*, Phys. Rev. A **27**, 3135-3145 (1983).

49. Yu.S. Kivshar and B.A. Malomed,*Dynamics of solitons in nearly integrable systems*, Rev. Mod. Phys. **61**, 761-915 (1989).

50. V.I. Karpman and E.M. Maslov,*Perturbation theory for solitons*, Zh. Eksp. Teor. Fiz. **73**, 537-543 (1977) [Sov. Phys. JETP **46**, 281-292 (1977).

51. V.I. Karpman,*Soliton system under perturbations. Oscillatory shock waves*, Zh. Eksp. Teor. Fiz. **77**, 114-123 (1979) [Sov. Phys. JETP **50**, 58-63 (1979)].

52. A. Hasegawa and Y. Kodama,*Signal transmission by optical solitons in monomode fiber*, Proc. IEEE **69**, 1145-1150 (1981).

53. P.K.A. Wai, C.R. Menyuk, Y.C. Lee, and H.H. Chen, *Nonlinear pulse propagation in the neighborhood of the zero-dispersion wavelength of monomode optical fiber*, Opt. Lett. **11**, 464-466 (1986);*Soliton at the zero-dispersion wavelength of single-mode fiber*, Opt. Lett. **12**, 628-630, (1987).

54. M.Desaix, D. Anderson, and M. Lisak,*Soliton emerging from pulses launched in the vicinity of the zero-dispersion point in a single-mode optical fiber*, Opt. Lett. **15**, 18-20 (1990).

55. P.K.A. Wai, H.H. Chen, and Y.C. Lee,*Radiations by solitonsat the zero-group-dispersion wavelength of a single-mode optical fiber*, Phys. Rev. A **41**, 426-439 (1990).

56. J.P. Gordon,*Interaction forces among solitons in optical fibers*, Opt. Lett. **8**, 596-598 (1983).

57. V.V. Afanasjev, Yu. S. Kivshar, V.V. Konotop, and V.N. Serkin,*Dynamics of coupled dark and bright optical solitons*, Opt. Lett. **41**, 805-807 (1989).

58. T. Ueda and W.L. Kath,*Dynamics of coupled solitons in nonlinear optical fibers*, Phys. Rev. A **42**, 563-571 (1990).

59. B.J. Hong, C.C. Yang, and L. Wang,*Generating dark solitons through cross-phase modulation in optical fibers*, J. Opt. Soc. Am. B **8**, 464-470 (1991).

60. S.Trillo, S. Wabnitz, E.M. Wright, and G.I. Stegeman, *Optical solitary waves induced by cross-phase modulation in optical fibers*, Opt. Lett. **13** 871-873 (1988).

61. V.V. Afanasyev, E.M. Dianov, and V.N. Serkin,*Nonlinear pairing of short bright and dark solitons*, Sov. Phys. JETP Lett. (Pisma Zh. Eksp. Teor. Fiz.) **48**, 638-640 (1988).

62. V.V. Afanasjev, E.M. Dianov, and V.N. Serkin,*Nonlinear pairing of short bright and dark solitons by phase cross modulation*, IEEE J. Quantum Electron., **25**, 2656-2664 (1989).

63. L. Wang and C.C. Yang,*Existence of solitary waves in optical fibers due to the mutual support between bright and dark pulses*, Opt. Lett., **15**, 474-476 (1990).

64. M. Lisak, A. Höök, and D. Anderson,*Symbiotic solitary-wave pairs sustained by cross-phase modulation in optical fibers*, J. Opt. Soc. Am. B **7**, 810-814 (1990).

65. Yu.S. Kivshar, D. Anderson, A. Höök, A.A. Afanasjev, and V.N. Serkin,*Symbiotic optical solitons and modulational instability*, Physica Scripta **44**, 195-202 (1991).

66. J.S. Aitchison, A.M. Weiner, Y. Silberger, M.K. Oliver, J.L. Jackel, D.E. Leaird, E.M. Vogel, and P.W.E. Smith,*Observation of spatial optcal solitons in a nonlinear glass waveguide*, Opt. Lett. **15**, 471-473 (1990).

67. F. Reynaud and A. Barthelemy*Optically controlled interaction between two fundamental soliton beams*, Eurohys. Lett. **12**, 401-405 (1990).

68. P.L. Kelly,*Self-focusing of optical beams*, Phys. Rev. Lett. **15**, 1005-1008 (1964).

69. M. Sheik-bahae, D.J. Hagan, and E.W. VanStryland, In: *Intern. Quantum Electron. Conf.*, Vol. 8 of The 1990 OSA Technical Digest Series (OSA, Washington, D.C., 1990), paper QTUM3.

70. H. Jerominek, R. Tremblay, and C. Delisle,*Optical branching effect in photorefractive sensitive Ti:LiNbO$_3$ slab waveguides*, IEEE/OSA J. Lightwave Technol. **3**, 1105-1107 (1985).

71. H. Jerominek, C. Delisle, and R. Tremblay,*Optical branching effect in Ti:LiNbO$_3$ waveguides: near field pattern studies*, Appl. Opt. **25**, 732-736 (1986),

72. P.A. Belanger and P. Mathieu,*Dark soliton in a Kerr defocusing medium*, Appl. Opt. **26**, 111-113 (1987).

Networks

Some experimental results on asymmetric Boltzmann machines.

Bruno Apolloni *
Angela Armelloni*
Giuseppe Bollani*
Diego de Falco**

*Dipartimento di Scienze dell' Informazione,Università di Milano, Italy
** Dipartimento di Matematica, Politecnico di Milano, Italy

ABSTRACT

We present some experimental results on the asymmetric block-sequential Boltzmann machine and, for comparison purposes, on the more conventional sequential symmetric Boltzmann machine.

1. Introduction

With reference to the geometric setup of figure 1 , we report in figures 2,3,4 the results[1,2] of training sessions aimed at learning by examples the elementary Boolean functions AND, OR, XOR, on a <u>symmetric</u> Boltzmann machine (BM) [3].

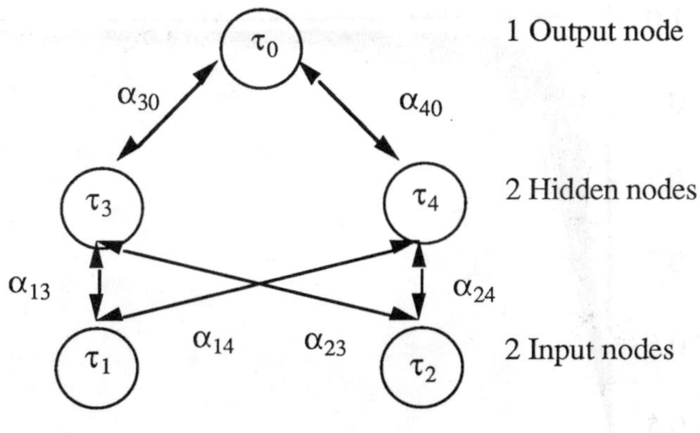

Fig. 1: A 2-2-1 multilayer perceptron (MLP); $\tau_i \in \{0,1\}$;
the double arrows indicate the symmetry of the coupling constants $\alpha_{ij} = \alpha_{ji}$.

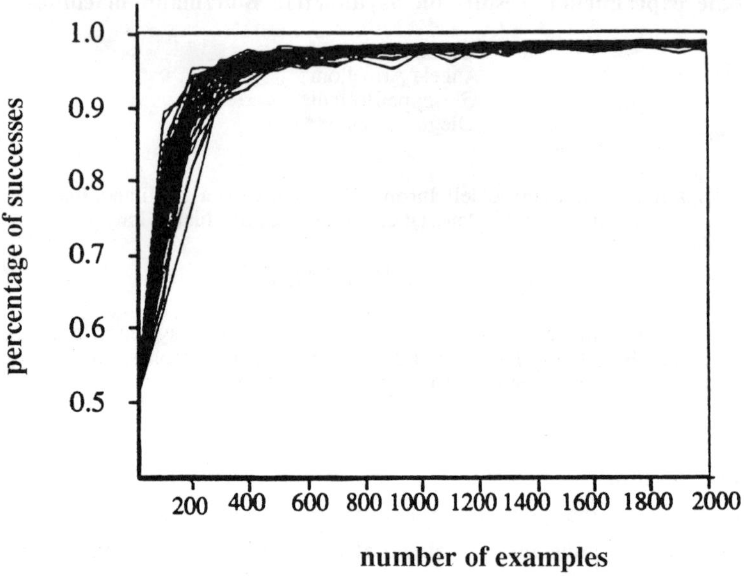

Fig 2: Training, by BM, the 2-2-1 symmetric network of Fig 1 to learn the AND function ($\tau_0 = \tau_1 \wedge \tau_2$).

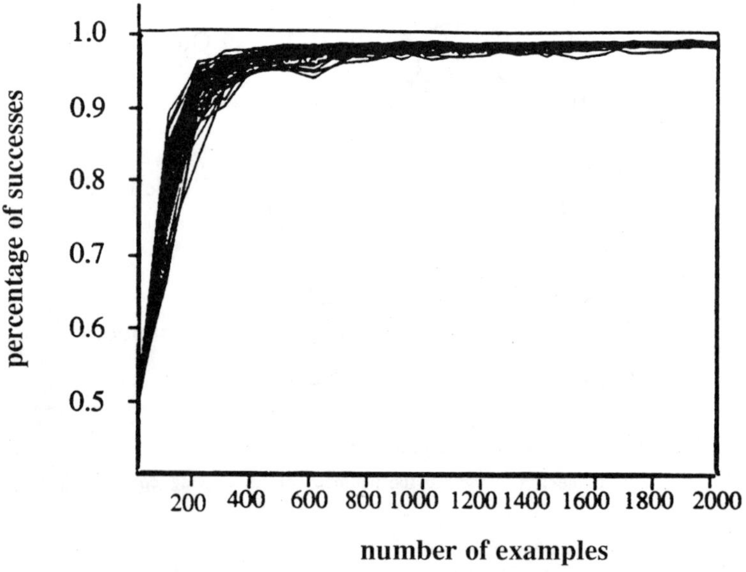

Fig 3: Training a 2-2-1 symmetric network to learn the OR function ($\tau_0 = \tau_1 \vee \tau_2$).

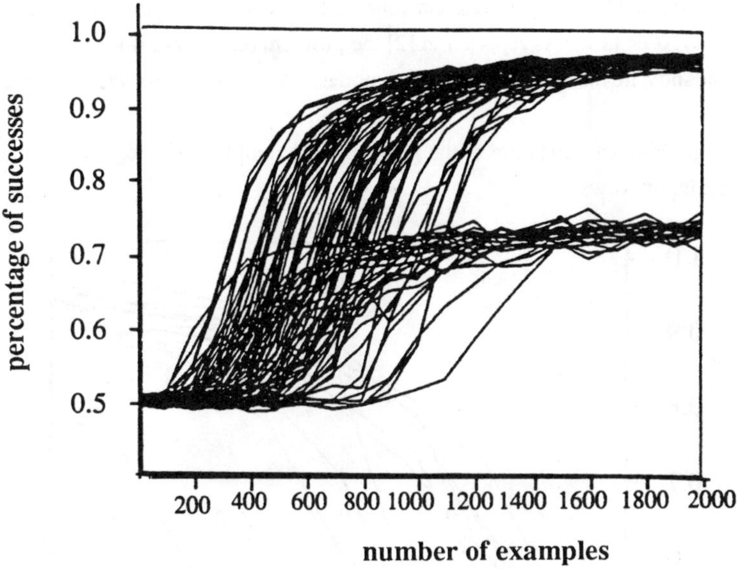

Fig 4: Training a 2-2-1 <u>symmetric</u> network to learn the XOR function ($\tau_0 = (\tau_1 \vee \tau_2) \wedge \overline{(\tau_1 \wedge \tau_2)}$); percentage of higher trajectories = 74%.

By a heat bath algorithm the network is, with the current symmetric matrix α of coupling constants and the current vector $\underline{\gamma}$ of local constants, driven to equilibrium at a Gibbs state $\mu(\underline{\tau},\alpha) \sim \exp(-\beta\,(1/2\,\underline{\tau}\cdot\alpha\,\underline{\tau} + \underline{\gamma}\cdot\underline{\tau}))$. On the basis of a training set of correctly associated pairs ($((\tau_1,\tau_2),\tau_0)$, the size n of which is reported in abscissa, the coupling constants α are updated by minimizing an estimated risk function (relative entropy) which measures the discrepancy between the joint distibution of (τ_0,τ_1,τ_2) estimated from the sample (training set) and the one in the Gibbs state attained with the current α. The percentage of successes (correctly associated pairs $((\tau_1,\tau_2),\tau_0)$ exhibited by the machine in its free evolution) reported in ordinate shows how the evolution $\alpha(n)$ of the coupling matrix with the sample size drives the support of $\mu(\underline{\tau},\alpha)$ toward the locus of correctly associated input-output pairs.

Each of the trajectories in the figures reports the progress of an individual network out of a set of 100 networks (All initialized with $\alpha = 0$, "tabula rasa"). It is interesting to relate the steepness of the trajectories to the difficulty of the task (technically speaking AND and OR are linearly separable functions, while XOR isn't). The fact that a percentage ($\sim$ 25 %) of the networks considered do not learn the XOR task is a remnant of the perceptron problem [4] and of the fact that the geometric setup of figure 1 is the minimal one, without direct connection between input and output, able to perform the XOR task. As reported in [2] the problem disappears if we allow for some redundancy in the number of hidden nodes, adopting for instance a 2-4-1 geometry.

Figure 5 might give an indication of the growth of the sample size n required as the length of the input grows.

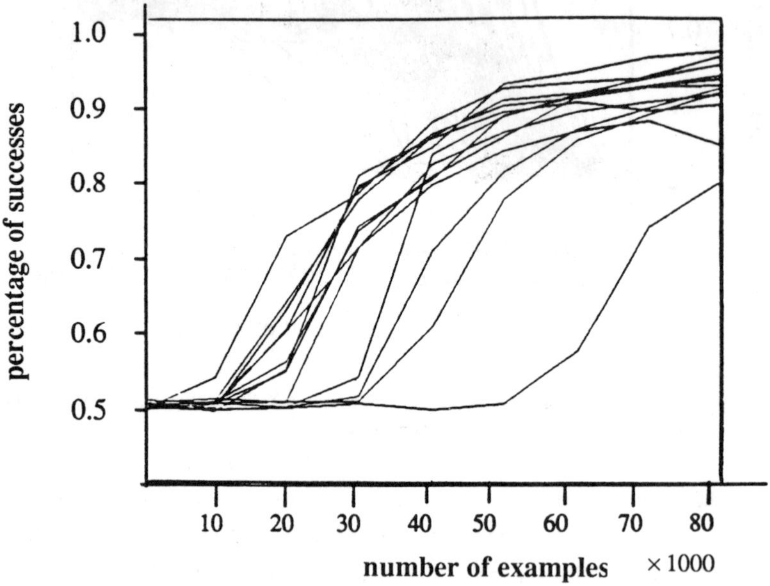

Fig 5: Training a 7-9-1 symmetric network to learn the PARITY function ($\tau_0 = ((-1)^{\Sigma_{input}\tau_i} + 1)/2$; percentage of higher trajectories = 87%.

Inspection of figure 1 shows a quite unnatural feature of the symmetric models considered up to now: if the issue of the learning task is the input-output relation between (τ_1 ,τ_2) and τ_0 (as encoded in the conditional distribution of τ_0 given (τ_1,τ_2)) the double arrows (symmetric coupling constants) by which the output node reacts back on the input nodes seem redundant.

We have given some thought to the idea of releasing the assumption of symmetric coupling constants, which, while quite natural in a statistical mechanical context (in which "spins exchange forces") is quite unnatural in a cybernetical context (in which "neurons exchange information"). In [5] we introduced an entropic learning rule for parallel asymmetric Boltzmann machines. In [6] we analyzed the capabilties of parallel asymmetric Boltzmann machines. In [6] we analyzed the capabilties of asymmetric machines from the point of view of the computational complexity of the tasks they can perform. Here we restrict our considerations to the peculiar asymmetry present in a feedforward architecture (in which, in particular, if $\alpha_{i,j} \neq 0$ then $\alpha_{j,i} = 0$) and report some experimental results on the specialization and implementation of the general learning algorithm of [5] on such an architecture.

2. The model

We refer to a multilayer perceptron architecture as in fig. 6.

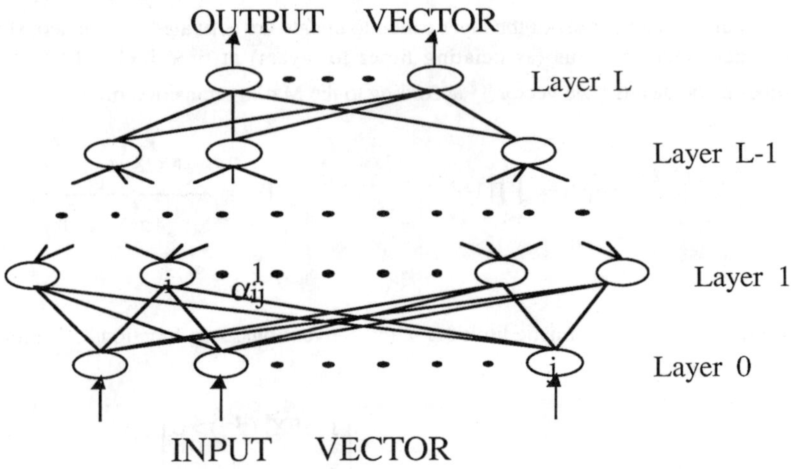

Fig 6. A multilayer perceptron

The neurons are arranged in layers, numbered from 0 to L. The ℓ-th layer has n_ℓ neurons; its state is described by a Boolean vector $\tau^\ell = (\tau^\ell_1, ..., \tau^\ell_{n_\ell}) \in \{0,1\}^{n_\ell}$; α^ℓ_{ij} is the synaptic weight of the connection from neuron j of layer ℓ-1 to neuron i of layer ℓ.

The signal present on neuron i of layer ℓ is

$$X_i^\ell(\tau^{\ell-1}) = \sum_{j=1}^{n_{\ell-1}} \alpha_{ij}^\ell \tau_j^{\ell-1} + \gamma_i^\ell$$

where $-\underline{\gamma}^{\ell} \in \mathbb{R}^{n_\ell}$ is the vector of thresholds for layer ℓ.

Based on this signal, neuron i of layer ℓ randomly updates its state T^{ℓ}_i according to the usual stochastic rule of the Boltzmann machine [3], namely

$$E(T_i^{\ell}(t+1)|\underline{T}^{\ell-1}(t)) = \frac{1}{1+e^{-\beta X_i^{\ell}(\underline{T}^{\ell-1}(t))}}$$

where E(|) indicates conditional expectation. The iteration mode is block sequential; all the neurons of the layer called to update are activated simultaneously and independently. Thus (associating times to layers) at time $\ell=1,\ldots,L$ layer number ℓ updates its state vector $\underline{T}^{\ell}$ according to the Markov transition rule:

$$P(\underline{T}^{\ell}=\underline{\tau}^{\ell}|\underline{T}^{\ell-1}=\underline{\tau}^{\ell-1}) = p_{\underline{\tau}^{\ell-1},\underline{\tau}^{\ell}} = \prod_{i=1}^{n_\ell}[1+e^{\beta(1-2\tau_i^{\ell})\sum_{j=1}^{n_{\ell-1}}\alpha_{ij}^{\ell}\tau_j^{\ell-1}+\gamma_i^{\ell}}]^{-1} = \frac{e^{\beta\,\underline{\tau}^{\ell}\cdot\underline{X}^{\ell}(\underline{\tau}^{\ell-1})}}{\sum_{\underline{\sigma}^{\ell}} e^{\beta\,\underline{\sigma}^{\ell}\cdot\underline{X}^{\ell}(\underline{\tau}^{\ell-1})}}$$

Notice that, in the deterministic limit $\beta\to\infty$, the conventional step function activation rule is recovered:

$$\lim_{\beta\to\infty} P(T_i^{\ell}=1|\underline{T}^{\ell-1}=\underline{\tau}^{\ell-1}) = \begin{cases} 1 & \text{if } X_i^{\ell}(\underline{\tau}^{\ell-1})>0 \\ \frac{1}{2} & \text{if } X_i^{\ell}(\underline{\tau}^{\ell-1})=0 \\ 0 & \text{if } X_i^{\ell}(\underline{\tau}^{\ell-1})<0 \end{cases}$$

3. The learning strategy

According to the usual schemes of supervised learning, the training agent initializes layer 0 with an input vector $\underline{\tau}^0 \in \{0,1\}^{n_0}$, and proposes also a desired output vector $\underline{\tau}^L \in \{0,1\}^{n_\ell}$.

The input-output pairs ($\underline{\tau}^0$, $\underline{\tau}^L$) are proposed according to a joint probability law $\varphi(\underline{\tau}^0,\underline{\tau}^L)$ which is part of the training instance.

The state $\underline{T}^L$ exhibited by the output layer L is a random variable both because $\underline{T}^0$ is randomly chosen, with probability law

$$\psi(\overset{0}{\underline{\tau}}) = \sum_{\underline{\tau}^L} \varphi(\overset{0}{\underline{\tau}},\overset{L}{\underline{\tau}}),$$

and because the transitions of the successive layers are random.

The input-output relation defined by a given assignment of weights and thresholds is described by the joint probability law

$$\pi_\beta(\overset{0}{\underline{\tau}},\overset{L}{\underline{\tau}}) = \psi(\overset{0}{\underline{\tau}}) \sum_{\underline{\tau}^1,\ldots,\underline{\tau}^{L-1}} p_{\underline{\tau}^0,\underline{\tau}^1} p_{\underline{\tau}^1,\underline{\tau}^2} \cdots p_{\underline{\tau}^{L-1},\underline{\tau}^L}$$

The learning task considered here is that of adjusting the parameters α and $\underline{\gamma}$ in such a way that the joint law π_β defined by the machine get close, in relative entropy to the joint law φ defined by the training agent. Namely, the task is that of minimizing the relative entropy :

$$D = \sum_{\underline{\tau}^0,\underline{\tau}^L} \varphi(\overset{0}{\underline{\tau}},\overset{L}{\underline{\tau}}) \ln \frac{\varphi(\overset{0}{\underline{\tau}},\overset{L}{\underline{\tau}})}{\pi_\beta(\overset{0}{\underline{\tau}},\overset{L}{\underline{\tau}})}$$

The direction of gradient descent motion in the parameter space is determined by the quite explicit form of the derivatives of D, expressible in terms of conditional expectations as:

$$\frac{\partial D}{\partial \alpha^{\ell}_{ij}} = \beta \sum_{\underline{\tau}^0,\underline{\tau}^L} \varphi(\overset{0}{\underline{\tau}},\overset{L}{\underline{\tau}}) E\Big(E(T^{\ell}_i T^{\ell-1}_j \mid \underline{T}^{\ell-1}) - T^{\ell}_i T^{\ell-1}_j \;\Big|\; \underline{T}^0 = \overset{0}{\underline{\tau}};\, \underline{T}^L = \overset{L}{\underline{\tau}} \Big)$$

$$\frac{\partial D}{\partial \gamma^{\ell}_{i}} = \beta \sum_{\underline{\tau}^0,\underline{\tau}^L} \varphi(\overset{0}{\underline{\tau}},\overset{L}{\underline{\tau}}) E\Big(E(T^{\ell}_i \mid \underline{T}^{\ell-1}) - T^{\ell}_i \;\Big|\; \underline{T}^0 = \overset{0}{\underline{\tau}};\, \underline{T}^L = \overset{L}{\underline{\tau}} \Big)$$

The Hebbian rule which emerges from such a structure is the following: among the input-output pairs presented as final state by the machine, those (to which we restrict by conditioning the expectations in the above formulae) which are significantly represented in the training set (which we take into account by weighing the above sums with φ), for short the "right answers", determine the direction of motion in parameter space.

This learning rule is to be compared with the conventional one for the sequential symmetric Boltzmann machine [3], and its variant for the parallel symmetric machine [7], in which the direction of motion in parameter space is determined by the unconditioned two point functions of the system at equilibrium.

4. The CELLINI algorithm

The theoretical considerations of the previous section motivate the following learning algorithm for feed-forward Boltzmann machines, which we call CELLINI since it is evocative of the pedagogical philosophy (the trainer intervenes to confirm the right answer of the trainee) which emerges from a passage of the autobiography of the Italian Renaissance artist..

The instance we consider is one in which the input-output pairs are generated by a function $g:\underline{\tau}^0 \in \{0,1\}^{n_0} \rightarrow \underline{\tau}^L \in \{0,1\}^{n_L}$.

The steps of the algorithm are the following

INITIALIZE: defines the geometrical structure of the feed-forward network, under user's control, and randomly sets the initial weights and thresholds.

EXAMPLES: generates a training set of points ($\underline{\tau}^0$,g($\underline{\tau}^0$)), with $\underline{\tau}^0$ distributed according to a user assigned law.

WALKING(k): For $\underline{T}^o$ set to the value of the k-th input vector $\underline{\tau}^o(k)$, iterates the random block sequential activation of section 2 until a sample path with $\underline{T}^L$=g($\underline{\tau}^0$(k)) is obtained.

STATISTICS: For the "right" trajectories singled out by WALKING cumulates in acc[i,j], for each link (i,j), the quantity

$$[\tau_j^{\ell-1} \tau_i^{\ell} - \tau_j^{\ell-1} \frac{1}{1+e^{-\beta X_i^{\ell}(\underline{\tau}^{\ell-1})}}] \quad (*)$$

(*) This and the following statistics will be referred to the translation of the above derivatives in {-1,1} domain, as the relative entropy landscape in these coordinates leads to an increase in success percentage.

UPDATING(label): On label=0, updates the weights according to $\alpha_{ij}+\delta\cdot acc[i,j]$, then sets acc[i,j]=0.

The dynamics of the above steps is ruled by the following MAIN:

```
MAIN (Nexamples,Samplesize)
    external STOPPING RULE
    INITIALIZE
    EXAMPLES
    k=0
    REPEAT
        FOR I=1 TO Nexamples DO
                        BEGIN
                        WALKING(I)
                        STATISTICS
                        END
        k=k+1
        UPDATING(mod(k,Samplesize))
    UNTIL (STOPPING RULE is TRUE)
END
```

For Samplesize large enough, we accumulate in acc[i,j] a good estimate of

$$E[T_j^{\ell-1} T_i^{\ell} - T_j^{\ell-1} \frac{1}{1+e^{-\beta X_i^{\ell}(\underline{T}^{\ell-1})}} \mid \underline{T}^0=\underline{\tau}^0 ; \underline{T}^L=\underline{\tau}^L]$$

The cyclic presentation of the training set, in turn, estimates the expectation of the above quantity with respect to the environmental distribution φ.

5. Numerical results

Figures 7 to 15 summarize our experimental work on the Cellini algorithm.
Comparison between figures 4 and 7 suggests the following considerations:

a. Quite independently of the algorithm adopted, learning proceeds by a transition from an indifferentiated phase, to two distinct plateaux.

b. Cellini is a little slower than BM in terms of number of learning cycles (actually 1 learning cycle is made of 4 examples),but more frequently successful: more trajectories reach the higher plateau in Cellini (92%) than in BM (75%).

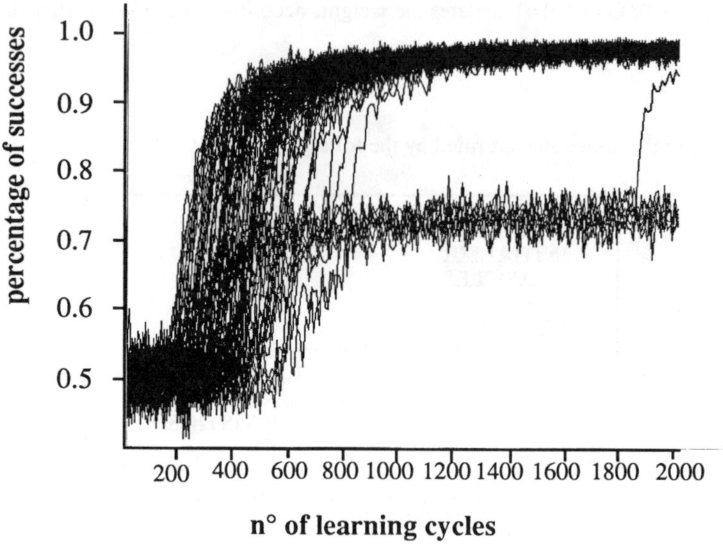

Fig 7: Training, by Cellini, a 2-2-1 feedforward network to learn the XOR function percentage of higher trajectories = 92%.

c.Inspection of the evolution of the weights shows that in most trajectories the plateau on which a given trajectory will land is determined very early in the learning process; Cellini shows a higher capability than BM in bringing trajectories from the lower to the higher plateau in later stages: this accounts for a large part of the difference in the success rate between the two methods.

Figure 8 shows, that, after an initial transient, the progress in the learning process is stable in terms of the number of right answers that a given network is able to give under deterministic evolution with the current weights attained along its individual history.

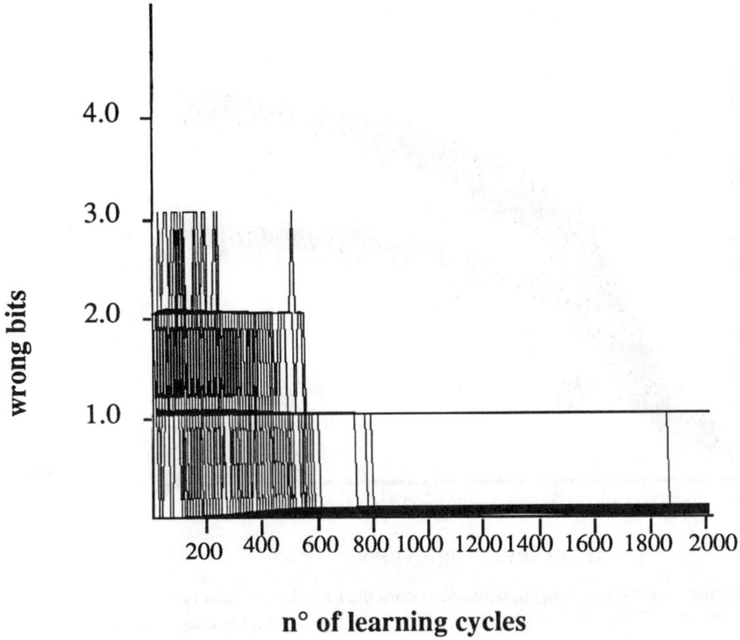

Fig 8: For the task of figure 7, the number of wrong answers (for deterministic evolution with the weights attained by Cellini) is plotted against the number of examples shown.

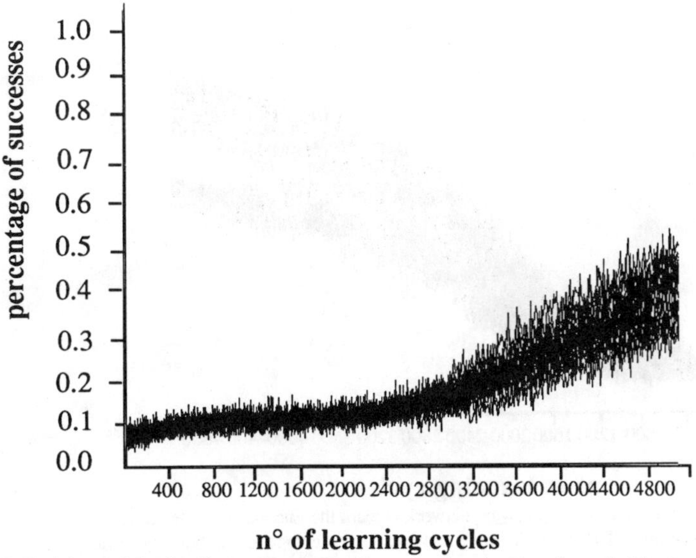

Fig 9: Training a 4-2-4 <u>feedforward</u> network to learn the Encoder-Decoder task: identity on (0,0,0,1), (0,0,1,0), (0,1,0,0), (1,0,0,0). Temperature $T \equiv 1/\beta = 2$; learning rate $\eta = .0001$;

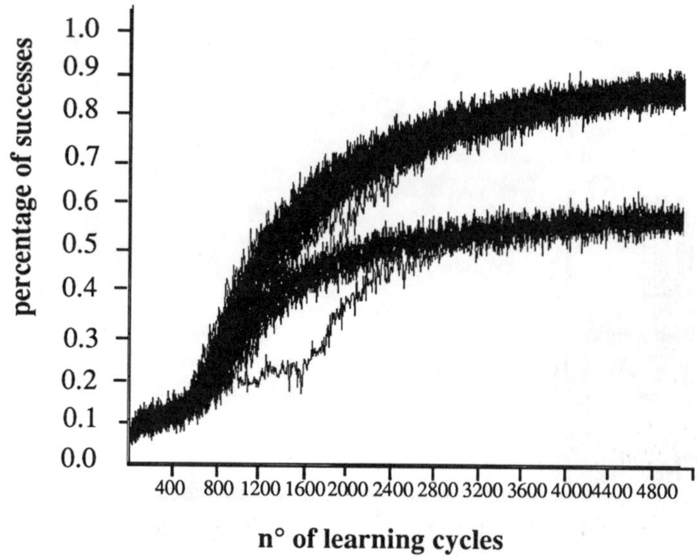

Fig 10: Training a 4-2-4 feedforward network to learn the Encoder-Decoder task. Temperature $T \equiv 1/\beta = 5$; learning rate $\eta = .001$; percentage of higher trajectories = 77%.

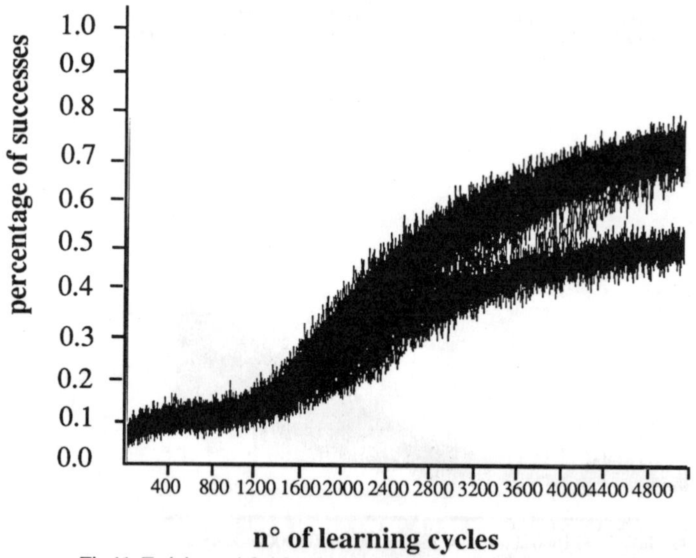

Fig 11: Training a 4-2-4 feedforward network to learn the Encoder-Decoder task. Temperature $T \equiv 1/\beta = 10$; learning rate $\eta = .001$; percentage of higher trajectories = 75%.

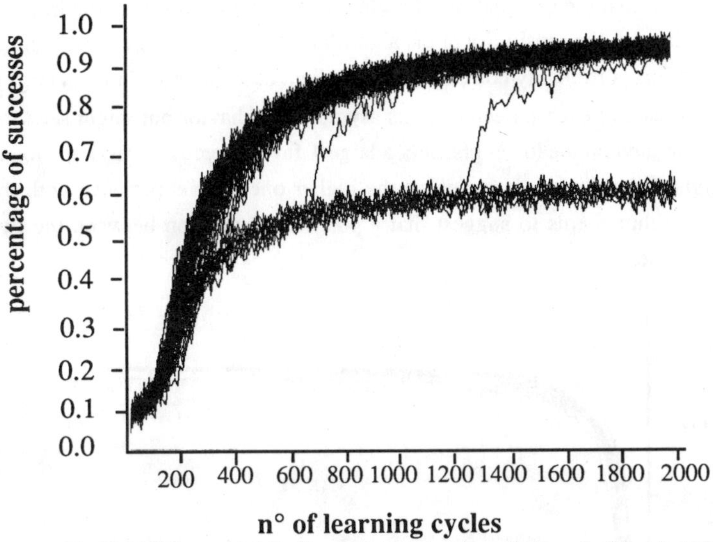

Fig 12: Training a 4-2-4 <u>feedforward</u> network to learn the Encoder-Decoder task. Temperature $T \equiv 1/\beta = 10$; learning rate $\eta = .01$; percentage of higher trajectories = 89%.

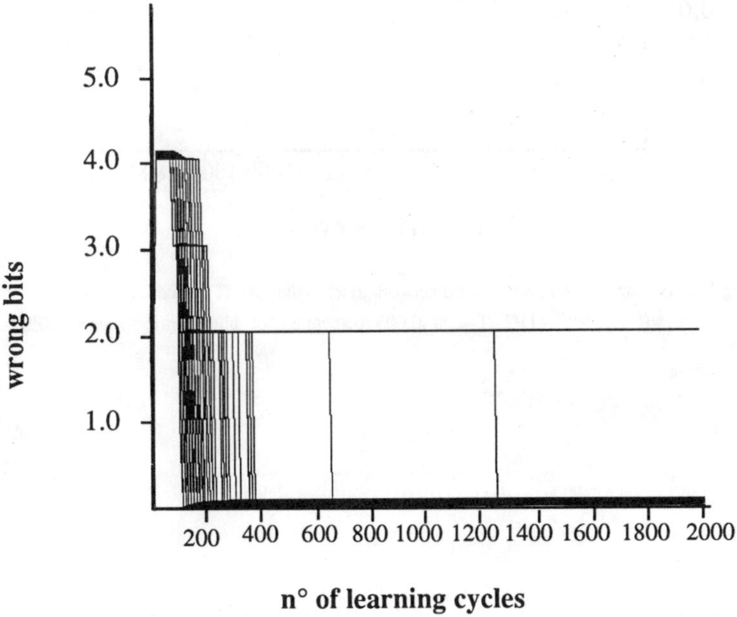

Fig 13: For the task of figure 12, the number of wrong answers (for deterministic evolution with the weights attained by Cellini) is plotted against the number of examples shown.

Figures 9 to 13, show that the two-plateaux structure is quite independent of the learning task, and try to point out the sensitivity of the algorithm to the temperature $T=1/\beta$ and to the learning rate η=proportionality constant in the gradient descent procedure: a large η favours a fast approach of each trajectory to its asymptotic behavior but might send too many trajectories to land on the lower plateau; a large T favours recovery from the lower plateau, but might destabilise trajectories on the higher one. Our experience with the tuning of the algorithm seems to suggest that a positive correlation between the two parameters is convenient.

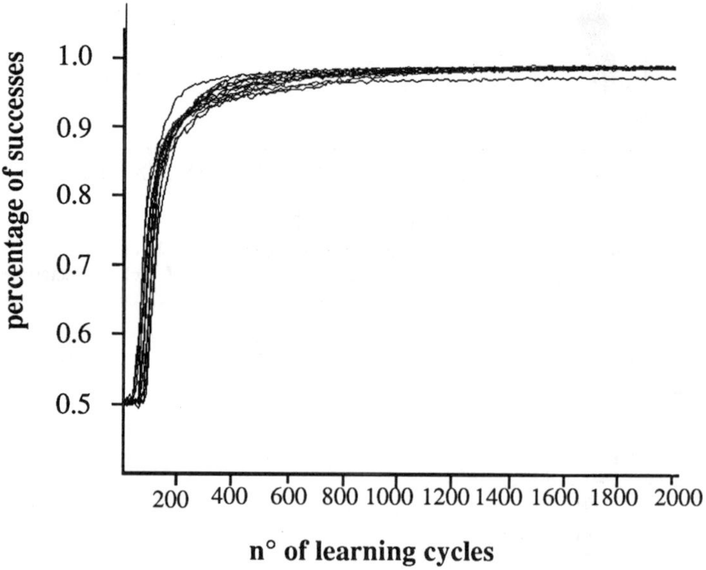

Fig 14: Training a 7-9-1 feedforward network to learn the PARITY function: $\tau_o = ((-1)^{\sum_{input}\tau_i}+1)/2$. T=2, η=0.005 ;percentage of higher trajectories = 100%.

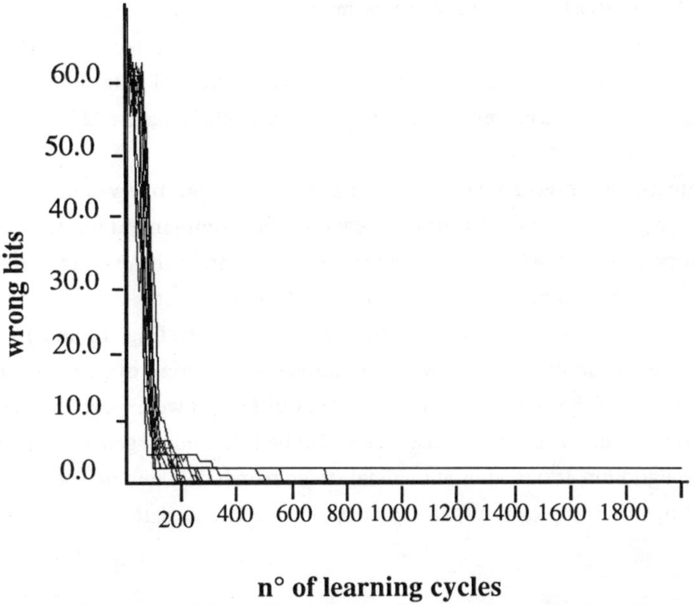

Fig 15: For the task of figure 14, the number of wrong answers (for deterministic evolution with the weights attained by Cellini) is plotted against the number of examples shown.

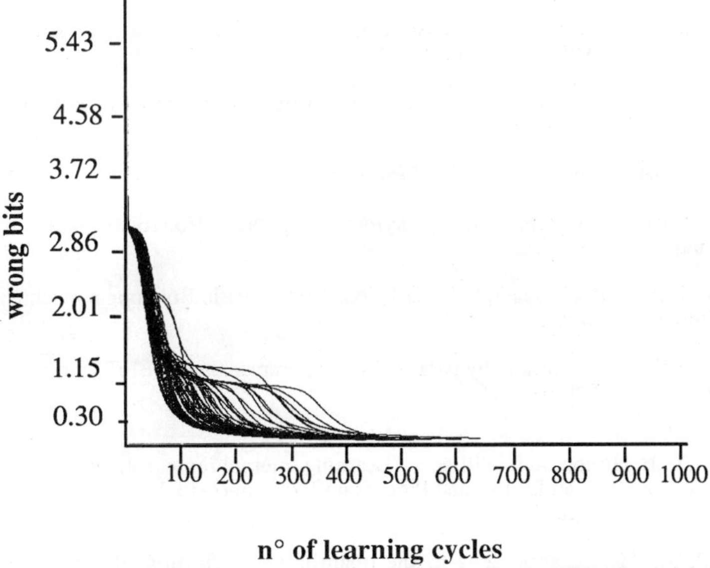

Fig 16: For the task of figure 12, the number of "wrong answers" (for deterministic evolution with the weights attained by Error backpropagation) is plotted against the number of examples shown.

Figures 14 and 15 reports the result of some preliminary runs of Cellini on the Parity function on 7 bit words. Some redundancy is allowed for in the number of hidden nodes, of which Cellini seems to benefit much more than BM in terms of both learning speed and success frequency. Note however that here a learning cycle is made of 128 examples.

As a final remark we observe that, in its trying to make more likely the trajectories underlying "right answers", Cellini appears to be complementary to the error backpropagation rule (BP)[8] which draws the error signal from those input-output pairs which are "wrong" compared to what is exhibited in the training set.
A comparison between our algorithm and BP error curves (see figg. 13 and 16) shows equivalence in error descent speed with the number of learning cycles. Actually the computational cost of the single cycle is heavier in Cellini, but we expect that this cost, possibly of no relevance with parallel machines, will be balanced by generalization skills, as an entropic algorithm as our aims at minimal structure, possible minimal program in the sense of Kolmogorov complexity [9] , able to describe the sample data.

References:

1.R. Pisano: "Reti neuronali asimmetriche" Thesis, Milano (1991).

2.B. Apolloni, D. de Falco, R. Pisano: "The Boltzmann Machine: Experimental results and theoretical perspectives", Milano preprint, 1991.

3.D.H. Ackley, G.E. Hinton, T.J. Sejnowski: "A learning algorithm for Boltzmann machines" Cognitive Science 9 (1985) 147.

4. M. Minsky, S. Papert : "Perceptrons", MIT Press 1968.

5. B. Apolloni, D. de Falco: "Learning by asymmetric parallel Boltzmann machines" Neural Computation 3 (1991) 402.

6.B. Apolloni, A. Bertoni, P. Campadelli, D. de Falco: "Asymmetric Boltzmann machines" to appear in Biological Cybernetics (1992).

7. B. Apolloni, D. de Falco: "Learning by parallel Boltzmann machines" IEEE Transactions on Information Theory 37 (1991) 1162.

8.D.E. Rumelhart, G.E. Hinton, R.J. Williams, "Learning internal representations by error backpropagation", in "Parallel Distributed Processing:Exploration in the Microstructures of Cognition" Vol I, MIT Press (1986).

9.Kolmogorov A.N., "Three approachs to the quantitative definition of information problems." J. Inform. Transm.,1,4-7, (1965).

INFORMATION-SPECIFIC DESIGNS OF MULTI-CONNECTED NEURAL NETWORK MODELS

Karl E. Kürten

Institut für Theoretische Physik, Johannes-Kepler-Universität, A-4040 Linz, Austria
and
Institut für Experimentalphysik, Universität Wien, A-1090 Wien, Austria

ABSTRACT

Designing application-specific architectures in artificial neural network models is a problem of high theoretical and practical interest. We illuminate various ideas which allow a system to evolve to a quasi-optimal interaction and connectivity structure such that information can be embedded as deeply and efficiently as possible. We present specific applications which include memorization and retrieval of highly structured patterns of information, "seven-segment display" patterns and finally "Hadamard" patterns characterized by group-theoretical internal symmetries. We also present concrete evidence that the existence of such symmetries and correlations can be exploited such that essential regularities of a pattern set becomes fully integrated into the network design, while the connectivity structure is optimally adapted to the information the network is asked to memorize.

INTRODUCTION

The last decade has shown a remarkable renaissance of artificial neural network research. This revival is now attracting the attention of scientists across a broad field of traditional disciplines such as neurobiology, computer science, artificial intelligence and theoretical physics [1,2]. In general, one can distinguish several main objectives persued by the neural network community:

• Deepening the understanding of neural structures and illuminating neural responses observed in living biological systems

• Developping algorithms for so-called "hard" problems in cognitive science (optimization, classification, pattern recognition and association)

• Designing a future generation of biological computers inspired by neurophysiological principles

• Studying artificial networks as complex statistical systems (disorder, frustration, collective phenomena)

In particular, for the theoretical physicist, these novel "particles", called neurons, can be viewed as entities of a complex, highly nonlinear dynamical many-body system characterized by disorder and frustration. The "atoms" or units of the system are considered as relatively simple mathematical objects, obeying individual dynamical laws, which govern the time evolution of the "activity" variables describing the current states of the neurons. Note however, that the interactions are in general asymmetric, i.e., Newton's third Law fails, and consequently there is no appropriate energy function which never increases under the dynamics and opens the way to straightforward applications of statistical physics. Furthermore, the interactions are time-dependent, i.e., the synaptic couplings underly changes due to the past history the network has experienced.

The connectivity structure of biological neurons is extremely intricate showing a high degree of order in a background of randomness. However, according to Braitenberg's picture about the interconnectivity of the cortical pyramidal cells [3], one observes short-range interactions with basal dendrites supplemented by fibers making long-range interactions with apical dendrites. Thus, models involving homogeneous connectivity or nearest-neighbour interactions are not realistic. Moreover, neural network models designed for visual pattern recognition or other cognitive tasks are expected to be flexible and adaptive. However, most of the currently popular models are based on fixed orders of the multi-neuron interactions as well as prescribed network connectivity structures. By contrast, living networks combine flexible architectures and learning strategies reminiscent of the growth of new synaptic connections or degeneration of old ones, a phenomenon, which could also be considered as neural development. Hence, it is our intention to concentrate on exploring design or learning strategies aiming at optimization of architectures as it might also happen in real biological systems.

1. THE MULTI-CONNECTED NETWORK MODEL

The model network consists of a set of N interacting binary threshold units σ_i which are only capable to take the value +1 or -1 corresponding to the states of "high" or "low" activity, respectively. We assume that each neuron i can interact with K_i other units of the network with $1 \leq K_i \leq N-1$, while self-interactions are excluded. The state of neuron i at time t then evolves in discrete time according to the deterministic threshold rule

$$\sigma_i(t+1) = \mathrm{sgn}[h_i(t)] \qquad i = 1, ..., N \quad . \tag{1.1}$$

The net internal stimulus or local field $h_i(t)$, felt by neuron i, is given in terms of a polynomial K_i-th order expansion of the commonly used linear superposition of the weighted input states

$$h_i(t) \quad = \quad [c_i \quad + \quad \sum_{(j_1)} c_{ij_1}\sigma_{j_1}(t) + \ldots + \sum_{(j_1)<\ldots<(j_{K_i})} c_{ij_1\ldots j_{K_i}}\sigma_{j_1}(t)\ldots\sigma_{j_{K_i}}(t)] / \parallel c_i \parallel_2 \tag{1.2}$$

with the spherical normalization

$$\parallel c_i \parallel_2 = \{c_{i0}^2 + \sum_{(j_1)} c_{ij_1}^2 + \ldots + \sum_{(j_1)<\ldots<(j_{K_i})} c_{ij_1\ldots j_{K_i}}^2\}^{\frac{1}{2}} \quad . \tag{1.3}$$

The sums in (1.2) and (1.3) are only taken over those K_i neurons that are supposed to interact with neuron i. We stress that each individual neuron i is allowed to receive stimuli from an arbitrary number K_i of other neurons of the network via an arbitrary order which might be determined by the designer, by self-organization or learning.

It is commonly known that the number of random unbiased patterns a network is able to memorize, is roughly porportional to the number of synaptic coefficients which describe the degrees of freedom of the system. The critical capacity parameter depends strongly on the learning algorithm. However, correlations among the different patterns can lead to much higher capacities, which crucially depend on the internal structure of the information the network is asked to capture. Hence, high-order networks offering impressive storage capacities have often been believed to be impractical due to the rapid proliferation of high-order contributions. As demonstrated in ref. [4-6] this problem can be successfully attacked by admitting a strongly reduced pattern-specific interconnectivity or by employing suitable summation techniques based on group-theoretical methods [7].

2. HOW TO DESIGN A NETWORK FOR A SPECIFIC TASK

Successful performance of a particular network model depends crucially on the designer's ability to choose a suitable subset of all possible 2^K high-order terms in (1.2) which adequately match the complexity of the problem [8]. Moreover, tayloring connectivity structures, the arrangement of nodes and patterns of connections , adapted to context-dependent tasks is of fundamental importance. Finally, a suitable learning rule for synaptic plasticity has to be chosen with great care. In summary, we distinguish three hierarchic levels of plasticity:

- i) **Variation of the order of the interactions**

- ii) **Variation of the connectivity structure**

- iii) **Variation of the synaptic weights**

According to Minsky and Papert [8], the order of the interactions has to match the predicate order of the problem in the sense that a set of k-order units is able to solve a k-order problem. There are a variety of problems, where the exact order is known, i.e. the "majority" problem is linearly separable, i.e. of predicate order one, the "one" problem is of order two, whereas the K-parity problem is of predicate order K [8]. However, for many classes of so called "hard" problems such as visual pattern recognition or classification tasks, the order of the problem is not known at all and has to be estimated. In the case of visual pattern recognition, studying correlations among the set of patterns the network has to memorize might be a usefull starting point.

Strong evidence has been presented in refs. [9-11] that generalization properties as well as the storage capacity of a network can be largely improved if the connectivity is adapted to the specific information the network is asked to capture. The network design can be thought of as a combinatorial optimization problem, where for each cell i a suitable set of input neurons of adequate order have to be carefully selected such that a suitable cost function measuring the quality of storage is optimized. Since an exhaustive neighbour search is not possible for networks of respectable size one might adopt trial and error schemes supplemented by refined *Monte Carlo* procedures or highly efficient deterministic strategies [9,10]. According to various experiments which indicate that low-efficacy synapses degenerate, and hence hardly contribute to the stability, substantial improvement can been achieved by weighting the connections with the magnitudes of their synaptic efficiencies such that the probability of being selected increases with decreasing magnitude of the synaptic weight.

Finally, the optimization of the synaptic weights could be performed with the aid of an arbitrary, not necessarily optimal learning procedure. One significant result of our studies suggested strongly that it is far more important to aim at appropriate connectivity structures than to concentrate on optimal learning procedures for the modification of the coupling coefficients. Hence, we tend to adopt simple Hebbian "one-shot" learning rules via

$$c_{ij_1\ldots j_s} = \sum_{\mu=1}^{p} S_i^\mu S_{j_1}^\mu \ldots S_{j_s}^\mu \quad . \tag{2.1}$$

Thus, the synaptic weights are given explicitly and need not be determined by iterative, time consuming learning procedures.

3. STABILITY OF A PRESCRIBED SET OF MEMORIES

In order to store p prescribed patterns $\underline{S}^1,\ldots,\underline{S}^p$ as fixed points of the dynamics (1.1) a suitable set of coupling coefficients has to be chosen such that *all* stability parameters $\kappa_{i\mu}$ defined by

$$\kappa_{i\mu} = S_i^\mu h_i(S_1^\mu,\ldots,S_N^\mu) = S_i^\mu \sum_{j_1<\ldots<j_s} c_{ij_1\ldots j_s} S_{j_1}^\mu \ldots S_{j_s}^\mu / \parallel c_i \parallel_2 \quad , \tag{3.1}$$

be nonnegative. Furthermore, their magnitude, commonly considered as a measure of the quality of memorization, is required to be as large as possible in order to ensure reasonably large basins of attraction for the individual patterns.

In order to gain more insight into the origin of stability we decompose $\kappa_{i\mu}$, defined for a fully connected network, into its individual contributions

$$\kappa_{i\mu}^{(j_1...j_s)} = \frac{S_i^\mu c_{ij_1...j_s} S_{j_1}^\mu ... S_{j_s}^\mu}{\| c_i \|_2} \tag{3.2}$$

and define the stability tensor

$$\mathbf{K}_i^{(j_1...j_s)} = (\kappa_{i\mu}^{(j_1...j_s)})_{\mu=1,...,p}^{j_1 \neq ... \neq j_s \neq i} \tag{3.3}$$

Here, the tensor element $\kappa_{i\mu}^{(j_1...j_s)}$ specifies the stability contribution for the i-th component of pattern μ due to the synaptic s-order connection from neurons $j_1...j_s$. We stress that only a positive value of $\kappa_{i\mu}^{(j_1...j_s)}$ favours embedding of the information. Taking the average over all p patterns produces

$$\frac{1}{p} \sum_{\mu=1}^{p} \kappa_{i\mu}^{(j)} = \frac{1}{p \| c_i \|_2} c_{ij_1...j_s} c_{ij_1...j_s}^H \; . \tag{3.4}$$

The sum rule (3.4) implies that if and only if the actual coupling $c_{ij_1...j_s}$ is chosen according to the Hebbian sign constraint

$$\mathrm{sgn}(c_{ij_1...j_s}) = \mathrm{sgn}(c_{ij_1...j_s}^H) \tag{3.5}$$

the arithmetic stability mean $\overline{\kappa}_i^{(j_1...j_s)}$ represented by the left hand side of eq. (3.4) is strictly positive. Averaging over the whole set of stability constributions $\kappa_{i\mu}^{(j_1...j_s)}$ gives the arithmetic mean

$$\overline{\kappa}_i = \frac{1}{p} \sum_{j_1<...<j_s} \sum_{\mu=1}^{p} \kappa_{i\mu}^{(j)} = \frac{1}{p \| c_i \|_2} \sum_{j_1<...<j_s} c_{ij_1...j_s} c_{ij_1...j_s}^H \quad , \tag{3.6}$$

which for Hebbian couplings reduces to

$$\overline{\kappa}_i = \frac{1}{p} \sqrt{\sum_{j_1<...<j_s} (c_{ij_1...j_s}^H)^2} = \frac{1}{p} \| c_i^H \|_2 \quad . \tag{3.7}$$

Hence, a substantial portion of the couplings should obey the Hebbian sign constraint (3.5) in order to allow for a positive mean. However, an *absolute* stability measure of the embedded information is given by the value of the minimal $\kappa_{i\mu}$, where the minimum is taken over *all* patterns.

In judging the quality of memorization one might also consider the number of positive and negative entries in column $j_1...j_s$ of the stability tensor $\mathbf{K}_i$ defined in (3.8), denoted by $n_{ij_1...j_s}^+$ and $n_{ij_1...j_s}^-$, respectively. These quantities obey the relations

$$n_{ij_1...j_s}^+ = \frac{1}{2}[p + c_{ij_1...j_s}^H \mathrm{sgn}(c_{ij_1...j_s})] \quad \text{and} \quad n_{ij_1...j_s}^- = \frac{1}{2}[p - c_{ij_1...j_s}^H \mathrm{sgn}(c_{ij_1...j_s})] \quad . \tag{3.8}$$

If the coupling $c_{ij_1...j_s}$ is chosen under the Hebbian sign constraint (3.5), $n^{+}_{ij_1...j_s}$ assumes its minimum value $\frac{1}{2}$ for $c^{H}_{ij_1...j_s} = 0$ and reaches its maximum value p for $|c^{H}_{ij_1...j_s}| = p$, where $|c^{H}_{ij_1...j_s}|$ takes its maximal value. Thus, the fraction of positive stability entries in $\mathbf{K}_i^{(j_1...j_s)}$ is always larger than $\frac{1}{2}$ increasing linearly with increasing magnitude of the Hebbian coupling. Accordingly, as is the case in biological networks, it may be inferred that large Hebbian efficiencies play a crucial role in optimized network architectures.

4. INFORMATION-SPECIFIC CONNECTIVITY STRUCTURES

4.1 Selective connectivity construction in first order

Basically, there are two completely different, though equivalent approaches for economizing at the number of connections. One might start from a completely connected network and selectively delete those weights which disfavour the stabilization of the information to be engraved. Alternatively we suggest to begin with a completely disconnected network and add only specific connections which are most effective in increasing a suitable cost function defined in terms of a minimum stability measure. Thus, starting from a "zero" network, we let synapses grow until a desired degree of stability is achieved. This procedure has the distinct advantage that a modification of the connectivity is not performed before but during the learning phase.

The design of a first-order network can be thought of as a combinatorial optimization problem, where each cell i has $\binom{N-1}{K_i}$ different ways to choose its interacting neighbours. Ideally, one would have to select K_i $(1 \leq K_i \leq N-1)$ columns of the stability tensor defined in eq.(3.3) such that the minimal stability parameter defined in eq.(3.2) attains its maximum value. Once an optimal set of connections has been found for each K_i , one might also try to minimize the number of connections with respect to a desired stability measure. The result might be a network with greatest stability and minimial resources.

The combinatorial optimization problem can be attacked by adapting Branch-and-Bound (B&B) strategies widely applied to the solution of optimization problems of high complexity [12]. These algorithms can be considered as general strategies to search for optimal solutions in a combinatorially large configuration space, in our case, the space of all possible connectivities. The procedure avoids an explicit enumeration of all possible configurations by restricting the search to certain subspaces of the connectivity space. The exclusion of those connections which might not yield near-optimal answers from further competition is based upon the computation of a lower bound defined in terms of a suitable cost function, in our case, a suitable stability measure. More details about an explicit procedure for building quasi-optimal connectivity structures with the aid of (B&B) strategies can be found in ref. [10]. Note that the strategy can directly be applied to arbitrary orders of the neuron interactions.

4.2 The maximal correlation network

As already indicated by Giles and Maxwell [6] prior knowledge of correlations among different patterns within the pattern set can be extremely useful for optimal network designs. In fact, there is an intimate connection between a Hebbian coupling $c_{ij_1...j_s}$ defined in eq. (2.1) and the strength of many-pixel correlations among *all* individual patterns. More precisely, the $(s+1)$-order coupling $c_{ij_1...j_s}$ captures $(s+1)$-order correlations between the components $j_1...j_s$ and the target component i. Hence, the quantitiy $C^{(s+1)}$ defined by the magnitude of the normalized s-order coupling $c_{ij_1...j_s}$

$$C^{(s+1)}_{(c_{ij_1...j_s})} = \frac{1}{p}\left|\sum_{\mu=1}^{p} S_i^\mu S_{j_1}^\mu ... S_{j_s}^\mu\right| \tag{4.1}$$

measures the strength of correlations between pixel $j_1, ..., j_s$ and i through multiple product operations summed over *all* patterns. There are no correlations present, if $C^{(s+1)}_{(c_{ij_1...j_s})} = 0$ and the pixel set $\{i, j_1, ..., j_s\}$ is maximally correlated if $C^{s+1}(c_{ij_1...j_s}) = 1$.

In chapter **3** we have shown that synaptic coefficients with $|c_{ij_1...j_p}| = p$ give exclusively positive contributions in the corresponding entries of the stability tensor $\mathbf{K}_i^{(j_1...j_s)}$ defined in (3.3) Hence "maximal" correlations with $C^{s+1}_{(c_{ij_1...j_s})} = 1$ should be exploited for the construction of a network architecture. In fact, admitting only connections incorporating "maximal" correlations, leads to stabilities being all identical with the value $\kappa_i^\mu = \sqrt{K_i}$, where K_i is the number of eligible contributions in the polynomial expansion (1.2). However, we point out that the commonly used stability measure is only a necessary condition for the quality of performance, since the shapes of the basins of attraction are usually far from being radial.

5. APPLICATIONS

5.1 First-order network for artificially structured information

In real-world applications stored information is never completely random but contains correlated structures. Not surprisingly, correlations among the patterns turn out to be a serious obstacle for a satisfactory performance of a fully connected network model based on the outer product rule (2.1). Evidently, in the absence of high-order interactions the Hebb rule treats the individual patterns equally and considers each pattern as a new piece of information. Thus, there is a tendency to enhance overlapping parts of information. On the other hand, important details which distinguish the patterns are not well captured by the network. This failure as well as massive interference effects result in the appearance of numerous spurious attractors which deteriorate the network performance considerably. However, we will demonstrate that a selectively constructed first-order network consisting of $N = 100$ neurons interacting in terms of *ternary* coupling coefficients $c_{ij} \in \{0, +1, -1\}$ shows excellent performance in processing 29 strongly structured patterns presented in figure 1.

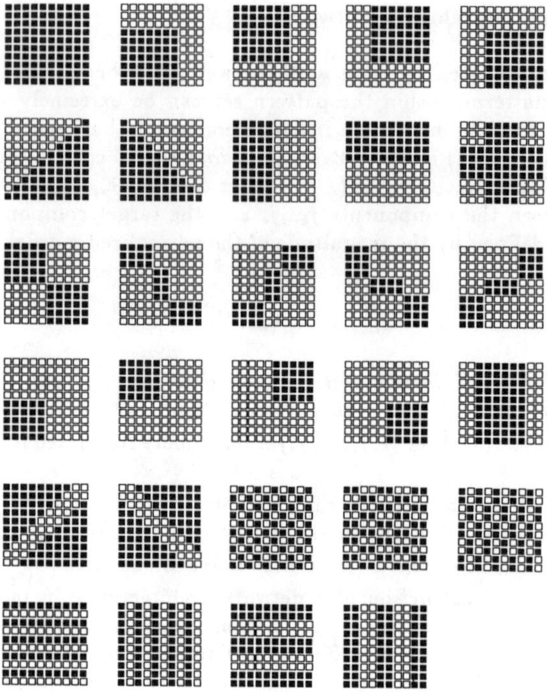

Fig.1: Samples the network is to memorize

We observe that 62% of the couplings have been set to zero, while 26% and 12% take the value "+1" and "-1", respectively. The minimal stability parameter turns out to be about 0.5. Note that a fully connected Hopfield network is not able to stabilize this strongly correlated training set, since the differences in the patterns cannot be well separated by the simple learning rule and interference effects are too strong. However, our selectively constructed network with ternary coupling coefficients can still recall the information almost completely even if up to 10% of the pixels have been damaged at random. For specific corruptions we find results comparable to those observed in refs. [13-14].

5.2 Memorizing "Seven-segment display" patterns

"Seven-segment displays" as shown in figure 2 are attractive for applications such as integration into electronic instrumentation, computer peripherals, desk top calculators or digital clocks. The basic elements are seven segments $s_1, s_2, ..., s_7$ allowing excellent readability dispite of their relatively small size and aesthetic visualization of alphanumeric characters for viewing distances up to several meters.

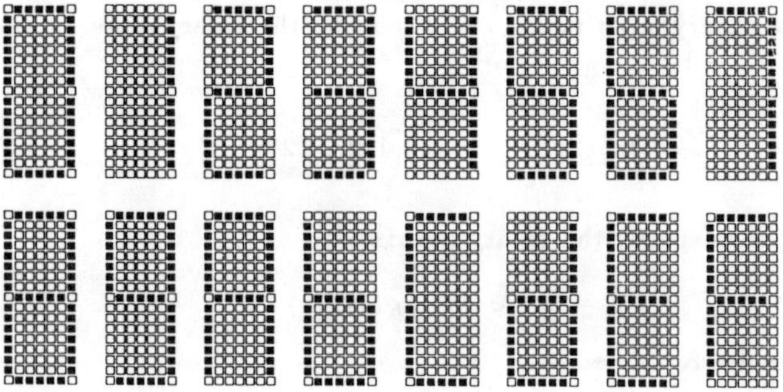

Fig.2: Sixteen alphanumeric characters in seven-segment display

Let us assume that each horizontal segment consists of n_1 pixels, while the four vertical segments are supposed to contain n_2 pixels, respectively. The information is obviously burried in only $3n_1 + 4n_2$ bits, whereas the remaining $2n_1n_2 + 6$ bits merely serve as a uniform background identical in all displays. Although there are 2^7 possible arrangements of seven segments , we restrict our choice to a subset $\mathcal{M} = \{0, 1, 2, 3, 4, 5, 6, 7, 8, 9, A, b, C, d, E, F\}$ used in hexadecimal coding.

Memorizing an arbitrary subset of the 2^7 possible seven-segment display patterns can be shown to be a **first** order problem. It is easy to verify that the maximality condition (4.1) $|c_{ij}| = p$ is satisfied if and only if pixel i and pixel j are elements of the same segment, while the corresponding zero-order condition $|c_i| = p$ holds if pixel i and j belong to the uniform background. Thus, the dynamical time evolution is given by

$$\sigma_i(t+1) = \text{sgn}(\sum_{(j)} \sigma_j(t)) \quad \text{for} \quad i, j \in s_k \quad k = 1, 2, ..., 7 \quad . \tag{5.1}$$

Otherwise, $\sigma_i(t+1)$ takes the value -1, if pixel i belongs to the uniform background.

The network divides into seven independently operating fully connected subnets consisting of n_1 or n_2 neurons, respectively. The couplings are all equal to one such that each neuron obeys a "majority" rule and each subnet behaves as a ferromagnet. The basins of attraction are all known exactly; starting from a stored pattern one may flip the states of half of the pixels belonging to one segment. Retrieval is obviously performed in one single time step.

5.3 "Hadamard" patterns

Hadamard matrices have been widely used in technology, in particular, encoding and transmission of huge amounts of data. One of their most prominent applications was realized in 1971, when they served as codewords of the Mars Mariner 9 code to send photographs of Mars back to earth.

Let us first present the group-theoretical background. We assume that an output neuron is fed by K binary input neurons and consider the binary representation of the numbers $0, 1, 2, ..., 2^K - 1$:

$$i = \sum_{l=0}^{K-1} \epsilon_i^l 2^{K-l} \quad (i = 0, 1, 2, ..., 2^K - 1) \tag{5.2}$$

The coefficients ϵ_i^l then specify the binary matrix

$$E_K = (\epsilon_i^j) \in \mathcal{M}_{2^K,K}(\{0,1\}) \quad , \tag{5.3}$$

which, for example, for K=3 reads

$$E_3 = \begin{pmatrix} 0 & 0 & 0 \\ 0 & 0 & 1 \\ 0 & 1 & 0 \\ 0 & 1 & 1 \\ 1 & 0 & 0 \\ 1 & 0 & 1 \\ 1 & 1 & 0 \\ 1 & 1 & 1 \end{pmatrix} \tag{5.4}$$

The row vectors ϵ_i of E_K are obviously a complete enumeration of *all* possible 2^K binary input combinations $(\epsilon_i^0, ..., \epsilon_i^{K-1}) \in \{0,1\}^K$ for $i = 0, 1, ..., 2^K - 1$. The K column vectors ϵ^j of E_K generate a well defined 2^K-dimensional subspace $\mathcal{H} \subset \{0,1\}^{2^K}$ such that an arbitrary vector $h^j \in \mathcal{H}$ can be naturally specified by a linear combination of the K basis vectors $\epsilon^0, ..., \epsilon^{K-1}$

$$h^j = \sum_{l=0}^{K-1} \epsilon_j^l \epsilon^l \quad , \tag{5.5}$$

where $h^{2^i} = \epsilon^i$ for $i = 0, 1, ..., K - 1$. Note that in eq.(5.5) the matrix element ϵ_i^j of E_K also plays the role of a coefficient and the summation is performed modulo two. Thus, the matrix H_K, defined by the matrix elements

$$h_i^j = \sum_{l=0}^{K-1} \epsilon_i^l \epsilon_j^l \quad , \tag{5.6}$$

is the *Gramsch* matrix specified by the scalar products of *all* possible input combinations represented by the row vectors of E_K. The matrix H_K is symmetric and for $K = 3$, e.g., H_K takes the form

$$H_3 = \begin{pmatrix} 0 & 0 & 0 & 0 & 0 & 0 & 0 & 0 \\ 0 & 1 & 0 & 1 & 0 & 1 & 0 & 1 \\ 0 & 0 & 1 & 1 & 0 & 0 & 1 & 1 \\ 0 & 1 & 1 & 0 & 0 & 1 & 1 & 0 \\ 0 & 0 & 0 & 0 & 1 & 1 & 1 & 1 \\ 0 & 1 & 0 & 1 & 1 & 0 & 1 & 0 \\ 0 & 0 & 1 & 1 & 1 & 1 & 0 & 0 \\ 0 & 1 & 1 & 0 & 1 & 0 & 0 & 1 \end{pmatrix} \tag{5.7}$$

Eventually , the transformation $\{0,1\} \Rightarrow \{1,-1\}$ gives the standard Hadamard description commonly found in literature [15]. The internal symmetries of Hadamard matrices are visualized in figure 3, where full squares and open squares correspond to the $\{0,1\}$ or $\{+1,-1\}$ notation, respectively.

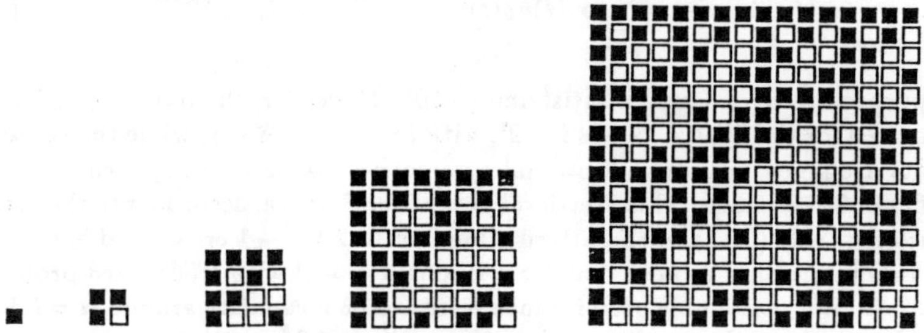

Fig.3: Hadamard matrices for K=0,1,2,3,4 (from left to right).

An equivalent design for Hadamard matrices of order 2^K with entries $h_i^j \in \{-1,+1\}$ or $h_i^j \in \{0,1\}$ is well known from the construction principle for a maximal set of 2^K pairwise orthogonal 2^K-dimensional vectors:

$$H_{l+1} = \begin{pmatrix} H_l & H_l \\ H_l & -H_l \end{pmatrix} \quad \text{for} \quad l = 0, 1, ..., K-1 \quad . \tag{5.8}$$

According to the iteration procedure, apart from the first row h_0, each row h_i of H_K contains 2^{K-1} entries with value "-1" as well as 2^{K-1} entries with value " $+1$". Thus, for two arbitrary row vectors h_i and h_j, we find $\sum_{l=0}^{2^K-1} h_i^l h_j^l = 0$ and since $\sum_{l=0}^{2^K-1} (h_i^l)^2 = 2^K$ for all i, we have

$$H_K H_K^T = 2^K I \tag{5.9}$$

such that H_K is Hadamard.

Let us now perform the summation modulo two of two arbitrary Hadamard vectors h^i and h^j . In the $(0,1)$- representation we have

$$h^i \oplus h^j = \sum_{l=0}^{K-1} \epsilon_i^l \epsilon^l \oplus \sum_{l=0}^{K-1} \epsilon_j^l \epsilon^l = \sum_{l=0}^{K-1} (\epsilon_i^l \oplus \epsilon_j^l) \epsilon^l \quad . \tag{5.10}$$

Obviously, the sum vector is also Hadamard and, in particular, $h^i \oplus h^j = h^0 = \{0\}$ for $i = j$. Hence, $\mathcal{H}$ is an Abelian group of order 2^K under the operation $\oplus$ with zero element h^0, generated by the K basis vectors $\epsilon^0, ..., \epsilon^{K-1}$ defined in (6.3). Thus, the identity

$$h^i \oplus h^j \oplus h^k = h^0 = \{0\} \qquad i \neq j \neq k \neq 0 \tag{5.11}$$

holds, if $h^i \oplus h^j = h^k$. This trivially proves that for fixed $i \neq 0$ there exist exactly $2^K - 2$ "maximal" three-pixel correlations introduced in **4.2**. Note that half of them are symmetric with respect to pixel exchange of j and k. Hence, the construction of an "optimally" designed network is straightforward. For the dynamical time evolution we have

$$\sigma_i(t+1) = \mathrm{sgn}(\sum_{j<k} \sigma_j(t)\sigma_k(t)) \quad , \quad i = 1, 2, 4, ..., 2^{K-1} \tag{5.12}$$

with an appropriate set i, j and k satisfying (6.10). Moreover, the dynamics have only to be performed for the "basis" pixels $i = 2^l$, with $l = 0, 1, ..., K-1$, while the remaining $2^K - K$ bits are given in terms of internal symmetries. Note that any two Hadamard vectors differ in 2^{K-1} pixels. Thus, retrieval is "optimal", since, according to the **second** order "majority" rule (6.11), each pattern allows up to $2^{K-2} - 1$ errors and is recovered perfectly. It is interesting to note that for a Hopfield model with Hadamard prototypes a natural limit for a reasonable performance seems to be a capacity around $\alpha = 0.5$ [16]. Although, due to the orthogonality of the information all 2^K patterns can be stored as fixed points of the dynamics, the retrieval is rather unsatisfactory and even impossible for α close to one. By contrast, the "maximal correlation" network, equipped with a minimal set of **second**-order interactions, is able to retrieve *all* 2^K patterns perfectly, dividing the 2^{2^K}-dimensional space into 2^K equally sized basins of attraction for the 2^K Hadamard vectors. Moreover, we find no spurious states, and retrieval is performed in one single time step.

CONCLUSION

We have demonstrated that an appropriate choice of the order of the neuron interactions as well as a carefully selected connectivity structure is of crucial importance for the functioning of a neural network model. Moreover, incorporating prior knowledge about correlations into the network design can be of substantial advantage, i.e., decreased architectural complexity by an efficient organization of the network structure as well as a dramatically improved quality of performance. In view of these results we suggest strongly that it is highly beneficial to study *economic* high-order designs which have been neglected too much in recent years.

Acknowledgements

The author acknowledges support from the German Science Foundation under contract number Se 251/32-1. I thank J.W. Clark, R. Folk and A. Schinner for numerous stimulating and helpful discussions. In particular, I gratefully thank N. Klingen for helpful advice with respect to group-theoretical aspects of Hadamard designs. Part of the computer experiments were carried out on the CRAY YMP of the HLRZ Jülich.

REFERENCES

1. J.W. Clark, 1989, in: *Nonlinear Phenomena in Complex Systems*, A.N. Proto, ed. Elsevier Science Publishers B.V. (North-Holland) "*Introduction to neural networks*"

2. J.J. Hopfield,1982, Proc.Nat.Acad.Sci.,**79**,2554-2558 "*Neural networks and physical systems with emergent collective computational abilities*"

3. V. Braitenberg,1977, *On the Texture of Brains*: "An Introcution to Neuroanatomy for the Cybernetically Minded" (Springer Verlag, Berlin)

4. K.E. Kürten,1990, Lecture Notes in Physics,**368** ,(Springer-Verlag) ,p.461-466, Statistical Mechanics of Neural Networks, "*Higher-order memories in optimally structured neural networks*"

5. D. Psalits, C.H. Park and J.Hong,1988,Neural Networks,**1**, 149-163 "*Higher order associative memories and their optical implementations*"

6. C.L. Giles and T. Maxwell,1987, Applied Optics **26**,23, 4972-4978 "*Learning, invariance, and generalization in high-order neural networks*"

7. K.E. Kürten, 1992, in: Neural Networks: From Biology to High-Energy Physics, O. Benhar, C. Bosio, P. del Giudice, and E. Tabet, eds. (Giardini, Pisa), in press. "*Many-body cluster theory in high-order neural networks*",

8. M.L. Minsky and S. Papert, 1969, Perceptrons, (MIT Press, Cambridge,MA

9. K.E. Kürten,1990, J. Phys. France **51**,1585-1594 "*Quasi-optimized memorization and retrieval dynamics in sparsely connected neural network models*"

10. K.E. Kürten,1992, J. Phys.**I** France,**2**,615-624, "*Adaptive architectures for Hebbian network models*"

11. Y.L. Le Cun, J.S. Denker and S.A. Solla, Neural Information Processing Systems, Vol.**2**, D. Touretzky ed. "*Optimal brain damage*"

12. C. Roucairol, 1989, INRIA report 962, "*Parallel branch and bound algorithms - an overview*"

13. K.E. Kürten,1990, Parallel Processing in Neural Systems and Computers,World Scientific, ed. by R.Eckmiller, G.Hartmann and G. Hauske, 191-194 "*Dynamical learning in networks with sparse connectivity*"

14. R. Folk, K.E. Kürten and H. Schweng, in *Complex Dynamics in Neural Networks*, E. Caianiello, J. W. Clark, R. M. J. Cotterill, and J. G. Taylor, eds. (Springer-Verlag, Berlin, 1992), in press. "*Visual comparison of information storage in various neural network models*"

15. W.K. Pratt, J. Kane and H.C. Andrees,1969, Proceedings of the IEEE, **57**,1,58-69 "*Hadamard transform image coding*"

16. O. Krisement,1990,Z.Phys.B,**80**,415-421, "*A Hopfield model with Hadamard prototypes*"

OPTIMIZATION WITH POTTS NEURAL NETWORKS

BO SÖDERBERG[1]
Department of Theoretical Physics, University of Lund
Sölvegatan 14A, S-22362 Lund, Sweden

ABSTRACT

The *Potts Neural Network* approach to non-binary discrete optimization problems is described. It applies to problems that can be described as a set of elementary '*multiple choice*' options. Instead of the conventional binary (Ising) neurons, mean field *Potts neurons*, having several available states, are used to describe the elementary degrees of freedom of such problems. The dynamics consists of iterating the mean field equations with annealing until convergence.

Due to its deterministic character, the method is quite fast. When applied to problems of Graph Partition and scheduling types, it produces very good solutions also for problems of considerable size.

1. Introduction

The idea of using artificial (binary) neural networks for optimization is due to Tank and Hopfield[1], and was first applied to the Travelling Salesman problem (TSP), but with limited success. It was subsequently applied by Anderson and Peterson to the Graph Bisection (GB) problem[2], with good results. They also applied it, using neuron multiplexing, to the general K-fold Graph Partition (GP) problem[3], but with less success. The reason for the difference in performance seems to be that GB has a natural formulation in terms of binary (Ising) degrees of freedom, while TSP and GP have not – there, a formulation in terms of a Potts glass[4] is more natural.

A Potts variable is simply a discrete variable with a fixed number of available states, not necessarily two as for the binary (Ising) case of conventional neurons. The generalization of the Hopfield-Tank approach to Potts systems was put forward by the author together with Peterson, and the additional refinement of adding self-couplings to secure stability was made. This approach was applied to TSP and GP[5], and performed considerably better than the binary predecessors. It has subsequently been tested on *scheduling* type problems, with excellent results[6].

Plan of Paper

In Sec. 2, the NN approach for the binary case is described for comparison. The basic formalism for Potts systems is presented in Sec. 3, with GP as a reference example, and a generic algorithm is given. Some aspects of the dynamics are discussed

[1]bs@thep.lu.se

in Sec. 4. Numerical results for some applications are discussed in Sec. 5, and the author's conclusions are given in Sec. 6.

2. The Binary Case

To see the analogy between the binary and Potts cases, the binary approach to the GB problem will first be discussed. The problem is defined as follows: Given a random graph of N nodes described by a (symmetric) connection matrix $C_{ij} = 0, 1$, divide the nodes in two equally large parts such that the number of connections between the two halves is minimal. This covers e.g. the problem of optimally dividing a given electric circuit in two parts.

Ising Formulation

The GB problem can be formulated as that of finding the lowest energy state of an Ising system. Assign to every node i an Ising (± 1) variable s_i, the value of which decodes which half the node winds up in. Then the *cutsize* "energy" is proportional to

$$E_0 = \frac{1}{2} \sum_{i,j} C_{ij}(1 + s_i)(1 - s_j), \tag{1}$$

to be minimized under the equal-size constraint

$$\sum_i s_i = 0. \tag{2}$$

This constraint can be added to the energy as a penalty term

$$E_c = \frac{\alpha}{2} (\sum_i s_i)^2, \tag{3}$$

where α is a parameter that can be adjusted according to the importance of exact equipartition. The total energy $E = E_0 + E_c$ to be minimized is, apart from a trivial constant, equal to

$$E[s] = \frac{1}{2} \sum_{ij} (\alpha - C_{ij}) s_i s_j, \tag{4}$$

where $[s]$ denotes the whole set of binary variables.

Mean Field Equations

The MF equations of this system can be derived in the following way. Introduce a "temperature" T, and consider a thermal distribution, given by the *partition function*

$$Z = \sum_{[s]} e^{-E[s]/T}. \tag{5}$$

Embedding the Ising spins in a linear space ($\mathcal{R}$), introduce a new set of variables v_i living in this space, one for each spin, and set them equal to the spins with a Dirac delta-function. Then we can express the energy in terms of the v_i:

$$Z = \sum_{[s]} \int d[v] e^{-E[v]/T} \prod_i \delta(s_i - v_i). \tag{6}$$

Now, Fourier-expand the delta-functions, introducing a set of conjugate variables u_i:

$$Z = \sum_{[s]} \int d[v] \int d[u] e^{-E[v]/T} \prod_i e^{u_i(s_i - v_i)}. \tag{7}$$

Then carry out the original sum over $[s]$:

$$Z \propto \int d[v] \int d[u] e^{-E[v]/T - \sum_i u_i v_i + \sum_i \log\cosh(u_i)}. \tag{8}$$

The original partition function is now rewritten entirely in terms of the new variables $[u, v]$, with an effective energy in the exponent.

The MF approximation to the expectation value of s_i is now given by the value of v_i at a saddle-point of the effective energy, equations for which are obtained by differentiation w.r.t. $[u]$ and $[v]$. This results in the MF equations:

$$u_i = -\partial E[v] \partial v_i / T, \tag{9}$$
$$v_i = \tanh(u_i). \tag{10}$$

Neural Network

For the GB problem, we obtain

$$v_i = \tanh(u_i) = \tanh(\frac{1}{T} \sum_j (C_{ij} - \alpha) v_j). \tag{11}$$

But these are precisely the equations of a Hopfield type (feedback) neural network, if we interpret v_i as the output state of a (± 1)-neuron and Tu_i as its input, with $C_{ij} - \alpha$ the (symmetric) coupling between neurons i and j, and $1/T$ as a gain factor. The dynamics consists in iterating the MF equations. When the network has settled, a spin solution is extracted from $s_i = \text{sgn}(v_i)$. For suitable values of the parameters T and α, it has been reported[2] to produce good solutions, starting from a random initial neuron configuration with all the v_i close to zero.

3. Potts systems

Graph Partitions in General

What if an N-node graph is to be partitioned into $K > 2$ parts of equal size ($=N/K$)? Attempts to encode such a problem with conventional neurons (neuron multiplexing) have been made[3], with somewhat discouraging results.

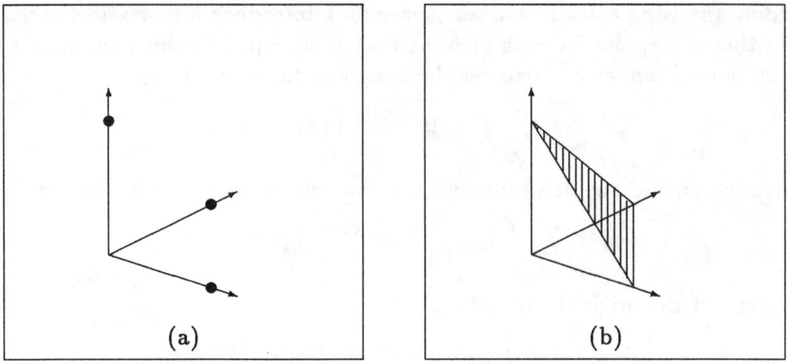

Figure 1: **(a)** The three allowed states for a ($K = 3$) Potts spin $\vec{S}_i$. **(b)** The accessible states (shaded area) for the corresponding mean field $\vec{V}_i$.

Potts Formulation

This problem is more naturally encoded in terms of a system of K-state Potts spins, each having K allowed states, in the following way: A K-state Potts spin is assigned to every node of the graph, such that the state a of a spin encodes which part of the partition the node winds up in.

The key idea of ref. 5 was to define a generalized neural network of Potts neurons, with a dynamics based on the proper MF equations for a Potts system, along the lines of the preceding section.

To explain how this is done, we choose a K-dimensional vector representation of the Potts spin states: The a:th possible state of the Potts spin vector $\vec{S}_i$ associated with the node i is represented by the unit vector in the a-direction. In this way, the K different states, $(1, 0 \ldots 0)$, $(0, 1 \ldots 0) \ldots (0, 0 \ldots 1)$, are all normalized and mutually orthogonal (c.f. Fig. 1a).

With this formulation, and with C_{ij} as before being the connection matrix of the graph, the number of interconnections can be written as

$$E_0 = \frac{1}{2} \sum_{i,j} C_{ij}(1 - \vec{S}_i \cdot \vec{S}_j), \tag{12}$$

which is to be minimized under the equipartition constraint, in component form:

$$\sum_i S_i^a = N/K, a = 1, \ldots, K. \tag{13}$$

As in the previous section, the constraint is implemented in a soft way as a penalty term,

$$E_c = \frac{\alpha}{2} \sum_{i,j} \vec{S}_i \cdot \vec{S}_j. \tag{14}$$

Removing trivial constants, we are left with the following energy to minimize (c.f. Eq. 4):

$$E[\vec{S}] = \frac{1}{2}\sum_{i,j}(\alpha - C_{ij})\vec{S}_i \cdot \vec{S}_j. \tag{15}$$

Potts Neural Network

Given a K-state Potts system, and an energy $E[\vec{S}]$, introduce a temperature T, and define a thermal partition function:

$$Z = \sum_{[\vec{S}]} e^{-E[\vec{S}]/T}. \tag{16}$$

Proceeding in analogy to what was done in the binary case in the previous section, the partition function is transformed into

$$Z = \int d[\vec{U}]d[\vec{V}] \exp(-E[\vec{V}]/T - \sum_i \vec{U}_i \cdot \vec{V}_i) \sum_{[\vec{S}]} e^{\sum_i \vec{U}_i \cdot \vec{S}_i}. \tag{17}$$

Performing the original sum, we are left with

$$Z = \int d[\vec{U}]d[\vec{V}] \exp(-E[\vec{V}]/T - \sum_i \vec{U}_i \cdot \vec{V}_i + \sum_i \log \sum_a e^{U_{ia}}). \tag{18}$$

The MF approximation to $< \vec{S}_i >$ is now given by the value of $\vec{V}_i$ at a saddle-point of the effective energy of $[\vec{U}, \vec{V}]$ in the exponent of Eq. 18. Differentiating, we obtain the Potts MF equations (c.f. Eqs. 9,10):

$$U_{ia} = -\frac{\partial E[\vec{V}]}{\partial V_{ia}}/T, \tag{19}$$

$$V_{ia} = \frac{e^{U_{ia}}}{\sum_b e^{U_{ib}}}, \tag{20}$$

where a are the vector components.

From the latter follows, that

$$\sum_a V_{ia} = 1. \tag{21}$$

One can thus think of the mean field V_{ia} as the probability for the Potts spin i to be in state a (*fuzzy logic*). The set of values available to $\vec{V}_i$ is the inside of a simplex in a $(K-1)$-dimensional hyperplane, with its corners on the allowed states for the Potts spin $\vec{S}_i$. For $K = 3$ this is a triangle (c.f. Fig. 1b). For $K = 2$, we recover the formalism of the Ising case, provided that we define v_i as $V_{i1} - V_{i2}$, etc. Thus, we have a natural generalization of the MF approach from $K = 2$ to general Potts systems.

In analogy with the binary case, we now regard $[\vec{V}]$ as a network of generalized *neurons* $\vec{V}_i$, with inputs $T\vec{U}_i$ and a common gain factor $1/T$. The dynamics of the network consists of iterating the MF equations.

Refinements

For small GP problems, one obtains good solutions for a range of parameter (T, α) values, starting from a random configuration close to the central point $\vec{V}_i = (1, 1 \ldots 1)/K, i = 1, N$. For larger problems, the performance improves considerably with the following refinements:

- Add a self-coupling term $-\beta \sum_i \vec{S}_i^2$ to the energy, with β adjusted to secure stability.
- Anneal: Instead of using a fixed temperature, start at a high temperature T_0, and decrease it slowly in between updates.

For most problems, the proper values of β and T_0 can be calculated or estimated in advance (see below).

Generic Algorithm

The complete neural network algorithm for a generic K-state Potts system will then look as follows:

1. Formulate an energy, the minimization of which solves the problem.
2. Add a self-coupling term, and calculate the proper values for β and T_0.
3. Initialize the system with a random configuration of $\vec{V}_i$ in the neighbourhood of the central point $(1, 1 \ldots, 1)/K$.
4. Update the network, one neuron at a time, according to Eqs. 19,20.
5. Decrease the temperature T by a suitable factor (in most cases, 0.95 will do).
6. Until the system has settled (all $\vec{V}_i^2 \approx 1$), repeat steps 4 and 5.
7. Extract solution: For each neuron, pick the nearest allowed Potts spin state.

4. Analysis Of The Dynamics

High Temperature

At a high temperature, the input of all neurons will be small, and they will converge to a *trivial fixpoint* close to the central point $(1, 1 \ldots, 1)/K$ – the network is in a *mixed state* of all possible decisions. Below some critical temperature T_c, this fixpoint will turn unstable, and the network will begin evolving towards a non-trivial fixpoint. The value of T_c can be computed by making a spectral analysis of the linearized dynamics in the neighbourhood of the trivial fixpoint. For some systems, this can be done exactly, but it suffices with an estimate.

The value of the critical temperature depends on the selfcoupling β, and on precisely how the updating is done. In *serial mode*, neurons are updated one by one, using fresh values of previously updated neurons. In *synchronous mode*, all neurons are updated in parallel, using only old values as input.

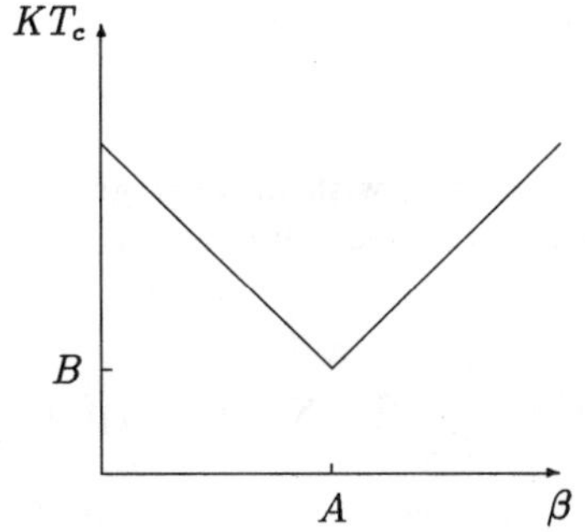

Figure 2: Schematic β dependence of the critical temperature T_c.

In both cases, the critical temperature depends on β in the simple fashion:

$$KT_c = B + |\beta - A| \tag{22}$$

with the constants A and B depending on the problem and the updating mode[5]. A graph of this critical line (c.f. Fig. 4.) has two straight line branches meeting at $\beta = A$. To obtain a smooth evolution (in particular for synchronous mode) when T_c is decreased across the critical line, the right branch must be used ($\beta > A$)[5]. Without a self-coupling term, this is not always the case, which explains the need for it. On the other hand, with too large a β value, the solution quality will decrease. A suitable β value is thus somewhere above, but not too far from, the turning point A. The initial temperature T_0 is then chosen close to the critical line.

Low Temperature

At low temperature, any difference between the components of a neuron's input will be strongly magnified. As a result, we have a *winner-take-all* situation, where one neuron component will be almost one, and the others close to zero. The dynamics turns discrete, and a kind of local optimization results as $T \to 0$. Eventually the network should settle at a fixpoint close to an allowed state of the Potts spin system, representing a (possibly approximate) solution to the optimization problem.

Also here, the β value has an impact: A higher value tends to stabilize also bad decisions (positive feed-back), while a lower value destabilizes even a good solution, and can lead to cyclic or even chaotic behaviour.

For serial mode, there is never a problem to choose a β that is satisfactory in both temperature limits[5]. This is often not the case for synchronous mode, so serial mode will be understood where not otherwise stated.

5. Numerical Explorations

Graph Partition

For the GP problem defined above, with an average of 10 connections per node, using the energy of Eq. 15 with an added self-coupling term, the MF equations read

$$\vec{U}_i = \frac{1}{T}(\beta \vec{V}_i + \sum_j (C_{ij} - \alpha)\vec{V}_j), \quad (23)$$

$$V_{ia} = \frac{e^{U_{ia}}}{\sum_b e^{U_{ib}}}. \quad (24)$$

With the simplest choice $\alpha = 1$, and proper estimated values for T_0 and β, the above generic algorithm has been applied to GP problems with $K = 4, 10$ and $N = 100$, supplemented with a *greedy heuristic* to clean up any slight imbalance that might remain in the final configuration. The resulting performance compares well in solution quality with standard stochastic methods (simulated annealing, SA), with considerably less CPU-consumption[5].

In the limit of large problem size N (with fixed K), the CPU needed per update sweep will scale as $O(N^2)$ on a sequential machine, while the number of sweeps needed is believed (and is empirically seen) to be approximately constant (≈ 20). On a parallel machine, the CPU will scale as $O(N)$ for serial update mode (and as $O(1)$ for synchronous mode!).

Travelling Salesman

Also this problem can be defined in terms of minimizing the energy of a Potts system, with $K = N$. In this case, the Potts NN algorithm does not fully reach the solution quality of SA, but it is still a considerably faster method for medium sized problems ($N < 200$). For large problem size, the CPU of the algorithm is expected to scale as $O(N^3)$ on a sequential machine. On a parallel machine, this will reduce to $O(N)$ for serial update mode (and to $O(1)$ for synchronous mode).

Scheduling Problems

The Potts neural network algorithm is ideal for scheduling problems - these have a natural formulation in terms of Potts systems, and the ground state energy is known – thus, one can identify exact solutions by inspection.

Numerical explorations have been made on artificial problems of school scheduling type – i.e. all combinations of N_1 teachers and N_2 classes are to be scheduled in a system of N_3 rooms and N_4 time-slots, such that collisions of various types are avoided[6]. The results are nothing but excellent: For problems of size up to $9\times9\times9\times9$, in almost all cases, a legal solution is found in the first attempt. Even when soft constraints are added, such as a preference for classes to stay in the same room in consecutive lectures, the rate of legal solutions stays close to one, while the soft constraints are to a high degree fulfilled.

Experiments have been made also with more realistic school scheduling problems, of sizes corresponding to about $50 \times 50 \times 50 \times 50$ with similar good results[7].

6. Conclusions

The Potts NN algorithm can be used on any discrete optimization problem that can be expressed as a set of elementary multiple-choice problems. This includes e.g. GP and scheduling type problems, where the method has been successfully tested.

A characteristic feature of the method is that it smears out the discrete nature of a problem, opening up shortcuts in configuration space. As the temperature parameter is decreased, the discrete character of the problem is slowly recovered.

When the problem size increases, the performance of the algorithm can be expected to decrease, at least for difficult problems. In such cases, only approximate solutions will result. In many practical applications, this is all that is needed. In other cases, the algorithm can be used to provide a reasonable starting point for more elaborate algorithms.

An advantage of the method is the speed: it produces a deterministic and highly parallelizable algorithm, that empirically produces good approximations to the solution in a short time. A disadvantage for large K is the need for memory: Every neuron needs K *floats*, as opposed to one *int* for a discrete Potts spin.

Another important feature is the control over parameters. Suitable values for the self-coupling and the initial temperature can be estimated in advance. No high sensitivity to parameter values has been observed.

So far, the algorithm has been run in serial update mode. The speed will be even further increased, if synchronous updating can be made feasible.

The author's conclusion is, that the Potts NN algorithm is a good alternative to more elaborate methods, wherever good approximative solutions to discrete optimization problems are desired.

References

[1] J. J. Hopfield and D. W. Tank, *Biol. Cybernetics* **52** (1985) 141.

[2] C. Peterson and J. R. Anderson, *Complex Systems* **2** (1988) 59.

[3] C. Peterson and J. R. Anderson, preprint MCC-ACA-ST-064-88.

[4] I. Kanter and H. Sompolinski, *J. Phys.* **A 20** (1987) L673.

[5] C. Peterson and B. Söderberg, *Int. J. Neur. Syst.* **1** (1989) 3.

[6] L. Gislén, C. Peterson and B. Söderberg, *Int. J. Neur. Syst.* **2** (1989) 167.

[7] L. Gislén, C. Peterson and B. Söderberg, preprint LU TP 91-21 (submitted to Neural Computation).

Vision

Early Vision with Cellular Automata Fields

Ricardo Alberto Marques Pereira
Dipartimento di Informatica, Università di Trento
Via Inama 7, Trento TN 38100, ITALIA

Abstract

We introduce a ***new approach***, based on the natural ***clique structure*** of diffusive networks, to both the standard and the graded ***weak membrane*** models of ***image reconstruction from noisy sparse data.*** The new clique-based formulation leads to a single discontinuity field, provides a natural solution to the long-standing localization problem concerning the gradient field construction, and yields an exact paradigm of ***cellular automata field.*** The ***standard weak membrane model*** of image reconstruction assumes weakly continuous images corrupted by white gaussian noise and operates through an iterative process of constrained anisotropic diffusion. The anisotropy is controlled by the ***line process*** l describing the discontinuity contours, while the constraint enforces closeness to the image data. The ***graded weak membrane model*** includes an extra field, the ***sparse process*** s, which computes a 0 to 1 visual significance map and excludes the less significant data from the image reconstruction scheme. In this mechanism the sparse process s couples with a local noise estimate and, in the interactive model, with the line process l. The resulting [0,1] significance hierarchy of image sites can be regarded as the membership distribution of a ***fuzzy minimal image.***

Keywords: Computer Vision, Early Visual Processing, Image Reconstruction, Edge Detection, Markov Random Fields, Distributed Processing, Cellular Automata

1 Introduction

One of the most fruitful insights in computer vision research over the past decade has been the realization that the so-called edge detection module is in fact but one essential component of the more comprehensive early vision [1-15] module of *image reconstruction from noisy sparse data.* In contrast with the traditional approach, which viewed edge detection as a thresholding calculus on a previously filtered image, the new generation of image reconstruction schemes combines image smoothing and edge detection in a cooperative and highly more effective way.

The *weak membrane model* [26-45] is a crucial paradigm of modern image reconstruction. The algorithm, constructed on the basis of prior assumptions about universal image structure and expected noise corruption, operates through an iterative process of edge estimation and edge-preserving image smoothing. The construction incorporates some of the standard ideas and techniques from bayes probability theory, markov random fields and statistical field theory [46-65].

In this article we review the essential aspects of both the standard and the graded weak membrane models in a *new formulation* based on the natural nearest-neighbour *clique structure* of diffusive networks [66-68]. Relatively to the pixel-based approach described in the literature so far, ours yields a simpler and more transparent formalism, and avoids the technical difficulties associated with the pixel-based construction of the gradient field. By computing the brightness gradients directly on the natural edge locations - the binary cliques - the new formulation involves a single discontinuity field and provides a natural solution to the recurrent *localization problem* - see [29] and references therein.

The *standard weak membrane model* [26-40] of image reconstruction assumes weakly continuous images corrupted by white gaussian noise. The local non-linear dynamics of the model is based on a competitive interplay between a form of *initial data constraint* and a characteristic process of *anisotropic diffusion.* The anisotropy is controlled by the *line process* l describing the discontinuity contours.

The *graded weak membrane model* [41-45] includes an extra process field, the *sparse process* s, which computes a 0 to 1 *visual significance* map and excludes the less significant data from the image reconstruction scheme. In this mechanism the sparse process s couples with a local noise estimate and, in the interactive model, with the line process l. The resulting $[0,1]$ visual significance hierarchy of image sites can be regarded as the membership distribution of a *fuzzy minimal image* [45,69-71].

From the operational viewpoint, both the standard and the graded weak membrane models lead to massively parallel and highly non-linear algorithms of the cellular automata [72] type, based on iterative local processing. In the weak membrane case, however, the local updating law is inhomogeneous, in so far as it depends on the local configuration of an external field which we call the *reference field.* In image reconstruction models of the weak membrane type the *main field* f represents the changing image under the iterative reconstruction scheme and the *reference field* g corresponds to the given image data. The processing structure of the weak membrane algorithm is thus an example of a *cellular automata field.*

The paper is organized as follows. In the next two sections we describe the new clique-based formulation as applied to the standard and the graded weak membrane models, respectively. We begin with a general overview of the standard model which, for the most part, applies also to the graded extension. The concluding section points out briefly a number of additional issues regarding the weak membrane model and discusses some of the present research perspectives.

2 The standard weak membrane model

Grey level images constitute the raw material of computer vision. They are regular bidimensional arrays of pixels, each of which encoding a grey level measure of the local image brightness [6]. Spatial resolution and grey level range are the two basic parameters of grey level images. The intrinsically parallel structure of the weak membrane algorithm renders it particularly suited to the analysis of large image arrays. The ones we use in experimentation, for instance, have $512 \star 512$ pixels and a grey level range of 256 values, which means 262,144 bytes of information content.

Technically speaking, grey level images are described by *image fields*, usually denoted g or f. Matrix indices are used for the *image domain*: with respect to an image field f, say, f_{ij} denotes the grey level associated with pixel p_{ij}. In the weak membrane model, g is used for the *image data* and f for the *image field* being transformed by the iterative reconstruction scheme. In other words, the *main field* f can be regarded as a *dynamical system* $f = f(t)$ with initial configuration given by the *reference field* $g = f(0)$.

The image reconstruction scheme combines anisotropic diffusion - iterative edge estimation and edge-preserving image smoothing - with a type of elastic constraint keeping f close to the image data g. The non-linear smoothing acts primarily on the small brightness gradients - neighbouring grey level differences below the edge threshold - thereby emphasizing the large brightness gradients.

It turns out natural to think of the non-linear smoothing as resulting from a network of *diffusion channels* between neighbouring pixels, whose *conductivity* is in some way inversely proportional to the grey level difference between the two pixels. This *diffusive network* scenario motivates the present clique-based construction of the gradient field and constitutes the basis for the new formulation of the weak membrane model in terms of binary clique structure.

Consider the *image domain* equipped with the usual 1st order neighbourhood system, in which pixel p_{ij} has 4 neighbours $\{p_{i-1,j}, p_{i+1,j}, p_{i,j-1}, p_{i,j+1}\}$, with adjustments at the image boundaries. The four binary *cliques* which converge at each pixel p_{ij} are denoted after their associated positions as

$$\lambda_{north(ij)} \ , \ \lambda_{south(ij)} \ , \ \lambda_{west(ij)} \ , \ \lambda_{east(ij)} \ .$$

Clearly, this means that

$$\lambda_{north(ij)} = \lambda_{south(i-1,j)} \ , \ etc \ .$$

The set of binary cliques, which we call the *dual dual image domain*, is the effective operational ground of the weak membrane algorithm in the new formulation.

• *Overview*

The weak membrane model solves the problem of image reconstruction from noisy sparse data on the basis of two fundamental prior assumptions concerning the universal structure of images and the type of noise corruption they are susceptible to.

The resulting image reconstruction scheme is controlled by an *effective visual cost function* $V_g(f)$ defined on the set of image field configurations $\{f\}$. Given image data g, the effective visual cost function $V_g(f)$ assigns a visual cost value to each field configuration f. The optimal image field $\bar{f}$, relatively to the image data g, is then that which minimizes $V_g(f)$.

The effective visual cost function $V_g(f)$, in turn, derives from a *fundamental visual cost function* $U_g(f,l)$ depending not only on the image field f but also on the *line process* l. At this level, l is an independent binary field on the dual image domain, $l_\lambda = 0, 1$. The image field f and the line process l are coupled in a characteristic way: the cross term $f \star l$ in the visual cost $U_g(f,l)$ is designed so that the line process l responds to the edge contours in the image field f.

The visual cost function $U_g(f,l)$, constructed in a bayesian framework, disambiguates the image reconstruction problem on the basis of two fundamental prior assumptions. The first of these, the a priori *universal image model*, is based on the notion of *weak continuity* [27]. Roughly speaking, images are assumed to be piecewise continuous surfaces with a 'minimal' number of discontinuity contours. Images of this type are commonly referred to as *mondrians* [7]. The second prior incorporated in the visual cost function $U_g(f,l)$ regards the type of noise corruption which is expected to intervene in the acquisition process leading to the image data g. The standard assumption in the weak membrane model is simply *white gaussian noise.*

From the operational point of view, the weak continuity assumption leads to the crucial thresholding mechanism which controls the edge-preserving image smoothing process. The white gaussian noise assumption, on the other hand, yields a form of elastic constraint keeping f close to the image data g.

The fundamental visual cost function $U_g(f,l)$ generates a gibbs probability distribution $e^{-\beta U}/Z$ which, in accordance with the minimization principle, assigns higher probability to the (f,l) configurations with lower visual cost. The temperature paramater $\beta = 1/T$ defines the spread of the probability distribution and thus controls uncertainty in the model. The motivation for the gibbs construction lies in the fact that the mean field configuration $(\bar{f}, \bar{l})$ associated with the probability distribution $e^{-\beta U}/Z$ is both easier to compute and more robust than the absolute optimal configuration for $U_g(f,l)$ itself. In the zero temperature limit, moreover, the mean field configuration should coincide with the optimal one.

One of the crucial aspects of the weak membrane model is that the visual cost function $U_g(f,l)$ is fully reducible with respect to the line process field associated with the dual image domain. As a result, it is essentially straightforward to integrate over the line process configurations $\{l\}$ in order to compute the effective visual cost function $V_g(f)$ and the effective line process $\bar{l}(f)$. Contrary to the fundamental visual cost function $U_g(f,l)$, both $V_g(f)$ and $\bar{l}(f)$ depend intrinsically on the temperature parameter β. The effective line process $\bar{l}$, moreover, is no longer a binary field, and takes values in the interval $[0,1]$.

Given the effective visual cost function $V_g(f)$, the optimal image field $\bar{f}$ can be numerically approximated in several ways. In the deterministic mean field approach the iterative process of image reconstruction consists of a combination of gradient descent and annealing, starting from the given image data g and leading to a progressive approximation of the optimal image field f. At each iterative stage $t = 0, 1, 2, ...$ the effective line process $\bar{l}^t$ is estimated from the current image field f^t and then f itself is updated through an appropriate edge-preserving smoothing filter.

In this presentation we will concentrate on the new formulation based on clique structure and on the derivation of the cellular automata field algorithm which solves the optimization problem for the effective visual cost function $V_g(f)$. Partly for that reason, we will skip all the bayesian considerations concerning the construction of the visual cost function $U_g(f, l)$, as well as the more technical mean field theory arguments involved in the computation of the effective visual cost function $V_g(f)$. These issues are already extensively discussed in the literature, see for instance [28-29] and references therein.

• *The standard model*

The weak membrane model of image reconstruction, from image data g to reconstructed image f, is best described in terms of two derived field variables defined on the dual image domain. The two variables, denoted δ^2 and ∇^2, are the basic building blocks of the new clique-based formulation.

The *dual field variable* δ^2 measures the local discrepancy between the current image field f and the original image data g, at each stage of the iterative scheme. Its definition is

$$\delta^2_\lambda = \tfrac{1}{2}\left((f_p - g_p)^2 + (f_{p'} - g_{p'})^2 \right) \tag{1}$$

where p and p' are the two pixels forming the link λ.

The *dual field variable* ∇^2, on the other hand, refers solely to the image field f and plays the role of its gradient. It describes the *local geometry* of the image f and is defined as

$$\nabla^2_\lambda = (f_p - f_{p'})^2 \,. \tag{2}$$

In this way, both the gradient field ∇^2 and the line process l are *localized* on the natural edge sites - the binary cliques - and each ∇^2_λ correlates only two pixels.

In contrast, the pixel-based approach leads to a gradient field ∇^2_{ij} whose effective *localization* is at the centre of the 4-pixel cell including p_{ij} and $p_{i-1,j-1}$. This has the following drawbacks: first of all, the notation ∇^2_{ij} is misleading since the gradient field does not in fact refer to the pixel location (ij); secondly, the centres of the 4-pixel cells are not the natural edge locations, particularly in the diffusive scenario induced by the weak membrane model; and thirdly, each ∇^2_{ij} correlates the grey level information of four pixels instead of two, with the result that the conductivity of the binary diffusion channels will depend on the correlations among at least six pixels (in the pixel-based approach the conductivity of one binary diffusion channel involves at least the two gradients on the sides), instead of the sole two pixels which are directly concerned.

In the new formulation, the g and f dependence in the visual cost function $U_g(f,l)$ is entirely conveyed by the two derived field variables δ^2 and ∇^2. As a result, $U_g(f,l)$ is effectively a function on the dual image domain. The exact form of the *fundamental visual cost function* $U_g(f,l)$ in the standard weak membrane model is

$$U = \sum_{\lambda \in \Lambda} U_\lambda \tag{3}$$

$$U_\lambda = \phi_\lambda + \psi_\lambda (1 - l_\lambda) \tag{4}$$

where ϕ_λ stands for $\frac{1}{2}\delta_\lambda^2$, and ψ is the *threshold field*

$$\psi_\lambda = \mu \nabla_\lambda^2 - \gamma_\lambda \tag{5}$$

whose role, roughly speaking, is that of differentiating large brightness gradients ($\psi_\lambda > 0$) from small brightness gradients ($\psi_\lambda < 0$).

The dividing line results from the specification of the free parameters μ and γ_λ, all positive real numbers. The *local edge threshold* $(\gamma_\lambda/\mu)^{\frac{1}{2}}$ depends essentially on the *parameter field* γ, whereas the *global parameter* μ controls the relative strength between the diffusive mechanism and the initial data constraint, within the process of image reconstruction.

Notice that ϕ is not a threshold field, only a name function adopted for notational convenience. In the graded model, however, the two field variables δ^2 and ∇^2 will be treated on the same ground and a true threshold field, associated to δ^2, will take the place of ϕ.

Consider now the local visual cost U_λ as in (4). It can be written as

$$U_\lambda = U_\lambda^I + U_\lambda^H \tag{6}$$

where U_λ^I and U_λ^H correspond, respectively, to ϕ_λ and $\psi_\lambda(1-l_\lambda)$.

The *inhomogeneous term* U_λ^I involving the image data g is associated with the prior assumption of white gaussian noise. Its effect is that of an elastic constraint on the image field f enforcing closeness to the image data g. The *homogeneous term* U_λ^H, on the other hand, results from the prior assumption of weakly continuous images. It induces the 'edge detector' l to respond when the threshold field ψ is positive - $l = 1$ when ∇^2 is large - and enforces local diffusive smoothing when ψ is negative.

The *gibbs probability distribution* $P_g(f,l)$ generated by the fundamental visual cost function $U_g(f,l)$ is of the form

$$P = e^{-\beta U}/Z = \left(\prod_\lambda e^{-\beta U_\lambda}\right) / Z \tag{7}$$

where the *partition function* Z is given by

$$Z = \sum_{\{f\}\{l\}} e^{-\beta U} = \sum_{\{f\}\{l\}} \left(\prod_\lambda e^{-\beta U_\lambda}\right) . \tag{8}$$

Notice the the introduction of the free *temperature parameter* $\beta = 1/T$, extrinsic to the visual cost function $U_g(f,l)$. By varying β one can control the spread of the probability distribution $P_g(f,l)$ and thus decide on the amount of uncertainty one wishes the model to incorporate.

Given that the visual cost $U_g(f,l)$ is fully reducible with respect to the line process structure associated to the dual image domain,

$$U = \sum_{\lambda} U_\lambda(l_\lambda) \tag{9}$$

the summation $\sum_{\{l\}}$ in (8) can be performed analytically. As a result, the partition function can be rewritten as

$$Z = \sum_{\{f\}} Z(f) = \sum_{\{f\}} \Big(\prod_{\lambda} e^{-\beta\phi_\lambda} \left(1 + e^{-\beta\psi_\lambda}\right) \Big) . \tag{10}$$

For each image field configuration f, therefore, we can define an *effective visual cost function* $V_g(f)$,

$$V = -\frac{1}{\beta} \, lnZ(f) = \sum_{\lambda} V_\lambda \tag{11}$$

$$V_\lambda = \phi_\lambda - \frac{1}{\beta} \, ln \left(1 + e^{-\beta\psi_\lambda}\right) \tag{12}$$

and a *effective line process* $\bar{l}(f)$,

$$\bar{l}_\lambda = 1 - \frac{1}{\beta}\frac{\partial lnZ(f)}{\partial \gamma_\lambda} = \frac{1}{1 + e^{-\beta\psi_\lambda}} . \tag{13}$$

As we had anticipated earlier, both $V_g(f)$ and $\bar{l}(f)$ depend on the temperature parameter β. In the zero temperature limit, the effective line process $\bar{l}(f)$ behaves as a binary field,

$$\lim_{\beta\to\infty} \bar{l}_\lambda = \theta(\psi_\lambda) \tag{14}$$

$$\mu\nabla_\lambda^2 > \gamma_\lambda \quad \leadsto \quad \bar{l}_\lambda = 1 . \tag{15}$$

Notice in (14) the explicit role of the threshold field ψ_λ. At zero temperature, moreover, the optimal image field $\bar{f}$ with respect to $V_g(f)$ and the effective line process $\bar{l}(\bar{f})$ coincide with the true mean fields associated with the original gibbs probability distribution $P_g(f,l)$ introduced in (7).

Given the effective visual cost function $V_g(f)$, the optimal image field $\bar{f}$ can be numerically approximated by means of an iterative scheme of gradient descent and annealing. Starting from the image data $f(t{=}0) = g$, the iterative process updates the image field f according to

$$f_{ij}^{t+1} = f_{ij}^{t} - \kappa \frac{\partial V^t}{\partial f_{ij}^t} \qquad t = 0, 1, 2, ... \tag{16}$$

where κ is some appropriate convergence rate parameter.

The gradient descent formula (16) yields the following iteration rule

$$f_{ij}^{t+1} = f_{ij}^{t} - \kappa \left(\varepsilon_{ij}^{t} + \mu \left(\varepsilon_{north(ij)}^{t} + \varepsilon_{south(ij)}^{t} + \varepsilon_{west(ij)}^{t} + \varepsilon_{east(ij)}^{t} \right) \right) \quad (17)$$

where the *increment fields* are given by

$$\begin{aligned} \varepsilon_{ij} &= (f_{ij} - g_{ij}) \\ \varepsilon_{north(ij)} &= (1 - \bar{l}_{north(ij)})\,(f_{ij} - f_{i-1,j}) \\ \varepsilon_{south(ij)} &= (1 - \bar{l}_{south(ij)})\,(f_{ij} - f_{i+1,j}) \\ \varepsilon_{west(ij)} &= (1 - \bar{l}_{west(ij)})\,(f_{ij} - f_{i,j-1}) \\ \varepsilon_{east(ij)} &= (1 - \bar{l}_{east(ij)})\,(f_{ij} - f_{i,j+1})\,. \end{aligned} \quad (18)$$

Notice in (17) the explicit role of the parameter μ in controlling the relative strength between the inertial increment ε_{ij} and the remaining diffusive increments.

The updating law (17 - 18) satisfies the *cellular automata field* paradigm. In contrast with the usual cellular automata scheme, in which the updating formula is homogeneous in the *main field* f, the increment fields in (18) depend not only on the neighbouring values of f itself but also on those of the *reference field* g. In the standard weak membrane model only the central value g_{ij} intervenes. In the graded model, instead, the local updating law involves the full set of local data values $\{g_{i,j}, g_{i-1,j}, g_{i+1,j}, g_{i,j-1}, g_{i,j+1}\}$.

The iterative process can be superimposed with an annealing schedule on the temperature parameter $\beta = \beta(t)$, from small values of β - large temperatures - to large values of β - small temperatures. The effect is that of progressively polarizing the conductivity coefficients $(1 - \bar{l}_\lambda)$ around zero - the *edge* case - or one.

3 The graded weak membrane model

From the point of view of the new formulation based on clique structure, the graded weak membrane model constitutes a sort of natural covariant extension of the standard model. In addition to the geometrical threshold mechanism associated with the dual field variable ∇^2 and operated by the threshold field ψ and the line process l, the graded model incorporates a second threshold mechanism, associated to the dual field variable δ^2. This inertial threshold mechanism is operated by a true *threshold field* φ - the covariant extension of the name function ϕ in the standard model - and an extra process field, referred to as the *sparse process* s.

The sparse process s, the crucial element in the graded model, is originally a binary field on the dual image domain, in analogy with the line process l. The fundamental visual cost function $U_g(f, s, l)$ couples the sparse process to the field variable δ^2, which can be regarded as a local noise estimator. The coupling is such that the sparse process responds to 'good' data - $s = 1$ when δ^2 is small - and excludes the 'bad' data from the image reconstruction scheme. In the interacting model, s couples also to the line process l and is inhibited in the absence of edge contours - $s = 0$ when $l = 0$. The sparse process s, therefore, computes a local *visual significance map* indicating a low level of noise and, in the interacting case, the presence of edge contours.

• *The free model*

The *fundamental visual cost function* $U_g(f, s, l)$ of the non-interacting graded weak membrane model is given by

$$U = \sum_{\lambda} U_\lambda \tag{19}$$

$$U_\lambda = \varphi_\lambda \, s_\lambda + \psi_\lambda \, (1 - l_\lambda) \tag{20}$$

where φ and ψ are the two *threshold fields*

$$\varphi_\lambda = \tfrac{1}{2} \, \delta_\lambda^2 - \eta_\lambda \tag{21}$$

$$\psi_\lambda = \mu \nabla_\lambda^2 - \gamma_\lambda \, . \tag{22}$$

As before, the free parameters μ, η_λ and γ_λ are all positive real numbers. The new *parameter field* η controls the *local noise threshold* $(2\eta_\lambda)^{\frac{1}{2}}$.

The *partition function* Z associated with the gibbs probability distribution generated by $U_g(f, s, l)$ is

$$Z = \sum_{\{f\}\{s\}\{l\}} e^{-\beta U} = \sum_{\{f\}\{s\}\{l\}} \Big(\prod_{\lambda} e^{-\beta U_\lambda} \Big) \, . \tag{23}$$

Given that $U_g(f, s, l)$ is again fully reducible,

$$U = \sum_{\lambda} U_\lambda(s_\lambda, l_\lambda) \tag{24}$$

the summation $\sum_{\{s\}\{l\}}$ in (23) can be performed analytically. As a result, the partition function Z can be rewritten as

$$Z = \sum_{\{f\}} Z(f) = \sum_{\{f\}} \Big(\prod_{\lambda} (1 + e^{-\beta\varphi_\lambda}) \, (1 + e^{-\beta\psi_\lambda}) \Big) \, . \tag{25}$$

Accordingly, the *effective visual cost function* $V_g(f)$ is

$$V = -\frac{1}{\beta} \, lnZ(f) = \sum_{\lambda} V_\lambda \tag{26}$$

$$V_\lambda = -\frac{1}{\beta} \, \ln \, (1 + e^{-\beta\varphi_\lambda}) \, (1 + e^{-\beta\psi_\lambda}) \tag{27}$$

and the *effective processes* $\bar{s}(f)$ and $\bar{l}(f)$ are

$$\bar{s}_\lambda = \frac{1}{\beta} \frac{\partial lnZ(f)}{\partial \eta_\lambda} = \frac{1}{1 + e^{+\beta\varphi_\lambda}} \tag{28}$$

$$\bar{l}_\lambda = 1 - \frac{1}{\beta}\frac{\partial lnZ(f)}{\partial \gamma_\lambda} = \frac{1}{1+e^{-\beta\psi_\lambda}} . \tag{29}$$

Notice the way in which the non-interacting effective visual cost (27) factors out into two independent contributions associated with φ and ψ. The same happens with the effective processes $\bar{s}$ and $\bar{l}$. In the free model, each process depends exclusively on the corresponding threshold field. The effective line process $\bar{l}$, in particular, is identical to the one in the standard model.

The gradient descent optimization of the effective visual cost $V_g(f)$ leads to an iteration rule of type (17), with *increment fields*

$$\begin{aligned}
\varepsilon_{ij} &= \bar{s}_{ij}\,(f_{ij} - g_{ij}) \\
\varepsilon_{north(ij)} &= (1 - \bar{l}_{north(ij)})\,(f_{ij} - f_{i-1,j}) \\
\varepsilon_{south(ij)} &= (1 - \bar{l}_{south(ij)})\,(f_{ij} - f_{i+1,j}) \\
\varepsilon_{west(ij)} &= (1 - \bar{l}_{west(ij)})\,(f_{ij} - f_{i,j-1}) \\
\varepsilon_{east(ij)} &= (1 - \bar{l}_{east(ij)})\,(f_{ij} - f_{i,j+1})
\end{aligned} \tag{30}$$

where $\bar{s}_{ij}$ stands for the central average of the four surrounding sparse process values,

$$\bar{s}_{ij} = \frac{1}{4}\left(\, \bar{s}_{north(ij)} + \bar{s}_{south(ij)} + \bar{s}_{west(ij)} + \bar{s}_{east(ij)}\,\right) . \tag{31}$$

Clearly, the sole difference between this iteration rule and the one in the standard model concerns the form of the increment field ε_{ij} in the top line of (30). In the graded model, ε_{ij} is proportional to the visual significance value $\bar{s}_{ij} \in [0,1]$, meaning that the strength of the initial data constraint enforced by ε_{ij} is modulated by the local visual significance of the data. When $\bar{s}_{ij}$ is small the diffusive process is locally dominant in order to compensate for particularly noisy data.

The free graded model, with its two independent thresholding mechanisms, turns out to be essentially equivalent [42-43] to a recently proposed scheme of outlier detection and robust statistics [44] based on a different extension of the weak membrane model. In both cases, in effect, the outlier detection module depends on a given parameter field - η in the graded model - and the image reconstruction scheme automatically discards the outliers. The major difference is that the sparse process s in the free graded model is considered an independent field variable on its own, a crucial fact that opens the way to the construction of the interacting model.

• *The interacting model*

The interacting graded weak membrane model can be constructed in several different ways, not necessarily equivalent among each other. The version we describe here is essentially the simplest one, for it contains no additional parameters other than the parameter field η already present in the non-interacting case. Instead of adding an interaction term $s \star l$ to the free fundamental visual cost function $U_g(f, s, l)$, the coupling between the two processes s and l is introduced by the so-called *hard constraint* mechanism, which consists in constraining the statistical system to a subset of the full configuration space.

In this construction of the interacting model, therefore, the *fundamental visual cost function* $U_g(f, s, l)$ is the same as in the free case - see (19 - 22). The coupling between s and l is introduced by excluding $(s_\lambda = 1, l_\lambda = 0)$ from the admissible set of single-clique states. This has the primary effect of inhibiting the sparse process s when no edge contours are present - $l_\lambda = 0$ implies $s_\lambda = 0$. In other words, large visual significance values require the presence of edge contours.

The exclusion of the single-clique state $(s_\lambda = 1, l_\lambda = 0)$ from the summation $\sum_{\{s\}\{l\}}$ in (23) modifies the *partition function* Z, which becomes

$$Z = \sum_{\{f\}} Z(f) = \sum_{\{f\}} \Big(\prod_{\lambda} (1 + e^{-\beta\varphi_\lambda} + e^{-\beta\psi_\lambda}) \Big) \ . \tag{32}$$

As a result, the *effective visual cost function* $V_g(f)$ in the interacting model is

$$V = \sum_{\lambda} V_\lambda \tag{33}$$

$$V_\lambda = -\frac{1}{\beta} \, ln \, (1 + e^{-\beta\varphi_\lambda} + e^{-\beta\psi_\lambda}) \tag{34}$$

and the interacting *effective processes* $\bar{s}$ and $\bar{l}$ are

$$\bar{s}_\lambda = \frac{e^{-\beta\varphi_\lambda}}{1 + e^{-\beta\varphi_\lambda} + e^{-\beta\psi_\lambda}} \tag{35}$$

$$\bar{l}_\lambda = \frac{1 + e^{-\beta\varphi_\lambda}}{1 + e^{-\beta\varphi_\lambda} + e^{-\beta\psi_\lambda}} \ . \tag{36}$$

As expected, the new effective processes $\bar{s}$ and $\bar{l}$ depend on both threshold fields φ and ψ in a non-trivial manner. Relatively to the free model, we observe that

$$\bar{s}_\lambda < \bar{s}_\lambda^{free} \qquad\qquad \bar{l}_\lambda > \bar{l}_\lambda^{free} \tag{37}$$

where $\bar{s}_\lambda^{free}$ and $\bar{l}_\lambda^{free}$ are as in (28) and (29), respectively. The interacting $\bar{s}_\lambda$ is close to $\bar{s}_\lambda^{free}$ when $\psi_\lambda \gg \varphi_\lambda$ and, conversely, $\bar{l}_\lambda$ is close to $\bar{l}_\lambda^{free}$ when $\varphi_\lambda \gg \psi_\lambda$.

In the zero temperature limit, $\bar{l}$ and $\bar{s}$ behave as binary fields according to

$$\lim_{\beta\to\infty} \bar{s}_\lambda = \theta(-\varphi_\lambda)\theta(\psi_\lambda - \varphi_\lambda). \tag{38}$$

$$\lim_{\beta\to\infty} \bar{l}_\lambda = 1 - \theta(-\psi_\lambda)\theta(\varphi_\lambda - \psi_\lambda) \tag{39}$$

as can be seen from (36) and (35). At zero temperature, therefore, the effective line process signals and edge when $\psi_\lambda > 0$ *or* $\psi_\lambda > \varphi_\lambda$,

$$\mu\nabla_\lambda^2 > \gamma_\lambda \qquad or \qquad \mu\nabla_\lambda^2 - \gamma_\lambda > \tfrac{1}{2}\,\delta_\lambda^2 - \eta_\lambda \qquad \rightsquigarrow \qquad \bar{l}_\lambda = 1 \tag{40}$$

whereas the sparse process responds to significant data when $\varphi_\lambda < 0$ *and* $\varphi_\lambda < \psi_\lambda$,

$$\tfrac{1}{2}\,\delta_\lambda^2 < \eta_\lambda \qquad \textit{and} \qquad \tfrac{1}{2}\,\delta_\lambda^2 - \eta_\lambda < \mu\nabla_\lambda^2 - \gamma_\lambda \qquad \rightsquigarrow \qquad \bar{s}_\lambda = 1. \tag{41}$$

Each case involves two conditions, an absolute threshold and a relative threshold. The latter is $\varphi_\lambda < \psi_\lambda$ in both cases. The *edge* response requires that either one of the two conditions (40) be satisfied. The *visual significance* requirement is more selective and requires that both conditions (41) be fulfilled.

Given the effective visual cost function $V_g(f)$ as in (33-34), one can easily establish numerical bounds to the variation of the effective visual cost value during the iterative optimization process leading to the reconstructed image field $\bar{f}$. Clearly, the upper bound is simply $V_R = V_g(f = g)$, corresponding to the initial effective visual cost value in the reconstruction process. We call V_R the *reference visual cost value*, for it depends on the reference field g. Explicitly,

$$f = g$$

$$V_R = \sum_\lambda \; -\frac{1}{\beta}\, ln\,(1 + \chi_\lambda e^{\beta\gamma_\lambda} + e^{\beta\eta_\lambda}) \tag{42}$$

where $\chi_\lambda \in [0,1]$ stands for $e^{-\beta(\mu\nabla_\lambda^2)}$. The lower bound, instead, depends exclusively on the function $V_g(f)$ itself, without any particular assumption regarding g or f. Given that

$$\frac{\partial V}{\partial \delta_\lambda^2} = \tfrac{1}{2}\,\bar{s}_\lambda > 0 \tag{43}$$

$$\frac{\partial V}{\partial \nabla_\lambda^2} = \mu(1 - \bar{l}_\lambda) > 0 \tag{44}$$

the minimum possible value for the effective visual cost corresponds to the field configuration $\delta^2 = \nabla^2 = 0$. This means

$$f = g = \text{constant}$$

$$V_\emptyset = \sum_\lambda \; -\frac{1}{\beta}\, ln\,(1 + e^{\beta\gamma_\lambda} + e^{\beta\eta_\lambda})\,. \tag{45}$$

Notice that if g is not a constant field then the effective visual cost $V_g(f)$ is necessarily larger than $V_\emptyset$, whatever the choice of f.

The gradient descent optimization of the interacting effective visual cost $V_g(f)$ leads to an iteration rule of type (17), with increment fields as in (30-31). Relatively to the free graded model, the difference lies only in the form of the effective processes $\bar{s}$ and $\bar{l}$. As a result of (35) and (36), the local updating law in the interacting model depends on the full set of local g and f values, in accordance with the general *cellular automata field* paradigm.

4 Concluding Remarks

The present binary clique formulation of the weak membrane model of image reconstruction is a sort of minimal description and can be improved in a number of ways, most of which apply equally to the standard and the graded models.

The most immediate of these modifications is the extension of the neighbourhood system from 4 to 8 neighbours, so as to allow *diagonal interactions* in the diffusive network. To this purpose it suffices to include the diagonal binary cliques in every summation $\sum_\lambda$ while keeping all the definitions and formulas unchanged.

In most cases the parameter fields γ and η are globally assigned - the same values γ and η are used for all binary cliques. In other words, the image reconstruction scheme usually operates with *global thresholds* for edge and noise. Under certain circumstances, however, it may be convenient not to do so. In the graded model, for instance, there is evidence that the *texture quality* of the reconstructed image improves by choosing the parameter field η proportional to the brightness gradient of the image data [43].

The continuity and the definition of the edge contours can be improved by two mechanisms called *hysteresis* and *nonmaximum suppression* [3], both of which can be incorporated in the weak membrane model by appropriate modifications of the fundamental visual cost function U_g - see [28-29].

Both the standard and the graded weak membrane models can cope with sparse data. In that case the reconstruction scheme is extended in order to include a *filling routine* based on local averaging of the available neighbouring data. The filling routine is applied at each stage t and for each empty pixel p_{ij}, progressively expanding the available data during the reconstruction process.

The sparse process s renders the graded weak membrane model particularly suited to the treatment of *sparse data* [41-44] and *fuzzy sparse data* [45]. Conventional sparse data is associated with the binary values of the sparse process s at zero temperature, whereas fuzzy sparse data encodes sparse process values $s \in [0,1]$ at any finite temperature.

To conclude, one final comment regarding the *2nd order extension* of the weak membrane model [34]. It involves an extra process field, the *roof process* r, which couples to second derivatives of the image field f and to the line process l. The 2nd order model is constructed on the basis of ternary clique structure and leads to four different classes of edges, each one associated with an appropriate smoothing action.

References

The list of references that follows is organized by topic and, within each topic, by alphabetical order. We hope that this will provide a clearer and more direct guide to the literature, particularly for non-expert readers.

Early vision

[1] D.H. Ballard, C.M. Brown (1982). *Computer Vision.* Prentice-Hall, Englewood Cliffs, NJ.

[2] M. Bertero, T. Poggio, V. Torre (1988). Ill-posed problems in early vision. *Proceedings IEEE*, 76, 869-889.

[3] J.F. Canny (1986). A computational approach to edge detection. *IEEE Transactions on Pattern Analysis and Machine Intelligence*, PAMI 8, 679-698.

[4] E.B. Gamble, T. Poggio (1987). Visual integration and detection of discontinuities: the key role of intensity edges. AI Memo 970, Artificial Intelligence Laboratory, MIT, Cambridge, MA.

[5] D. Geiger, T. Poggio (1987). An optimal scale for edge detection. *Proc. International Joint Conference on Artificial Intelligence IJCAI*, August 87.

[6] B.K.P. Horn (1977). Understanding image intensities. *Artificial Intelligence*, 8, 201-231.

[7] B.K.P. Horn (1986). *Robot vision.* McGraw-Hill & MIT Press, Cambridge, MA.

[8] D. Marr (1976). Early processing of visual information. *Phil. Trans. Roy. Soc. Lond.*, B 275, 483-524.

[9] D. Marr (1982). *Vision.* Freeman, San Francisco, CA.

[10] D. Marr, E. Hildreth (1980). Theory of edge detection. *Proc. Roy. Soc. Lond.*, B 207, 187-217.

[11] T. Poggio, V. Torre, C. Koch (1985). Computational vision and regularization theory. *Nature*, 317, 314-319.

[12] T. Poggio, H. Voorhees, A.L. Yuille (1984). Regularizing edge detection. AI Memo 776, Artificial Intelligence Laboratory, MIT, Cambridge, MA.

[13] D. Terzopoulos (1986). Regularization of inverse visual problems involving discontinuities. *IEEE Transactions on Pattern Analysis and Machine Intelligence*, PAMI 8, 413-424.

[14] V. Torre, T. Poggio (1986). On edge detection. *IEEE Transactions on Pattern Analysis and Machine Intelligence*, PAMI 8, 147-163.

[15] A.P. Witkin (1983). Scale space filtering. *Proc. 8th International Joint Conference on Artificial Intelligence IJCAI*, Karlsruhe, Germany, pp. 1019-1022.

Image reconstruction before the weak membrane model

[16] R. Chellappa, R.L. Kashyap (1982). Digital image restoration using spatial interaction models. *IEEE Transactions on Acoust. Speech and Signal Processing*, ASSP 30, 461-472.

[17] W.E.L. Grimson (1982). A computational theory of visual surface interpolation. *Proc. Roy. Soc. Lond.*, B 298, 395-427.

[18] A. Habibi (1972). Two dimensional bayesian estimate of images. *Proceedings IEEE*, 60, 878-883.

[19] F.R. Hansen, H. Elliot (1982). Image segmentation using simple markov field models. *Computer Vision, Graphics and Image Processing*, CVGIP 20, 101-132.
[20] R.A. Hummel, S.W. Zucker (1983). On the foundations of relaxation labeling processes. *IEEE Transactions on Pattern Analysis and Machine Intelligence*, PAMI 5, 267-287.
[21] B.R. Hunt (1977). Bayesian methods in non-linear digital image restoration. *IEEE Transactions on Computers*, C 23, 219-229.
[22] R.L. Kashyap, R. Chellappa (1983). Estimation and choice of neighbors in spatial interaction models of images. *IEEE Transactions on Information Theory*, IT 29, 60-72.
[23] N.E. Nahi, T. Assefi (1972). Bayesian recursive image estimation. *IEEE Transactions on Computers*, C 21, 734-738.
[24] W.H. Richardson (1972). Bayesian-based iterative method of image restoration. *Journal Opt. Soc. Amer.*, 62, 55-59.
[25] H.J. Trussell (1980). The relationship between image restoration by the maximum a posteriori method and a maximum entropy method. *IEEE Transactions on Acoust. Speech and Signal Processing*, ASSP 28, 114-117.

Standard weak membrane model

[26] L. Ambrosio (1988). Variational problems in SBV. Technical Report Lecture Notes, MIT, Cambridge, MA.
[27] A. Blake, A. Zisserman (1987). *Visual Reconstruction.* MIT Press, Cambridge, MA.
[28] D. Geiger, F. Girosi (1991). Parallel and deterministic algorithms for markov random fields: surface reconstruction and integration. *IEEE Transactions on Pattern Analysis and Machine Intelligence*, PAMI 13, 401-412.
[29] D. Geiger, A.L. Yuille (1991). A common framework for image segmentation. *International Journal of Computer Vision*, IJCV 6, 227-243.
[30] S. Geman, D. Geman (1984). Stochastic relaxation, gibbs distributions and the bayesian restoration of images. *IEEE Transactions on Pattern Analysis and Machine Intelligence*, PAMI 6, 721-741.
[31] W.E.L. Grimson, T. Pavlidis (1985). Discontinuity detection for visual surface reconstruction. *Computer Vision, Graphics and Image Processing*, CVGIP 30, 316-330.
[32] C. Koch, J.L. Marroquin, A. Yuille (1985). Analog 'neuronal' networks in early vision. *Proc. Nat. Acad. Sci.*, 83, 4263-4267.
[33] D. Lee, T. Pavlidis (1987). On dimensional regularization with discontinuities. *Proc. IEEE (Computer Soc.) 1st International Conference on Computer Vision ICCV*, June 1987, London, England, pp. 572-577.
[34] R.A. Marques Pereira, A. Broggi, D. Geiger (1991). Beyond step edges: 2nd order image restoration. Technical Report 3.CS.91, Dipartimento di Ingegneria dell' Informazione, Università di Parma, Italy.

[35] J.L. Marroquin (1987). Deterministic bayesian estimation of markovian random fields with applications to computational vision. *Proc. IEEE (Computer Soc.) 1st International Conference on Computer Vision ICCV*, June 1987, London, England, pp. 597-601.
[36] J.L. Marroquin, S. Mitter, T. Poggio (1987). Probabilistic solution of ill-posed problems in computer vision. *Journal Amer. Statist. Assoc.*, 82, 76-89.
[37] D. Mumford, J. Shah (1985). Boundary detection by minimizing functionals. *Proc. IEEE Conference on Computer Vision and Pattern Recognition CVPR*, San Francisco, CA.
[38] T. Richardson (1990). Ph.D. Thesis, Dept. of Electrical Engineering and Computer Science, MIT, Cambridge, MA.
[39] A.L. Yuille (1989). Energy functions for early vision and analog networks. *Biological Cybernetics*, 61, 115-123.
[40] J. Zerubia, R. Chellappa (1990). Mean field annealing for edge detection and image restoration. *Proc. European Conference on Signal Processing*, September 1990, Barcelona, Spain.

Graded weak membrane model

[41] D. Geiger, R.A. Marques Pereira (1991). On 'intelligent' sparse data. *Proc. 7th Israeli Symposium on Artificial Intelligence and Computer Vision*, December 1990, Ramat Gan, Israel. Eds. Y.A. Feldman, A. Bruckstein, Elsevier Science Publishers, Amsterdam, The Netherlands, pp. 465-473.
[42] D. Geiger, R.A. Marques Pereira (1991). The outlier process. *Proc. IEEE (Signal Processing Soc.) Workshop on Neural Networks for Signal Processing*, Sept/October 1991, Princeton, NJ, pp. 60-69.
[43] D. Geiger, R.A. Marques Pereira (1991). Selecting minimal surface data. To appear in *Proc. 5th Conference on Neural Information Processing Systems NIPS*, December 1991, Denver, CO.
[44] F. Girosi, T. Poggio, B. Caprile (1990). Extensions of a theory of networks for approximation and learning: outliers and negative examples. Technical Report 9005-12, IRST, Trento, Italy.
[45] R.A. Marques Pereira, M. Fedrizzi (1991). Fuzzy image reconstruction. Technical Report 11.INF.91, Istituto di Informatica, Università di Trento, Italy.

Bayes probability theory, markov random fields, statistical field theory

[46] E. Aarts, J. Korst (1989). *Simulated annealing and Boltzmann machines.* John Wiley & Anchor Press, Tiptree, England.
[47] J. Besag (1974). Spatial interaction and the statistical analysis of lattice systems. *Journal Roy. Statist. Soc.*, B 36, 192-326.
[48] L. Breiman (1969). *Probability and stochastic processes.* Houghton Mifflin Company, Boston, MA.

[49] D. Griffeath (1976). Introduction to random fields. In *Denumerable markov chains*, eds. Kemeny, Knapp, Snell, Springer-Verlag, New York, NY.
[50] J.M. Hammersley, D.C. Handscomb (1964). *Monte Carlo methods.* Methuen, London, England.
[51] J.J. Hopfield (1984). Neurons with graded response have collective computational propreties like those of two-state neurons. *Proc. Nat. Acad. Sci.*, 81, 3088-3092.
[52] J.J. Hopfield, D. Tank (1985). Neural computation of decision in optimization problems. *Biological Cybernetics*, 52, 141-152.
[53] P.J. Huber (1981). *Robust Statistics.* John Wiley, New York, NY.
[54] V. Isham (1981). An introduction to spatial point processes and markov random fields. *Int. Statist. Rev.*, 49, 21-43.
[55] E.T. Jaynes (1968). Prior probabilities. *IEEE Trans. Syst. Sci. Cybern.*, SSC 4, 227-241.
[56] R. Kinderman, J.L. Snell (1980). *Markov random fields and their applications.* Amer. Math. Soc., Providence, RI.
[57] S. Kirkpatrick, C.D. Gellat, M.P. Vecchi (1983). Optimization by simulated annealing. *Science*, 220, 219-227.
[58] D.V. Lindley (1969). *Introduction to probability and statistics from a bayesian viewpoint.* Cambridge University Press, Cambridge, England.
[59] N. Metropolis, A.W. Rosenbluth, M.N. Rosenbluth, A.H. Teller, E. Teller (1953). Equation of state calculations by fast computing machines. *Journal Chem. Phys.*, 21, 1087-1091.
[60] V.A. Morozov (1984). *Methods for solving incorrectly posed problems.* Springer-Verlag, Berlin, Germany.
[61] G. Parisi (1988). *Statistical field theory.* Addison-Wesley, Reading, MA.
[62] M. Plischke, B. Bergersen (1989). *Equilibrium statistical physics.* Prentice-Hall, Englewood Cliffs, NJ.
[63] A.N. Tikhonov, V.Y. Arsenin (1977). *Solutions of ill-posed problems.* W.H. Winston, Washington, DC.
[64] E. Wasserstrom (1973). Numerical solutions by the continuation method. *SIAM Review*, 15, 89-119.
[65] J.W. Woods (1972). Two dimensional discrete markovian fields. *IEEE Transactions on Information Theory*, IT 18, 232-240.

Diffusive networks for early vision

[66] J. Harris, C. Koch, J. Luo (1990). A two dimensional analog VLSI circuit for detecting discontinuities in early vision. *Science*, 1209-1211.
[67] A. Lumsdaine, J. Wyatt, I. Elfadel (1990). Non-linear analog networks for image smoothing and segmentation. *VLSI Signal Processing*, to be published.
[68] P. Perona, J. Malik (1987). Scale space and edge detection using anisotropic diffusion. *Proc. IEEE (Computer Society) 5th Workshop on Computer Vision*, Miami, FL, pp. 16-22.

Fuzzy set theory

[69] D. Dubois, H. Prade (1980). *Fuzzy Sets and Systems: theory and applications.* Academic Press, New York, NY.

[70] L.A. Zadeh (1965). Fuzzy sets. *Information and Control*, 8, 338-353.

[71] H.J. Zimmermann (1991). *Fuzzy Set Theory* (2nd ed.). Kluwer Academic Publishers, Dordrecht, The Netherlands.

Cellular Automata

[72] T. Toffoli, N. Margolus (1987). *Cellular Automata Machines.* MIT Press, Cambridge, MA.

Complexity, Data Analysis
and
Prediction

ON THE USE OF THE BI-ORTHOGONAL DECOMPOSITION

Joaquim A. Dente
Depart. de Engenharia Electrotécnica e Computadores, Instituto Superior Técnico, Av. Rovisco Pais, 1, 1096 Lisboa Codex Portugal

Ricardo Lima
Centre de Physique Théorique, CNRS-Luminy Case 907
13288 MARSEILLE Cedex 9 - France

ABSTRACT

We describe how the bi-orthogonal decomposition can be used in the analysis as well as in the control of different kind of dynamics.

1. Spatio-temporal phenomena and multivariable signals

We are mainly interested in analizing the evolution of spatially extended systems. This can be either real physical systems or mathematical models. But in within case we assume a signal is recorded. This signal is a number (more generally a vector) depending on the spatial coordinate x (the "site") that can take either discrete or continum values and in one, two or more dimensions.

Clearly the signal depends also on a time variable t, and here also we consider discrete or continuous time.

We denote by $u(x,t)$ the signal - the value of a physical quantity or the issue of a numerical or analytical computation in a model - for a given site x and time t.

Related to the signal function u, it is natural to introduce two symetric objects. Namely, for each $t \epsilon T$ we can define the configuration ξ_t and, for each site $x \epsilon X$ the probe η_x. The former being functions of X and the latter functions of T. They are defined in the obvious way :

$$\xi_t(x) = u(x,t) = \eta_x(t) \ , \ t \epsilon T \ , \ x \epsilon X \tag{1.1}$$

This way of conceiving the signal u, completely symmetric in space and time is particularily adapted to our analysis.

It seems natural to describe a spatio-temporal system in this double way. From one side, as a "dynamical system", it evolves in some phase space. Here the phase space is the space (eventually infinite dimensional) of functions ξ of the spatial variable $x \epsilon X$, that we call <u>configurations</u>. For the sake of definitness we consider here the case where ξ is an element of $L^2(X)$, the space of complex square-integrable functions on X, even if the generalization to other spaces of functions is possible. In this context ξ_t can be viewed as an orbit of a dynamical system associated to the phenomena under consideration.

On the other hand, as an "experimental device" it is natural to consider each <u>probe</u> η as a function depending of time $t \epsilon T$, and here, again we may suppose $\eta \epsilon L^2(T)$.

We perform below the analysis of the spatio-temporal signal $u(x, t)$ via the bi-orthogonal decomposition. Before let us only mention that this type of signals can arise in many other circumstances. As it is clear in what follows the method can be used if, instead of a spatial-temporal signal, we consider any other function depending on two variables. Even if in these cases the dynamical scenario described above is meaningless it is possible, in many cases, to give a similar interpretation adapted to each problem.

For instance, by applying the bi-orthogonal decomposition to image analysis [D,L,L,V] we were faced to a signal $u(i, j)$ defining the brightness or gray level at each pixel (i, j). Here $i, i = 1, ...R$, stands for the row and $j, j = 1, ...C$, for the column of the considered corresponding image. Therefore, following the lines of the previous case, we can consider the space of the row vectors and the space of the column vectors. In that way, an image may be throught of as an orbit of a row vector in $l^2(J)$ as well as an orbit of a column vector in $l^2(I)$.

Before going to the details of the bi-orthogonal decomposition we review some of the relevant features of the system that we want to capture.

2. Spatio-temporal dynamics

For us, the important features of such systems are related to the great number of degrees of freedom of the underlying dynamics. As it is known, spatial extended systems present different transitions from one regime to another (laminar, turbulent for instance) when some external parameter (the Reynold number, for instance) varies. But it can also happen, and this is a peculiar feature of spatial-extended systems, that coexistence of different dynamical regimes appears; in this case complex patterns may arise, each spatial domain having its own peculiar dynamics and all the domains changing in time, see for instance [P] for a complete discretion of this situation from a different point of view.

The emergence of coherent structures is also a special touch of such systems.

These phenomena consists in the appearance of large scale correlations in space and in time out of a disorded system.

The precise meaning of such coherent structures in a general context is still an open question.

But, from one side, it is possible to relate such structures with statistical properties of the signal. They appear then as the most probable configuration during a lapse of time. We refer to Lumley [L] that first formulate this concept in his study in hydrodynamics.

From another point of view such coherent structures can be defined in terms of symetries of the orbits in the phase space of the underlying dynamical system. From this point of view, the most simpler structures are the periodic - in space as well as in time - configurations analized in [B,L,L].

The general purpose of the bi-orthogonal analysis, as it was recently introduced in [A,G,L,1], is to give an effective tool to relate these two complementary approachs and to study the different situations that may occur in this respect.

We refer to [A,G,L,1] for a complete description of the method that we now summarize.

3. Signal analysis

Associated to a signal u we define a linear operator acting from the space $L^2(X)$ to $L^2(T)$ by

$$(U\varphi)(t) = \int_x u(x,t)\varphi(x)dx \quad , \ \forall \varphi \epsilon L^2(X) \tag{3.1}$$

Since $L^2(X)$ as well as $L^2(T)$ are endorsed with scalar products, an adjoint operator $U^\star$ is automatically defined from $L^2(T)$ to $L^2(X)$.

Under suitable conditions on u, which are very often fulfiled in pratics, the spectral decomposition of U give a one to one correspondence between the elements of an orthonormal family of functions in $L^2(T)$ and an orthonormal family of functions in $L^2(X)$, fulfilling

$$U\varphi_k = \alpha_k \psi_k \quad , \quad k = 1, 2, \ldots \tag{3.2}$$

where the numbers α_k verifie

$$\alpha_1 \geq \alpha_2 \geq \ldots \geq \alpha_k \geq \ldots > 0 \tag{3.3}$$

with

$$\lim_k \alpha_k = 0 \tag{3.4}$$

From that immediatly follows, a canonical decomposition of the signal, that we call bi-orthogonal decomposition, namely :

$$u(x,t) = \sum_k \alpha_k \varphi_k(x) \psi_k(t) \tag{3.5}$$

The relation of this decomposition with the statistical feature of the signal cames from the fact that the operator $U^\star U$ is the Kernel operator defined by the spatial correlation function and the operator $UU^\star$ is the corresponding time kernel operator.

From the point of view of the pure analysis of the signal we want to stress that contrary to other signal analysis (Fourier, Wavelets, etc.), here the functional basis that decompose u is produced by u itself... with the advantages and inconvenients of this fact !

The relation of this analysis with the dynamics goes as follows.

The set of the topos $\varphi_k, k = 1\ldots$ generate a linear space $\chi(X)$ in $L^2(x)$. This space is also the orthogonal complement of the kernel of U, i.e. the range of $U^\star$. In a symmetric way the chronos $\psi_k, k = 1\ldots$ generate a linear space $\chi(T)$ in $L^2(T)$, which is defined in the same way with U replaced by $U^\star$.

We call this spaces the characteristic spaces of u, and by definition it common dimension, $dim(u)$, is called the dimension of the signal.

Now, it turns out that the characteristic space $\chi(X)$ is the smallest vectoriel space inside the phase space $L^2(X)$ that contains the orbit ξ_t of the system. Also $\chi(T)$ is the smallest vectoriel space in $L^2(T)$ containing all the probes η_x.

Therefore we end up with a complete localization of the dynamics inside a subspace of the phase space. The structure of this characteristic spaces, $\chi(X)$ and $\chi(T)$, as well as the way in which they are structurally tied, (3.2), contain important information about the dynamics of the system.

In [A,G,L,1] we describe, among other features, the possibility of detecting and describing, in this way, some of the bifurcations of the system. In the same way we can made the connection of the bi-orthogonal analysis with other know geometric characterizations of the atractors of the dynamics (Liapunov dimension, for instance).

Let also mention that in [A,G,L,1] we can give an estimation of the time spended by the dynamics in a bounded subsets of the characteristic space. It turns out, as expected, that this localization is more effective the less energetic is the signal.

There are cases for which it may be interesting to get a priori bounds for the dimension of the signal. For instance, if we want to compare an experimental or numerically computed signal with a theoretical one, in order to fix some parameter of a model which is intended to reproduce them, it is rather interesting to know what the maximum possible dimension D of each signal is. This is due to the fact that, in general, we can get a more accurate estimation of the parameter when only the first D components of the former are used in (3.5). Furthermore such bound may also be important to determine the time needed in the computation.

For instance, let us suppose that we can anticipate the following form for the signal u:

$$u(x,t) = \sum_{k=1}^{N} f_k(t) g_k(x) \tag{3.6}$$

then it follows that the dimension D of u satisfies $D \leq N$. The reason for having this bound is very simple if is thought as a linear combination, given by (3.6), of the N vectors $g_1, g_2, ... g_N$ in the characteristic space of u.

We will give later an example where we use this remark.

4. The dynamics of a complex system

The localization of the dynamics in a space $\chi(X)$ of smaller dimension is consistent with the idea that the effective number of degrees of freedom in some regimes can be considerable less than the corresponding dimension of the phase space. This is clear because, the latter may contain the orbits of all the possible dynamics, for any regime of the system, and the former is intended to be only able to contain a given orbit or a set of orbits of the same type, for instance for a given value of a parameter or for a given bassin.

In [B,L,L] we describe in a rigorous manner how to explicitely reduce the dynamics of a system of a coupled map lattice - a system with a large number of degrees of freedom - to a 2 dimensional dynamical system. Translating this result in our language is simply to say that $\chi(X)$ is a space of dimension 2.

In this case, this space is well defined, since it is the 2 dimensional plane of the periodic sequences of period two. Recall that x is here the finite set of sites in the lattice (the extension to the infinite lattice is a technical question).

Of course this picture is only valid inside an interval of values of the diffusion parameter ε of the model, see [B,L,L].

Now it is clear that also the position of $\chi(X)$ inside $L^2(X)$, or that of $\chi(T)$ in $L^2(T)$, may change when the value of the parameter change. For instance, in the case mentioned above, we noticed in [A,G,L,1] that the dimension of $\chi(X)$ remains 2 even after the critical value of ε, computed in [B,L,L], for which the periodic configuration of period 2 in space is no longer stable. Therefore after this value of $\varepsilon, \chi(X)$ can no more be the set of periodic sequences and starts to move inside $L^2(X)$.

On the other hand, the 2 dimensional dynamical system describing the motion below the critical value of ε, when expressed in two abstract variables (x_1, x_2) still exist for larger values of ε. It is not difficult to verify that this system still describe the dynamics of the coupled map lattice for this region of the parameter ; but now (x_1, x_2) are no longer the values of the odd and even sites but the coordinates in $\chi(X)$.

In other words the 2 dimensional dynamical system always leave in $\chi(X)$ but its interpretation as values of the signal for odd-even sites is only true if $\chi(X)$ is generated by two periodic topos φ_1 and φ_2. After that and as long as no external bifurcation occurs, i.e. a bifurcation concerning a change of dimension of $\chi(X)$, the dynamical system describe the motion of the lattice, but in the coordinates of $\chi(X)$. In order to translate this motion into the motion of each site of the lattice we need to use the explicit form of the topos φ_1 and φ_2 as a change of coordinates.

Also bifurcations of the dynamics of the system living in $\chi(X)$ give rise to bifurcations of the corresponding dynamics of the large system. On the contrary the displacement of $\chi(X)$ inside $L^2(X)$, when the parameter changes, does not give rise by itself to any topological modification of the dynamics and therefore no bifurcation can arise if only this displacement is present.

Nevertheless we can imagine the extreme complexity of the dynamics in the spatial extended system in this case, since the motion of the small dynamical system in $\chi(X)$ is expressed in the global coordinates of $\chi(X)$, i.e. the topos φ_k, and these ought to be very complicated functions of the lattice. The two changing according to the values of the parameter, i.e. the change in the small dynamical system and the change of the characteristic space $\chi(X)$ in $L^2(X)$.

Finally since the position of $\chi(T)$ is also free, and the interval of stability is not the same of the one for $\chi(X)$, combining the two by (3.5), we can see various scenarios in the dynamics by varying the parameter. In [G,L] we describe some situations arising in coupled map lattices and we show that the rigidity of $\chi(X)$ is keeped as long as the change of the value of the parameter brings to a linear perturbation of the signal and is destabilized by non linear effects of this modification in the value of the parameter.

5. Dynamical model generators

In [D,L,M] we were able to use the capabilities of the bi-orthogonal analysis to generate dynamical models for control purposes in an automatic way.

The idea consisted in analysing the real data to continuously update the parameter value in a suitable family of dynamical generators. Conversely, we can also use such a tool to react on the experimental device in order to force it to follow a predetermined regime.

Two properties of the bi-orthogonal analysis are likely to be used for this purpose : first, its capability of noise-reduction that permits a more accurate comparison between the data and the model generator; second, the bi-orthogonal analysis optimizes the truncation procedure since it gives the best approximation for the signal for a given number of "modes", see [A,G,L,1], and this fact can be used to minimize time consuming computation allowing the possibility of a control "in line".

We have used an electromechanical actuator as an example for application. This system is based on an electric motor and its load, an electronic power supply converter and also a nonlinear speed controller.

In this study only simulation results are given which have been obtained using the simplified model presented below.

The motor is a separately excited *DC* motor and it is represented by equations (5.1), where i is the current, u is the current, u is the applied voltage, $T(t)$ is the load torque and ω the angular velocity

$$\begin{cases} \dfrac{di}{dt} = -\dfrac{R}{L}i - \dfrac{K_\phi}{L}\omega + \dfrac{u}{L} \\ \dfrac{d\omega}{dt} = \dfrac{K_\phi}{J}i - \dfrac{1}{J}T(t) \\ \dfrac{d\Theta}{dt} = \omega \end{cases} \tag{5.1}$$

This model is deliberately simplified. For instance, in practice the magnetic circuit is nonlinear and there are parameter variations.

The electric power is supplied to the motor throughout a DC/DC electronic converter. In this study the commutation of the power electronic devices is considered ideal. However in the simulation, the command and the speed control of the motor are taken into consideration.

As far as the electronic power supply is able to generate a high enough voltage, one can show that the resultant dynamic of the system approximately obeys the following equation [U], [D,M]

$$\frac{d\omega}{dt} = -\frac{K_\omega}{K_\gamma}\omega + \frac{K_\omega}{K_\gamma}R \tag{5.2}$$

Here R is a reference for the speed and K_ω, K_γ are the parameters to be adjusted.

One should notice that this equation does not include the commutation process which, however, is included in simulation results. Experimental recordings taken in systems like the one presented show other minor differences which are the consequence of phenomena not included in the simulation.

It is true that the controller attenuate this deviation, but there is always a limit for the achieved performances which is imposed by the uncertainties of the model used to represent the dynamic of the real system.

In this section we use the bi-orthogonal decomposition as a model generator. We hope that this technique using experimental data can identify in a simple way the structures that concentrate the dynamics.

To illustrate some questions related to the learning process using the biorthogonal decomposition we consider in a first approach that the system is represented by the equation (5.2). This means that there is a linear relation between the speed, the reference and the acceleration. The signal $\omega(\gamma, R)$ is considered and the bi-orthogonal decomposition is done using the theoretical values obtained by the relation (5.2). The available information about the system should be used at start - since it is commonly acknowledged that is better to start a learning process knowing something than to start it completely blind. However, we consider that there is an error in the parameter K_ω in the model. That is, at the start the value of this parameter used in the simulation is 70% of the value used to generate data for the decomposition.

In order to obtain a first result we compare the values of the speed (obtained by simulation of the global system ω and the one obtained by the bi-orthogonal decomposition $\omega^\star$) when the reference input R changes. Figure 1 presents a block diagram of this situation and figure (2) presents the evolution of the error between the two referred variables.

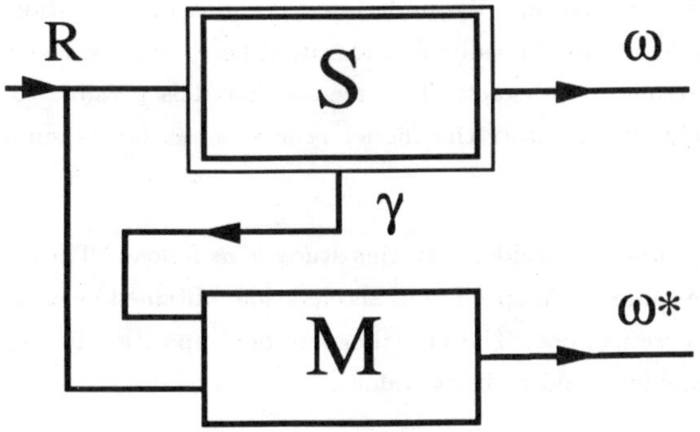

Fig. 1 Block S represents the global simulated model and block M is the one obtained by bi-orthogonal decomposition.

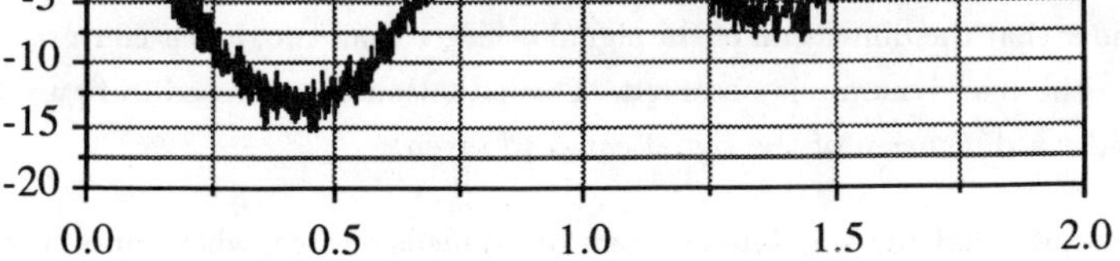

Fig. 2 Relative error evolution between speeds ω and $\omega^\star$.

One can see in figure 2 that the error presents some high frequency oscillations which appear only at the output signal $\omega^\star$. These oscillations are the result of the commutation process which is not represented by the simplified equation (5.2). In fact the system will be better represented by the equation (5.3) than by the equation (5.2)

$$\frac{d\omega}{dt} = -\frac{K_\omega}{K_\gamma}\omega + \frac{K_\omega}{K_\gamma}R + \varepsilon \tag{5.3}$$

The term referred above, ε, is not important as far as implementation is considered and can be attenuated by means of the updating of the model using measured values or using a low pass filter for $\omega^\star$.

If the oscillations are not taken into account it is easy to verify that the error results from the parameter deviation. We underline that in a real situation the parameter deviation will be less than the value consideration here, because the controller makes the resultant dynamic more exact. This means that it is possible to use theoretical values in a simple way to start the model generator, as far as similar systems are concerned.

The learning process considered in this study is as follows. The system input R is changing and the values of R, speed, and acceleration, obtained by simulation, are used to update the theoretical one. The bi-orthogonal decomposition is then applied to the signal $\omega(\gamma, R)$ combining old and new values.

Figure 3a presents the relative error evolution in a similar way as the one already presented in figure 2. However, the model has "learned something". This means that the bi-orthogonal decomposition is achieved using some values obtained directly by simulation. In this process we have considered that the signal has dimension two up to noisy terms. This means that the decomposition was computed only for the first two terms.

Figure 3a shows that the model generator presents a less significant error after some simulated data have been used in the bi-orthogonal decomposition. However, if we consider that the dimension of the signal is larger than two, a fast convergence of the model to the real dynamic is observed. This situation is presented in figure 3b where we consider a dimension of the signal equal to seven.

In despite that the simulated system has dimension two, when some true data are combined with theoretical ones, the signal considered for decomposition has in fact a dimension larger than two. As far as implementation is concerned, an estimation of the dimension of the signal must be performed in order to obtain a fast convergence. On another way, some limitation to this dimension need to be imposed in order to maintain the noise rejection capability of the bi-orthogonal decomposition. We will came to this problem below.

a
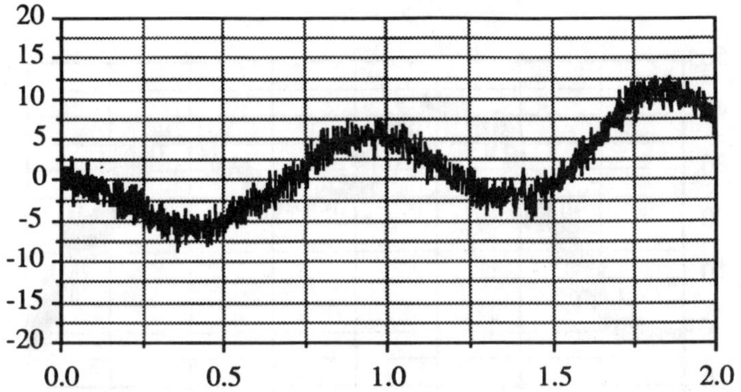

b
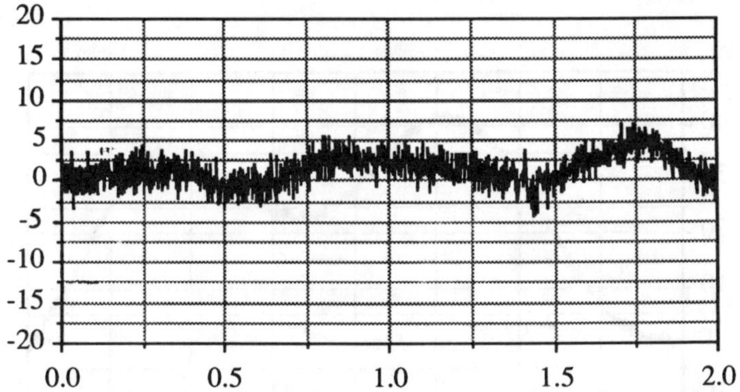

Fig. 3 Relative error evolution between speeds after a learning process. (a) with truncation at dimension 2 and, (b) at dimension 7.

In the figures 2 and 3 one can note that oscillations are not affected by the learning process.

The acquisition of the experimental values is disturbed by noise. The bi-orthogonal decomposition used as an algorithm can filter the signals. The theoretical results concerning signal dimension and all that was referred to above indicate how the algorithm works. For systems like the one presented the signal is linear and as a consequence only two terms of the decomposition will be significant. So to filter the values it is enough to consider these two terms.

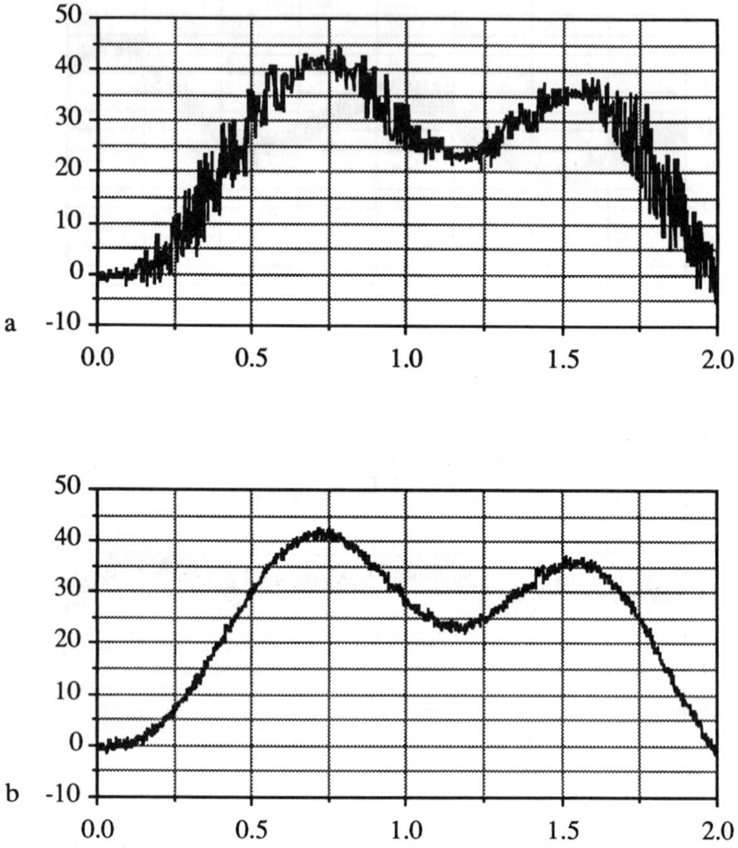

Fig. 4 (a) Time evolution of speed $\omega^{\star}$ and,
(b) same with a filtering by the bi-orthogonal decomposition.

In order to show this filter capability, figure 4 presents the evolution of the speed obtained by a model in two situations. In figure 4a the model is obtained using values given by the relation (5.4) where ξ is a random variable in $]-1,1[$ interval

$$\omega = -\frac{K_{\gamma}}{K_{\omega}}\gamma(1+0.5\xi)+R \tag{5.4}$$

In figure 4b the model is obtained using also the referred relation, but only two terms of the bi-orthogonal decomposition are considered.

Figure 4 shows that the signal obtained by the bi-orthogonal decomposition is free from noise. One must note that the visible oscillations in figure 4b have a different origin, as was already mentioned. They are a consequence of the commutation process and this entails some correlation between the variables considered to represent the dynamic of the system. However, the commutation depends also on other variables, such as the current or the voltage. In our opinion, it will be necessary to consider in the learning algorithm these values in order to represent the commutation effects.

If a model generator like the one presented is available, (i.e., an algorithm with self-learning capability to decode complex relations), some implementation in control area are made possible in an easy way. We present two of these applications as an example.

Full compensation

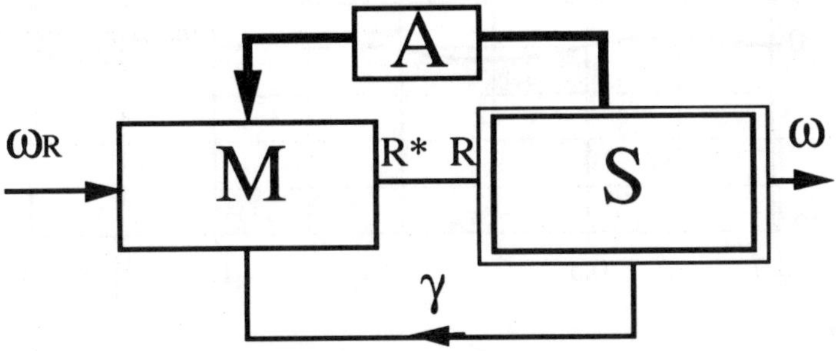

Fig. 5 Diagram of the compensation method.

Consider the block diagram presented in figure 5. Block S is the real system and block M is the model which is updated trough block A. Suppose that the system constrains the relation $\omega(R, \gamma)$ and the model generator is able to capture the signal $R(\gamma, \omega)$, that is the inverse model. Under this situation the output evolution becomes equal to ω as far as the input ω_r is compatible with physic constraints.

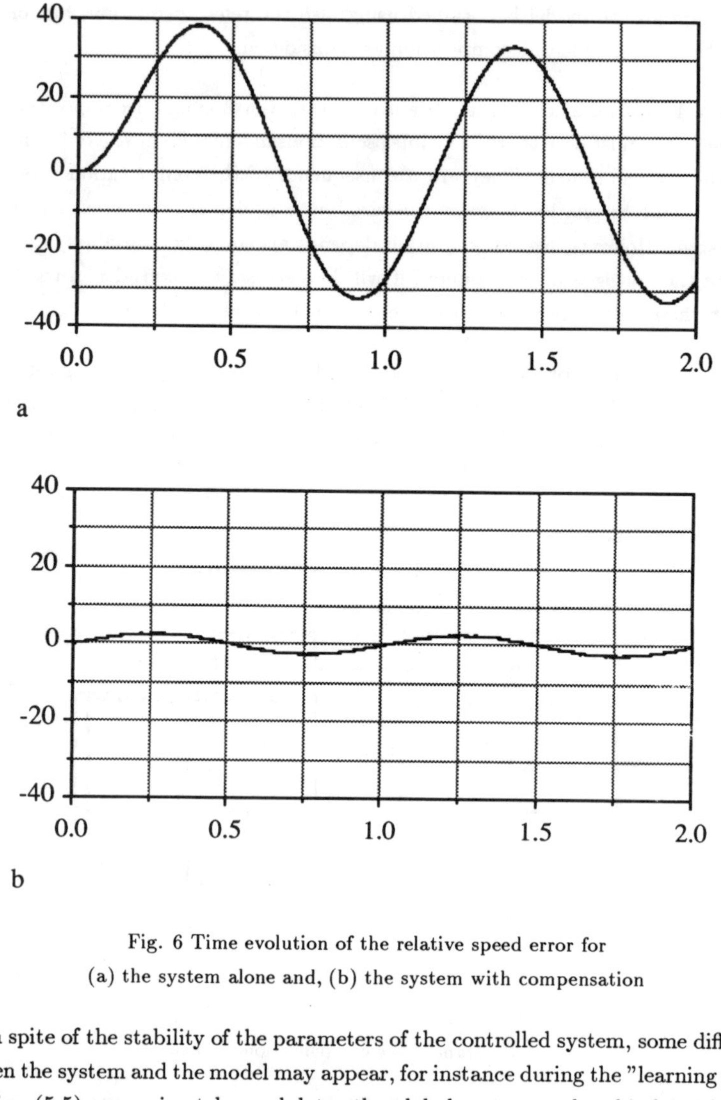

Fig. 6 Time evolution of the relative speed error for
(a) the system alone and, (b) the system with compensation

In spite of the stability of the parameters of the controlled system, some differences between the system and the model may appear, for instance during the "learning phase". Equation (5.5) approximately modulates the global system under this hypothesis. It is easy to verify that this form of compensation may give better performances to the system. Figure 6 present this case when an error in parameters is considered. In this figure one can see that, in spite of the error in parameter K_ω, the error at the output is lower than the error obtained in the case where the compensation is not performed.

We must note, using (5.5) that some instability may appear:

$$\frac{d\omega}{dt} = -\frac{K_\omega K_\omega^\star}{K_\omega^\star K_\gamma - K_\gamma^\star K_\omega}\omega + \frac{K_\omega K_\omega^\star}{K_\omega^\star K_\gamma - K_\gamma^\star K_\omega}\omega_r \tag{5.5}$$

Predictive control

Another typical application for the model generator is the predictive control. The control methodology is based in a preview capability and in a correction action. Usually the preview of the dynamic is made using simplified linear models which enable the predictive action in a very short time interval. The model generator based on bi-orthogonal decomposition has no limitation as far as the complexity of the dynamics is concerned. This fact is important, because it enables us a more reliable prediction in a longer time interval.

A better performance of the model may be achieved with a complementary process. For instance, according to the error between the output signal of the system and the output signal of the model the decomposition may use other variables or may consider a greater dimension of the signal. These tasks will be object of further investigations.

6. Global quantities

By analogy with the case for which the signal u is a velocity field, we define the energy by

$$E(u) = \int_X \int_T u(x,t)u(\bar{x},t)dxdt = \sum_{k=1}^{\infty} \alpha_k^2 \tag{6.1}$$

Notice that $E(u)$ can be expressed in terms of the operator U and its adjoint by :

$$E(u) = Tr(U^\star U) = Tr(UU^\star) \tag{6.2}$$

From (6.1) we see that each space-time structure, i.e. a pair (φ_k, ψ_k) participate independently to the energy of the signal.

Using (6.2) it is possible to control several properties of the global energy. In particular its robustness with respect to perturbations or change in the space or time definition set.

The normalized energy of each structure is

$$p_k = \frac{\alpha_k^2}{\sum_k \alpha_k^2} \tag{6.3}$$

and we define the global entropy of the signal by :

$$H(u) = - \lim_{N \to \infty} \frac{1}{log N} \sum_{k=1}^{N} p_k log\ p_k \tag{6.4}$$

Therefore, $H(u)$ is zero if and only if one eigenvalue is non zero and is one (its maximal possible value) if all the eigenvalues are equal.

In fig. 7 we show a plot of the global entropy for different values of the coupling constant ε in a system of coupled map lattice. Without going in details we notice that two regions of regular regime of the motion are detected by low values of the entropy. For very small ε a chaotic regime is present and for ε between 0.18 and 0.52 a more complex regime can be observed. Here we can see some local maxima of the entropy corresponding to the case where the values of the different eigenvalues are closed together ; in this case, we can sometimes follow two structures for which the corresponding energies present a crossing.

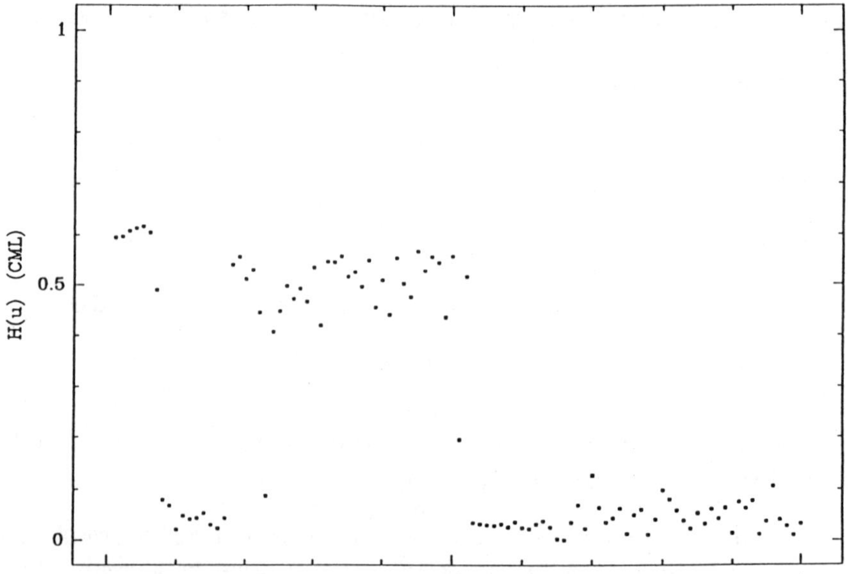

Fig. 7. The global entropy as function of the diffusion parameter ε in a model of coupled map lattice.

This situation is more clearly observed in fig. 8 where we present the global entropy as a function of Reynolds number in a rotating disk experiment. When this curve is compared with energy as a function of the same parameter ploted in fig. 9 we see that the local maxima of the entropy are located in the values of the parameter for which the first and second eigenvalue (sometimes the second and the third) present a cross. If we suppose that, closed to these points, the crossing eigenvalues have a linear dependence in the parameter and at the same time the remainder have only a small variation, then the parabolic shape of the entropy closed to the local maximum is easy to understand.

From this picture emerges a transition to the turbulent regime caracterized by a series of crossing in the energies of structures, where the new ones develope when old ones die. We notice that even if it is possible to use Fourier analysis in the first stage of the instability, it becomes useless as a thich packet of wave numbers gets activated. See [A,C,G] for more details.

The existence of a large Fourier spectrum may be present even in the case of low dimensional (two for instance) characteristic spaces. This is due, as mentioned in section 3, see also [G,L] for details, that these spaces, $\chi(X)$ and $\chi(T)$, can take different positions in phase and probe spaces and, in particular inside subspaces of arbitrary complexity in Fourier coordinates (arbitrary wave numbers or frequencies).

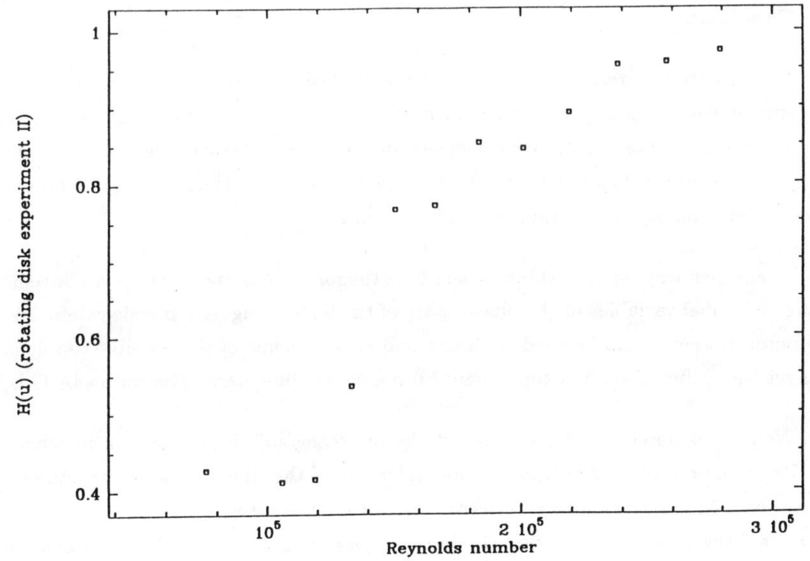

Fig. 8 The global entropy as function of Reynolds number in rotating disk experiment.

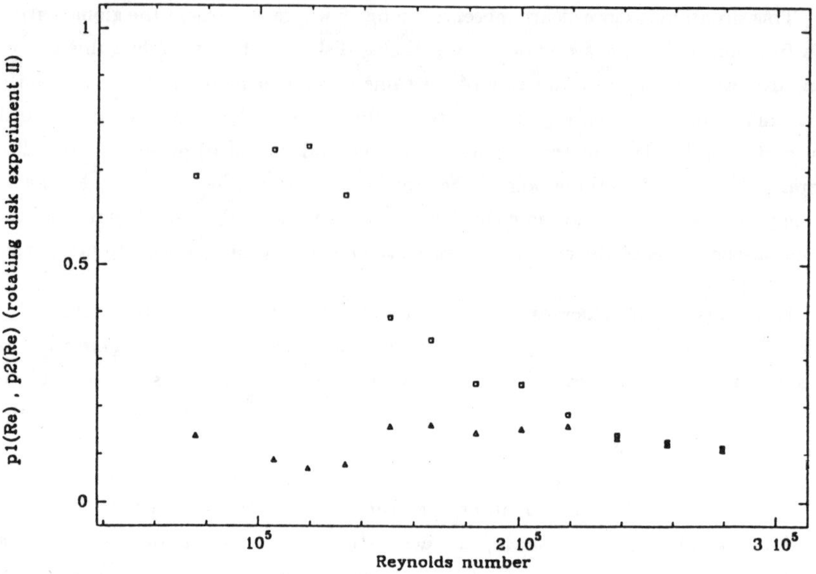

Fig. 9 Normalized energies as function of Reynolds number, in rotating disk experiment.

Conclusion

The essential ingredients which carry the analysis of a complex space-time signal in a simple way using the bi-orthogonal decomposition are : 1) the separation of time and space components, 2) the low dimension of the characteristic space inside the phase space where the signal effectively lies and 3) inside that space, the location of a bounded region where the dynamics stays most of the time.

In another way we can think of the bi-orthogonal decomposition as a convenient change to global variables in the phase space of the underlying dynamical system. From this point of view it can be used to detect and analyze some of the possible topological changes (i.e. bifurcations) in the dynamical regime as some parameter varies [A,G,L,2].

We also explored the capabilities of the bi-orthogonal decomposition in what reduction of noise level and optimal compactification of the signal relevant structures are concerned. An application to an electromechanical actuator was presented as an example to test the possibility of using the bi-orthogonal analysis as a dynamical generator for control purposes in an automatic way.

References

[A,C,G] N. Aubry, M.P. Chauve, R. Guyonnet ; Transition to Turbulence on a Rotating Flat Disk, Marseille (1990).

[A,G,L,1] N. Aubry, R. Guyonnet, R. Lima ; Spatio-Temporal Analysis of Complex Signals : Theory and Applications, J. Stat. Phys. 63 (1991) 683.

[A,G,L,2] N. Aubry, R. Guyonnet, R. Lima ; Spatio-Temporal Symmetries and Bifurcations via the Bi-orthogonal Decomposition, J. Nonlinear Sci., vol. 2, issue 2 (1992).

[A,G,L,3] N. Aubry, R. Guyonnet, R. Lima ; Turbulence Spectra, J. Stat. Phys. 67, issue 1/2 (1992) 203.

[B,L,L] L. Bunimovich, A. Lambert, R. Lima ; J. Stat. Phys. 1/2 (1990) 61.

[D,L,L,V] J.A. Dente, A. Lambert, R. Lima, R. Vilela-Mendes ; The Bi-orthogonal Decomposition and the Analysis of Complex Images, (1991) submitted.

[D,L,M] J.A. Dente, R. Lima, J.H. Maia, The Bi-orthogonal Decomposition as a Dynamical Model Generator, (1991), submitted.

[D,M] J.A. Dente, J.H. Maia, A Robust Position Controller for a DC Drive Using Sliding Mode : Some Implementation Requirements, EPE 91-Florence, vol. 3, (1991), 553.

[G,L] R. Guyonnet, R. Lima, Complex Dynamics in Coupled Map Lattices via the Bi-orthogonal Decomposition, Marseille (1991).

[L] J.L. Lumley, Stochastic Tools in Turbulence, Academic Press (1972).

[P] Y. Pommeau ; Topics in pattern formation problems and related questions, "Chaos, order and Patterns", NATO Advanced Intitute, Como, Italy.

[U] V.I. Utkin, Discontinuous Control Systems : State of Art in Theory and Applications, IFAC-Munich, (1987), 75-94.

EARTHQUAKE BEHAVIOUR OF STRUCTURES: FROM OVERABUNDANT COMPLEXITY TO USEFUL KNOWLEDGE THROUGH CONSTRUCTIVE PROBABILITIES

RICARDO TEIXEIRA DUARTE
National Laboratory for Civil Engineering (LNEC),
Av. do Brasil 101 P-1799 LISBOA CODEX

ABSTRACT

It is assumed that useful knowledge is knowledge about the probability of failure of a structure for earthquake occurrences. A mathematically based approach is presented, where probability distributions are constructed to represent knowledge and bayesian methods are employed to update it after new results are obtained. Applications to the earthquake design of bridge structures are described.

1. Introduction

The complexity and apparent randomness of earthquake vibrations and the richness and variety of the dynamic behaviour of structures strained far beyond the elastic range, coupled to the significance of the seismic hazard in large regions of the world, have given rise to powerful methods of analysis, both experimental and computational. Unhappily, in the present state of knowledge the uncertainty about the quantification of the input needed for the application of these methods is only balanced by the enormous volume of the information content of the results. The relationship between bending moments and curvatures, plotted in figure 1, of a single section of a single pier of a bridge acted by a single earthquake vibration illustrates the overabundant complexity and richness of detail of results obtained through nonlinear dynamic computations. Hence criteria and methods to sort out essential and useful information from irrelevant details and particularities is a prioritary endeavour. At LNEC, in the last years, the probability of failure has been used as the basic criterium in the assessment of structural performance and the concept of vulnerability function has provided the essential tool to compute, in an efficient manner, the value of that probability. The present paper mainly presents, for the first time in a comprehensive manner, the partials results published elsewhere. [1,2,3]

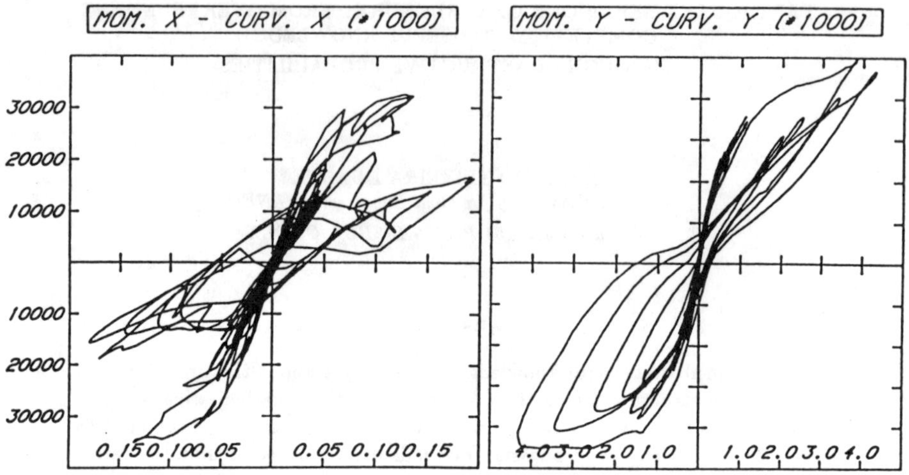

Figure 1: Moment-curvatures loops (major and minor inertia directions) of a reinforced concrete pier acted by bi-directional earthquake vibrations

2. Structural Models in Earthquake Engineering

Earthquake Engineering have to deal with structures strained far over the yelding limit, subjected to dynamic disturbances and undergoing large displacements; moreover, both the characteristics of the action and of the structure are difficult to quantify. Hence, Earthquake Engineering seldom can use 'exact' models and consequently, such relationship as may be established amoung models of different accuracy, are specially useful. Furthermore, the concept of an 'exact' model itself may be considered meaningless and advantageously substituted by the concept of reference model which is defined as a model without simplifying assumptions.

A model is defined by a set of hypothesis which idealize the relevant aspects of nature. In engineering, models are selected by their usefulness, not because they are 'exact'. Three aspects can be considered in the evaluation of models[4] : the logical consistency (syntactical aspect), the descriptive power (semantic aspect) and the practical usefulness (pragmatical aspect).

The syntactical aspect of models emphasizes models as theories relating a set of propositions to a set of assumptions. A model is sintactically correct when those relationships obey the rules of mathematics or of inductive or deductive logic. It should be noted that it is possible to conceptualize in a more advanced way this subject; in effect, in modern economics, a theory is obtained from a model, defined as a theoretical construct, by the addition of assumptions permiting to establish relationships between the variables in the model and real objects[5]. As the best example of a non-trivial sintactically correct model the classical field theories of mechanics deserve to be referred[6,7].

The semantic aspects of a model deals with the correspondence between model and that part of reality that the model purposes to describe; thus the semantics of a model is clearly to be established by experimental means.

The pragmatical aspects of a model are associated with its value for decision-making and are of fundamental importance in engineering, although this subject have not yet had the development it deserves; but, with the increasing availability of ever more sophisticated models, the search for optimal models could not be postponed much longer. A model is said to be optimal in the following sense: Let $M \in \mathcal{M}$ be an ensemble of models; define a cost functional $c(M)$ as the cost incurred in using model M and a penalty functional $p(M)$ as the penalty suffered because of the use of a wrong model. In general one of the models in $\mathcal{M}$ will be considered as 'exact' and will be designed as the reference model M^r; hence $p(M^r) = 0$. Now the optimal model is defined as the one which minimizes $c(M) + p(M)$. Obviously the main difficulty lies in establishing the cost and penalty functionals.

3. Vulnerability Functions

The vulnerability function expresses the relationship between the essential characteristics of the earthquake vibration and the essential characteristics of the corresponding structural response; thus the purpose of the vulnerability function is to eliminate the irrelevant detail in the description of vibration and response. Furthermore, two models very different in complexity, can have almost identical vulnerability functions; in consequence the simpler model may be used for practical problems; the vulnerability function then becomes an yardstick for the comparison of models.

The earthquake vibration will be considered to have several components and a finite duration T_1; hence, every possible earthquake vibration will be assumed to belong to the space L_s of vector valued functions Lebesgue integrable in absolute value. Conversely, every function belonging to L_s will be considered to be a generalized earthquake vibration and two functions equal almost everywhere will be considered to correspond to the same action; thus, L_s will be denoted by $\mathcal{A}_s$.

The response of the structure is assumed to be described by a finite number r of variables (displacements, accelerations, internal forces ...); the time histories of the responses to all generalized actions constitute an ensemble R_r which is contained in L_r ($\mathcal{R}_r$ is a linear space when the structure has linear behaviour); the duration of interest T_2 of the response time histories will be assumed to be finite and greater than T_1. In the following, to present concepts that apply to both the L_s and L_r spaces, a generic L_v space will be used; simillarly, a function $\mathbf{x}(t)$ may represent either a vibration $\mathbf{a}(t)$ or a response $\mathbf{r}(t)$.

The relationship between an element $\mathbf{a}(t)$ of $\mathcal{A}_s$ and the corresponding element $\mathbf{r}$ of $\mathcal{R}_r$ defines the structure operator:

$$\mathbf{r}(t_2) = \mathcal{E}(\mathbf{a}(t_1)) \quad t_1 \in [O, T_1] \quad t_2 \in [0, T_2] \tag{1}$$

The operator $\mathcal{E}$ will be assumed to be sufficiently continuous i.e. for two vibration histories $\mathbf{a}_1(t_1)$ and $\mathbf{a}_2(t_2)$ sufficiently close to one another, the corresponding response histories will also be close to one another. The Eq. 1 must be understood in the sense that it establishes a correspondence between the complete time history of the vibration $\mathbf{a}(t_1)$ and the complete time history of the response $\mathbf{r}(t_2)$; in the case of a structure with linear behaviour, Eq. 1 may be written as

$$\mathbf{r}(t_2) = \int_o^{t_2} \mathbf{H}(t_2 - t_1)\mathbf{a}(t_1)dt_1 \quad t_2 \in [0, T_2] \tag{2}$$

where $\mathbf{H}(t)$ is the unit impulse response matrix and the definition of $\mathbf{a}(t)$ is extended by setting $\mathbf{a}(t) = \mathbf{0}, T_1 < t$. The essential characteristics of the vibrations and responses may be expressed through the values of a number of functionals defined on the time histories $\mathbf{a}(t)$ and $\mathbf{r}(t)$, which will be referred to as descriptive functionals. If this number is finite, some information is necessarily lost; but if this number is infinite, all the information contained in the time histories may be preserved if the functionals are adequately selected. For illustration, consider the functionals $\mathbf{h}_k$ defined by $\mathbf{h}_k = (\int_0^{T_1} \mathbf{x}(t)e^{-2\pi ikt/T_1}\ dt)/T_1 \qquad k = 0, 1, -1, 2, -2, ...$ Knowledge of the values of those functionals is equivalent to the knowledge of the time histories themselves since they are their Fourier coefficients.

There are a large number of possibilities for selecting descriptive functionals. One possibility is to define a general functional, which is indifferently applied to every time history. For instance, the value of the supremum norm may be selected as the descriptive functional; the supremum norm $\| \ \|^\infty$ of a time history $x(t)$ defined in a time interval T is $\| x \|^\infty = \max \mid x(t) \mid \quad t \in T$. Hence, in this case, the descriptive functionals of the vector time histories $\mathbf{a}(t)$ and $\mathbf{r}(t)$ are the vectors $\| \mathbf{a} \|^\infty$ and $\| \mathbf{r} \|^\infty$ whose components are the peak values of the corresponding time histories. The number of descriptive functionals can be readily increased if the general functional is also applied to the time-derivatives or time-integrals of the histories of any order or to the non-integer order time derivatives or integrals[8].

Let $\mathbf{h}$ be the vector representing the values of the descriptive functionals $\mathbf{h}(\mathbf{a}(t))$ of the vibrations and $\mathbf{c}$ the vector representing the values of the descriptive functionals $\mathbf{c}(\mathbf{r}(t))$ of the responses (it should be noted that the same letter is used to denote the functional and its value; the first interpretation applies if and only if the letter is followed by an argument). Those last functionals will be called 'control variables'. Using Eq. 5 it is possible to write that

$$\mathbf{c} = \mathbf{c}(\mathcal{E}(\mathbf{a}(t_1))) \tag{3}$$

In this new and simplified form, the equation of motion represents a mapping from the space of the vibrations into the Euclidean vector space of the control variables $\mathcal{C}$; in general the range of this mapping will only be a subset of $\mathcal{C}$. The last step needed for the vulnerability function to appear is to substitute $\mathbf{a}(t)$ by $\mathbf{h}$. This substitution poses somes problems. In effect, it is possible to use the inverse image $\mathcal{F}(\mathbf{h})$ of $\mathbf{h}$ defined by

$$\mathcal{F}(\mathbf{h}) = \{\mathbf{x}(t) : \mathbf{h}(\mathbf{x}) = \mathbf{h}, \mathbf{x} \in \mathcal{A}_s\} \tag{4}$$

In general, $\mathcal{F}(\mathbf{h})$ will be constituted by a large number of time histories $\mathbf{x}(t)$; for instance if the descriptive functional is the peak value of acceleration, $\mathcal{F}(\mathbf{h})$ will be all acceleration time histories with peak value $\mathbf{h}$. Thus if the vulnerability function $\mathbf{V}(\mathbf{h})$ is defined as

$$\mathbf{V}(\mathbf{h}) = \mathbf{c}(\mathcal{E}(\mathcal{F}(\mathbf{h}))) \tag{5}$$

for a given vector $\mathbf{h}$ a large number of possible values for the control variables would result. The effective way to side-step this indetermination is to define as the value of the vulnerability function an weighted average of all the possible $\mathbf{c}$ values. The mathematical tool that permits the rigorous formulation of weighted averages is Measure Theory which, in pratical terms is applied to this problem under the guise of Probability Theory.

4. Stochastic Models

For the present purposes, a stochastic process is an weighted ensemble of functions of the time. In mathematical terms and considering a L_v space as the ensemble of functions, the definition of a measure in L_v needs the definition of a σ-field S of subsets of L_v; this σ-field is the σ-field generated by the cylinder subsets of L_v i.e. subsets comprising all functions $\mathbf{x}(t)$ for which $\mathbf{x}(t_1) \in \mathbf{X}_1$, $\mathbf{x}(t_2) \in \mathbf{X}_2, \ldots\ \mathbf{x}(t_N) \in \mathbf{X}_N$, where $t_1, t_2, \ldots t_N \in (0, T_1)$ and $(\mathbf{X}_1, \mathbf{X}_2, \ldots \mathbf{X}_N)$ are Borel sets in R^{vN}. Now a stochastic process is defined as a triplet (L_v, S, μ) with $\mu(L_v) = 1$; μ is a measure defined on the σ-field S. The specification of the measure μ defines the process; in general, a large number of functions will have a nonzero probability density of occurrence for a given process; in particular, for gaussian processes all functions $\mathbf{x}(t) \in L_v$ have nonzero probability density [9]. A function associated to a nonzero probability density will be called a realization of the process.

The measure defined in a L_v space may be used for the definition of functionals which will be descriptive of the complete space and not of a single realization. For instance it is now possible to define the mean peak value of a function $\mathbf{x}(t)$ as $mpv(\mathbf{x}) = \int_{L_v} \| \mathbf{x}(t) \|^\infty d\mu$ i.e. the mean peak value is the ensemble average of the greatest maximum (in absolute value) of each function in the ensemble. In a similar way the mean value of any functional $g(\mathbf{x}(t))$ is $E(g(\mathbf{x}(t))) = \int_{L_v} g(\mathbf{x}(t))d\mu$. The stochastic analysis of a structure signifies the characterization, in terms of probability concepts, of its response from the knowledge of the probability measure defined on the vibration space $\mathcal{A}_s$. In a general way, a stochastic model of the vibrations is a triplet $(\mathcal{A}_s, S_a, \mu_a)$ where $\mathcal{A}_s$ is the vibration space, S_a is a σ-field defined on $\mathcal{A}$ and μ_a is its probability measure; the corresponding stochastic model of the responses is a triplet $(\mathcal{R}_r, S_r, \mu_r)$ where $\mathcal{R}_r$ is the response space $\mathcal{R}_r = \mathcal{E}(\mathcal{A}_s)$, S_r is the σ-field generated by the $\mathcal{E}$-images of the subsets in S_a, and μ_r is the measure induced in $\mathcal{R}_r$ by $\mathcal{E}$ i.e. to each measurable set in $\mathcal{R}_r$ is attributed the measure of its inverse $\mathcal{E}$-image in $\mathcal{A}_s$. In general it is only needed or useful to know some functionals of the responses, as for instance the mean peak value:

$$mpv(\mathbf{r}) = \int_{\mathcal{R}_r} \| \mathbf{r}(t_2) \|^\infty d\mu_r = \int_{\mathcal{A}_s} \| \mathcal{E}(\mathbf{a}(t_1)) \|^\infty d\mu_a \tag{6}$$

This expression emphasizes that the responses may be characterized by performing the ensemble averaging with the probability measure defined on the vibration space.

It is now possible to define rigorously and in general terms a vulnerability function. Taking the expected value of Eq. 3:

$$E(\mathbf{c}) \equiv \int_{\mathcal{R}_r} \mathbf{c} d\mu_r = \int_{\mathcal{A}_s} \mathbf{c}(\mathcal{E}(\mathbf{a}(t_1)))d\mu_a = E(\mathbf{c}, \mu_a) \tag{7}$$

The expected value of a descriptive functional $\mathbf{h}(\mathbf{a}(t))$ of the actions is

$$E(\mathbf{h}) \equiv \int_{\mathcal{A}_s} \mathbf{h}(\mathbf{a}(t_1))d\mu_a = E(\mathbf{h}, \mu_a) \tag{8}$$

Both expected values depend on the measure μ_a; to explicit this fact the notation $E(\mathbf{c}, \mu_a)$ and $E(\mathbf{h}, \mu_a)$ was used. Let μ_{ai}, $i \in I$ be a family of probability measures, with I being a finite or countable or uncountable index set; this family is constructed in a such way that

$$E(\mathbf{h},\mu_{ai}) \neq E(\mathbf{h};\mu_{aj}) \quad i \neq j \tag{9}$$

Hence the mapping

$$E(\mathbf{h}) = E(\mathbf{h},\mu_{ai}) = \int_{\mathcal{A}_e} \mathbf{h}(\mathbf{a}(t_1))d\mu_{ai} \tag{10}$$

is a bijective mapping between i and $E(\mathbf{h})$ with domain $\mathcal{I}$ and range $\mathcal{H}$ defined by

$$\mathcal{H} = \{\ \mathbf{x} : \mathbf{x} = \int_{\mathcal{A}_e} \mathbf{h}(\mathbf{a}(t_1))d\mu_{ai}, i \in \mathcal{I}\} \tag{11}$$

The family of probability measures is supposed to be sufficiently diversified so that $\mathcal{H}$ is large enough for the applications; in general, as will be shown bellow, the family of probability measures will depend continuously on a given number k of independent parameters and thus $\mathcal{I} \equiv \mathcal{R}^k$ and in consequence $\mathcal{H}$ will be homeomorphic to $\mathcal{R}^k$; let $\mathcal{F}$ be the inverse mapping of the mapping defined by Eq. 10 i.e. $i = \mathcal{F}(E(\mathbf{h},\mu_{ai}))$. Then the vulnerability function is defined by the composition of the mapping $\mathcal{F}$ and the mapping defined by Eq. 7:

$$\mathbf{c} = E(\mathbf{c},\mu_{ai}) \qquad i = \mathcal{F}(\mathbf{h}) \ \ \mathbf{h} \in \mathcal{H} \tag{12}$$

where $\mathbf{c}$ and $\mathbf{h}$ are vectors of numerical values (and not values of the descriptive functionals); emphasizing the dependence of the probability measure on the index i by the notation $\mu_a(i) = \mu_{ai}$, the vulnerability function is written as:

$$\mathbf{c} = \mathbf{V}(\mathbf{h}) = E(\mathbf{c},\mu_a(\mathcal{F}(\mathbf{h}))) \tag{13}$$

To illustrate those concepts, the response of structures to a two component earthquake action is considered. Let $\mathbf{a} = [a_x(t), a_y(t)]^T$ where a_x and a_y are the time histories of acceleration in two orthogonal directions; a_x and a_y are time segments of realizations of two independent stationary gaussian stochastic processes defined by the power spectral densities functions $\alpha_x S_x(\omega)$ and $\alpha_y S_y(\omega)$, where α_x and α_y control the intensity and $S_x(\omega)$ and $S_y(\omega)$ are given spectral shapes. Let $(L, S, \mu(\alpha_x))$ and $(L, S, \mu(\alpha_y))$ be the stochastic processes defined by $\alpha_x S_x(\omega)$ and $\alpha_y S_y(\omega)$; then the stochastic process of the earthquake is $(L \times L, S \times S, \mu(\alpha_x, \alpha_y))$ where $\mu(\alpha_x, \alpha_y)$ is the product measure of $\mu(\alpha_x)$ and $\mu(\alpha_y)$. The family of measures $\mu_{ai} \equiv \mu(\alpha_x, \alpha_y), i \equiv (\alpha_x, \alpha_y) \in R^2$ obeys the condition expressed by Eq. 14 for almost all possible functionals $\mathbf{h}(\mathbf{a}(t))$; for instance if the earthquake intensity vector $\mathbf{h}$ represents the mean peak values of acceleration, $\alpha_x = (h_1/r_x\sigma_x)^2$ and $\alpha_y = (h_2/r_y\sigma_y)^2$ where σ_x and σ_y are the standard deviations and r_x and r_y are the mean peak factors for the S_x and S_y spectra. In this case, the vulnerability function relates the values of the peak acceleration to the values of the control variables computed for the corresponding power spectral functions.

Probabilities used in this section are completely artificial in the sense that they do not pretend to represent any physical random process. On the other hand they are the basis for the evaluation process of the state of knowledge described bellow, and they may be considered to be 'constructed' for that purpose.

5. Hazard Models

In the present approach seismicity is idealized by a probabilistic model which defines the probability of the occurrence of a given earthquake (defined by its source mechanism parameters e.g. focus and area of fault rupture) in a reference time interval, for example 50 years. It is usually reasonable, when assessing the effects of a strong earthquake, to disregard the decrease in resistance due to the prior occurrence of smaller earthquakes; it is also usual not to define hazard in terms of the histories of ground motion at the site but in terms of their descriptive functionals, namely peak acceleration or peak velocity. Hence, hazard models will be identified with a probability distribution in the space $\mathcal{H}$ of the earthquake intensity vector (Eq. 11) and the corresponding events will be the more severe earthquakes in the reference period. The concept of 'more severe earthquake' applies when the destructiveness of earthquake actions is evaluated by multi-parameter intensity measures[10] : for an ensemble of earthquakes motions the more severe earthquakes are those for which at least one of the descriptive functionals have the ensemble maximum value and the other functionals have values not smaller than those of the motions which also have the maximum value of the descriptive functional considered.

6. Probability of Failure and Limit States

Let μ^s be the probability measure defined in $\mathcal{H}$ by the hazard corresponding to the seismicity of the site. This probability defines two probability distributions of interest, one on the time histories of the response and other on the values of the control variables. The probability distribution of the time histories of the response is defined by the the composition of the mapping defined by the structure operator $\mathcal{E}$ (Eq. 1) with the inverse of the mapping between $\mathbf{a}(t)$ and $\mathbf{h}$; this last mapping is a probabilistic mapping, since to every value of $\mathbf{h} \in \mathcal{H}$ it associates a probability measure $\mu_a(i)$ in $\mathcal{A}_s$, with $i = \mathcal{F}(\mathrm{h})$ (Eq. 12). Let S_r be the σ-field generated in $\mathcal{R}_r$ by the $\mathcal{E}$-images of the Borel sets in $\mathcal{A}_s$ and $\mu_r(i)$ be the measure induced in $\mathcal{R}_r$ by $\mu_a(i)$; then for every set $X \in S_r$ is associated a probability:

$$P(X) = \int_{\mathcal{H}} \int_X d\mu_r(\mathcal{F}(\mathbf{h}))d\mu_h \equiv \int_X d\mu_S \tag{14}$$

where μ_h is the probability measure defined in the space of the actions by the hazard model and μ_S may be the probability measure in the space of time histories defined by the seismicity model and the stochastic model of the actions.

The probabilistic idealization of structural failure may be carried out in two ways. In the more rigorous approach, the structural behaviour is dependent on the actual values of the limit states; hence the vulnerability function is dependent on the probabilistic definition of the limit states and either elementary or averaged vulnerability functions may be considered; an elementary vulnerability function is the vulnerability function for a given realization of the limit states; the averaged vulnerability function is the probability-weighted average of the elementary vulnerability functions for all limit states realizations. In the simplified approach limit states and structural behaviour are mutually independent i.e. the vulnerability function is not dependent on the limit states values, and the limit states may be defined without reference to the vulnerability function; only this approach will be considered in the present paper and only with deterministic limit states.

The concept of limit state itself may be derived from a more fundamental viewpoint of the question. Let $\mathcal{Q}$ be the subset of the response space $\mathcal{R}_r$ which contains all acceptable time histories and let $\mathbf{c}(\mathbf{r}(t))$ be a vector of K control variables. The optimal values $\mathbf{c}_o^l$ for the limit states expressed in terms of those control variables may be defined in several ways and *inter alia* shall take into account the characteristics of the design processes. In the present framework it is reasonable to define $\mathbf{c}_o$ through the minimization of the errors in the probability of failure. The true probability of failure is

$$p_f^t = 1 - \int_{\mathcal{Q}} d\mu_S \tag{15}$$

where μ_s is the probability measure defined by expression (14). Let $\mathcal{R}^a(\mathbf{c}^l)$ be the ensemble of the response histories whose values of the descriptive functionals are smaller than $\mathbf{c}^l$: $\mathcal{R}(\mathbf{c}^l) = \{\mathbf{r}(t) : \quad \mathbf{c}(\mathbf{r}(t)) \leq \mathbf{c}^l\}$. A vector is considered smaller than or equal to another if every component is smaller than or equal to the corresponding component of the second vector. Let $\mathcal{W}(\mathbf{c}^l) = \mathcal{Q}\Delta\mathcal{R}^a(\mathbf{c}^l)$, where Δ means the symmetric difference, be the set of all response histories which are acceptable but whose values of the control variables are greater than $\mathbf{c}^l$ and of all response histories which are not acceptable but whose values of the control variables are smaller than or equal to $\mathbf{c}^l$. Then the error associated to an arbitrary limit states vector $\mathbf{c}^l$ may be represented by the probability of $\mathcal{W}$: hence the optimal limit states $\mathbf{c}_o^l$ are those which minimize this probability, and the corresponding probability p_f^e is a measure of the goodness of the ensemble of descriptive functionals for the assessment of the probabilities of failure:

$$p_f^e = \min \int_{\mathcal{W}(\mathbf{c}^l)} d\mu_S \qquad \mathbf{c}^l \in R^K \tag{16}$$

Hence for the optimum limit states vector $\mathbf{c}_o^l$ (which is assumed to be unique) $p_f^e = \int_{\mathcal{W}(\mathbf{c}_o^l)} d\mu_S$. The ensemble of descriptive functionals $\mathbf{c}(\mathbf{r}(t))$ gives a good description of the probability of failure whenever $p_f^e << p_f^t$.

7. Choice of Simplified Models

Since it is not practicable to use 'exact' models in earthquake engineering and moreover it is not justificable to use the more sophisticated model available in a large number of cases[11], it is important to investigate criteria to guide the choice of simplified model. Two situations may arise; in the first situation the control variables are identical to the control variables in the reference model; in the second situation the control variables are different, e.g. in the reference model the control variables are the relative vertical distances between contiguous storeys (on the assumption that total colapse is the limit state of interest) and in the simplified model the control variables are the maximum relative storey displacements.

When the reference and simplified models have the same control variables, the error in the probability of failure due to the use of the simplified model is $p_f^{es} = \int_{\mathcal{W}_s} d\mu_s$ where $\mathcal{W}_s = \mathcal{R}^r \Delta \mathcal{R}^s$ and $\mathcal{R}^r$ and $\mathcal{R}^s$ are the ensemble of response histories, computed by the reference model and by the simplified model respectively, which correspond to values of the descriptive functionals not higher than the limit states values; the probability measure μ_R is defined on the response histories by the mapping corresponding to the reference model. Hence, if several simplified models are available, the one with smaller probability error p_f^{es} should be selected.

When the reference and simplified models have different control variables a similar criterium may be followed; however a more interesting situation devellops if the values of the limit states are corrected to account for inaccuracies of the simplified model; hence the error due to the model is minimized by an optimized quantification of the values of its limit states. Let $\mathcal{R}^s(\mathbf{c}^{ls})$ be the subset of the response space which correspond to values of the simplified model descriptive functionals not higher than $\mathbf{c}^{ls}$ and let $\mathcal{W}_s(\mathbf{c}^{ls}) = \mathcal{R}^r \Delta \mathcal{R}^s(\mathbf{c}^{ls})$. Then, the optimal value of $\mathbf{c}^{ls}$ corresponds to the minimum of the probability

$$p_f^{ls} = \min \int_{\mathcal{W}_S(\mathbf{c}^{ls})} d\mu_R \tag{17}$$

The criteria presented for the choice of simplified model must be devellopped in two directions: a) by considering a loss function to be minimized[12] or an utility function to be maximized[13] associated to the probability of failure (which is an almost trivial devellopment): b) by considering that the optimization is carried out in regard to the choice of values for the 'decision variables' which quantify the design of the structure[14].

8. Probabilistic Representation of Knowledge

Assuming that the probability distributions of the earthquake vibrations and resistances are known, the evaluation of the probability of failure is dependent on the knowledge about the vulnerability function. The analysis process here presented has the purpose of estimating this function by a process which is optimal in regard to errors in the probability of failure. The estimation of the vulnerability function involves a probabilistic representation whose mean value will converge to the true value of the vulnerability function and whose variability will tend to zero as the analysis advances.

Presuming total absence of information about structural behaviour, the vulnerability function may be any nondecreasing function, since it is reasonable to expect that an increase of the load would not correspond to a decrease of the load effect. Hence a complete ensemble of nondecreasing functions is a first step to the representation of our knowledge (or its absence) about structural behaviour. Completness means that given the vulnerability function of any structure, there should be in the ensemble a function which is equal (under an appropriate norm) to that vulnerability function. From a practical viewpoint, it is not necessary, nor makes sense, to find the value of the probability of failure with great accuracy. Thus, it is not needed to identify the vulnerability function with much sharpness. Hence, completeness may be understood to be attained if in the ensemble there is always a function which is sufficiently close to any possible vulnerability function. Under this approximation and assuming that the domain and the range of the vulnerability functions are finite (since they represent the values of physical quantities) it follows that the number of ensemble functions that must be considered is finite. This fact suggests that a not very large number of ensemble functions V_i may be used ($i<1000$) if during the analysis process it is investigated if this number should be increased.

The robustness of the final estimates depends on the qualities of the functions in the ensemble. To control the robustness two partial ensembles are used[2]. The first one is composed of analytical functions of the form

$$c = \alpha h + \beta h^2 + \gamma h^3 \tag{18}$$

appropriate values of α, β and γ being chosen to generate the selected number of functions. The second partial ensemble is constituted by realizations of discrete

Markov processes where the value c_i of the control variable, corresponding to a value h_i of the load intensity, suitably discretized is given by

$$c_i = (1 + \delta x)c_{i-1} \tag{19}$$

where δ is a constant and x is a random variable with an uniforme distribution in the interval (0,1). This discrete function is transformed into a continuous function by a linear interpolation rule applied on a bilogarithmic plot. The values of δ are chosen to generate a partial ensemble with the desired characteristics. Results obtained by considering separately the partial ensembles allow to evaluate the robustness.

The ensemble of functions is probabilized by associating to each function V_i a probability value p_i. Each set of values $\mathbf{p} = \{p_1, p_2...\}^T$ represents a state of knowledge. There is a a good state of knowledge when all the ensemble functions are associated to very small probability values with the exception of those which are close to the true vulnerability function, which have much larger probability values.

Before any nonlinear analysis is performed there is no knowledge, i.e. the state of knowledge is non-informative. Absence of information must necessarily be understood in relation to some quantity, which in this case is considered to be the probability of failure. A non-informative state of knowledge is represented by a constant probability density in a logarithmic scale[15]. This means e.g. that the probability of the probability of failure lying in the interval $(10^{-3}, 10^{-4})$ is equal to the probability of it being comprised in the interval $(10^{-4}, 10^{-5})$. It should be noted that the total probability is infinite, as frequently happens with non-informative distributions. Probability values p_i may be easily computed to secure approximately a non-informative distribution[2]. The uncertainty associated to a state of knowledge may be quantified by the difference between the 5% and 95% fractiles of the probability distribution of the probability of failure.

9. Bayesian Analysis

The value of the vulnerability function is the mean value of the control variable. When a nonlinear computation is performed, necessarily using a realization of the model, the control variable value obtained is only an estimate of the true value of the vulnerability function. However, several realizations may be used and, as a result, a sample of control variables values are obtained. The sample mean value is obviously a better estimate than any individual value. If the number of sample elements is not small the probability distribution of the sample mean value is approximately a gaussian distribution (by the central limit theorem) with a mean value equal to the mean value of the response to one realization and with a variance equal to the variance of the response to one realization divided by the number n of sample elements. In the case of structural response to earthquakes it may be assumed that variance values corresponds to a coefficient of variation with a value of 0.3. This permits the definition of the conditional probability $P(r \mid V_i(h))$ of obtaining average response r for earthquake intensity h if $V_i(h)$ is the true probability function by

$$P(r \mid V_i(h)) = G(V_i(h), 0.09V_i(h)^2/n) \tag{20}$$

where $G(\mu, \sigma^2)$ represents a gaussian distribution with mean value μ and variance σ^2. Hence, after performing n computations and obtaining an average value r the *a posteriori* probabilities $P(V_i \mid r)$ may be computed by Bayes theorem:

$$P(V_i(h) \mid r) = p_i P(r \mid V_i(h)) / \sum_i p_i P(r \mid V_i(h)) \tag{21}$$

These probabilities represent the new state of knowledge.

The value h of the intensity of the earthquake vibration to be used in the nonlinear computations may be selected to provide an optimal increase in knowledge through a *preposterior* analysis. The idea behind this analysis is to evaluate for a large number of values of h what change in knowledge may be expected if the computations are performed. This evaluation is carried out by considering the probability distribution of obtaining a value r of the response, as may be computed from the probabilities associated to the ensemble functions[2] for each intensity. Then, the probability of the increase in knowledge due to that probability distribution is computed and the intensity corresponding to the larger expected increase in knowledge is selected.

10. Concluding Remarks

The present methodology has been proved very useful in the assessment of the probability of failure of structures behaving in the nonlinear range under earthquake load. Typical results are presented in figure 2 for a 420 m long bridge with 9 spans[2]. The left plot represents the probability distributions in the space of ensemble functions in the initial situation when the probabilities associated to the vulnerability functions originate a non-informative state of knowledge. The right plot represents the probability distributions obtained after performing 3 Bayesian analysis, optimized by *preposterior* analyses, with 4 elements in each sample. This corresponds to a sufficiently small confidence interval for the probability of failure.

The present methodology has also been sucessful used[16] in the quantification of the 'behaviour coefficients'which, in earthquake resistance codes, permit to take into account the influence of the nonlinear behaviour by 'correcting' the results of a linear analysis.

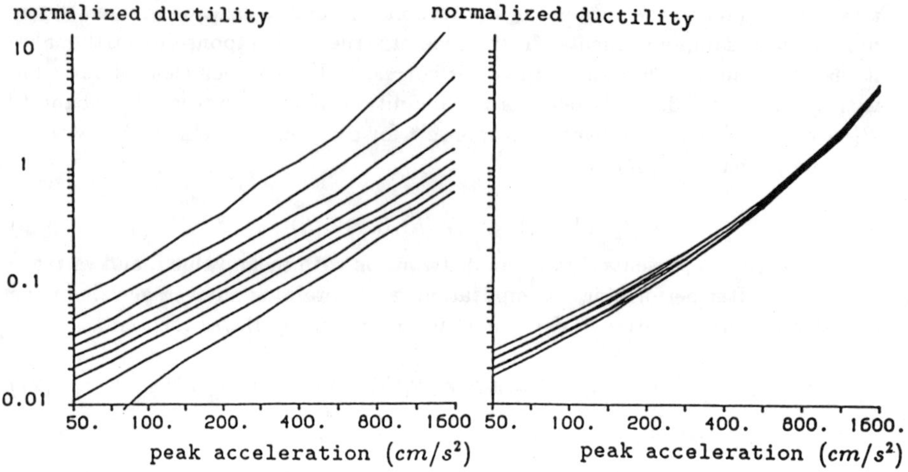

Figure 2: Initial (left plot) and final (right plot) state of knowledge expressed in terms of vulnerability functions. The 10%, 20%, ... 90% fractiles are plotted.

11. References

1. R.T. Duarte - *The Mathematical Foundations of Structural Safety Assessment for Earthquake Hazards,* Proceedings of the 9th European Conference on Earthquake Engineering, Moscow; 1990.

2. R.T. Duarte, M. Ritto-Corrêa, C.T. Vaz and A. Campos-Costa - *Bayesian Methodology for the Quantification of the Failure Probability and the Optimization of Analysis* (in portuguese), Proceedings of the 2s Jornadas Portuguesas de Engenharia de Estruturas, Lisbon; 1990.

3. R.T. Duarte, C.T. Vaz and M. Ritto-Corrêa - *Probabilistic Analysis of a Bridge under Seismic Actions - Portuguese State-of-Art,* International Workshop in Seismic Design and Retroffiting of Reinforced Concrete Bridge, 1991, Bormio, Italy.

4. D. Veneziano - *Basic Principles and Methods of Structural Safety,* CEB Bulletin d'Information no. 112, Paris, 1976.

5. Silberberg E. - *The Structure of Economics: A Mathematical Analysis*, McGraw-Hill, New York, 1978.

6. C. Truesdell and R. Toupin - *The Classical Field Theories*, in *Encyclopedia of Physics*, Springer Verlag, Berlin, 1960, Vol. III/1, pp. 226-793.

7. C. Truesdell and W. Noll - *The Nonlinear Field Theories of Mechanics*, in *Encyclopedia of Physics*, Springer Verlag, Berlin, 1965, Vol. III/3, pp. 1-579.

8. R.T. Duarte - *Generalized Derivative Analysis of Strong Ground Motion Earthquake Records*, European Earthquake Engineering, Vol. 1, No. 1, pp 44-54, 1987.

9. I. Ibrahimov and Y. Rozanov - *Processus Aleatoires Gaussiens*, Editions Mir, Moscow, 1974.

10. R.T. Duarte - *Definition of Seismic Actions with Multi-parameter Intensity Measures*, CSNI Specialist Meeting on Probabilistic Methods in Seismic Risk Assessment for Nuclear Power Plants, Lisbon, 1980, Vol. I, pp. 55-79.

11. E. Rosenblueth and H. Contreras - *Optimum Expenditures on Structural Engineering Services*, Proc. International Conference on Applied Numerical Modelling, Southampton, 1977.

12. R.T. Duarte - *Optimal Values for the Earthquake Reliability of Building Structures*, Proceedings of the 9th European on Earthquake Engineering, Moscow, 1990.

13. R.T. Duarte, A.G. Costa and A. Campos-Costa - *Definition and Quantification of Behaviour Coefficients Based on Structural Reliability Concepts*, Proceedings of the 9th European Conference on Earthquake Engineering, Moscow, 1990.

14. R.T. Duarte - *An Essay on Design Methods and Regulations in Earthquake Engineering*, I & D Report, LNEC, Lisbon, 1987.

15. J.O. Berger - *Statistical Decision Theory and Bayesian Analysis*, 2nd Ed., Springer-Verlag, New York, 1985.

16. R.T. Duarte, C.T. Vaz and M. Ritto-Corrêa - *Evaluation of Earthquake Behaviour of Reinforced Concrete Bridges, Quantification of Behaviour Coefficients*, Background Document to EC8, Part 2: Bridges, LNEC, 1990.

ASPECTS OF COMPLEXITY IN SLEEP ANALYSIS

José M. N. Leitão and Agostinho C. da Rosa

Centro de Análise e Processamento de Sinais, INIC

Dept. de Eng. Electrotécnica e de Computadores
Instituto Superior Técnico
Av. Rovisco Pais, Complexo I
1096 Lisboa Codex Portugal

ABSTRACT

The paper presents a selection of sleep analysis problems where some aspects and concepts of complexity come about. Emphasis is given to the electroencephalogram (EEG) as the most important sleep related variable.

The conception of the EEG as a message to be deciphered stresses the importance of the communication and information theories in this field. An optimal detector of K complexes and vertex sharp waves based on a stochastic model of sleep EEG is considered. Besides detecting, the algorithm is also able to follow the evolution of the basic ongoing activity.

It is shown that both the macrostructure and microstructure of sleep can be described in terms of symbols and interpreted as sentences of a language. Syntactic models and Markov chain representations play in this context an important role.

1. INTRODUCTION

Sleep is a complex dynamic process whose function and mechanisms are not clearly known. We take words of Webb [1] in the Encyclopaedia of Ignorance, as a choice of direction in sleep research: "From my perspective, anchored in my adaptive theory of efficient and final cause of sleep, we must, rather than learn the principal causes of sleep, learn the laws of sleep".

Learning the laws of sleep involves the development of models from observations and testing them with new observations. This paper describes the approach and some results of an effort of development of models and methods for automatic sleep analysis. Several aspects and concepts of complexity, as surveyed by Nicolis and Prigogine [2], emerge along this work.

In this section we briefly present the problematic under study and go through a selection of questions and definitions pertinent to the understanding of the next sections.

Figure 1 is a formal representation of the sleep analysis problem. Sleep is assumed as a process S(t) running inside the sleeper. This process itself cannot be observed; instead, it is assessed by a set of sleep related variables.

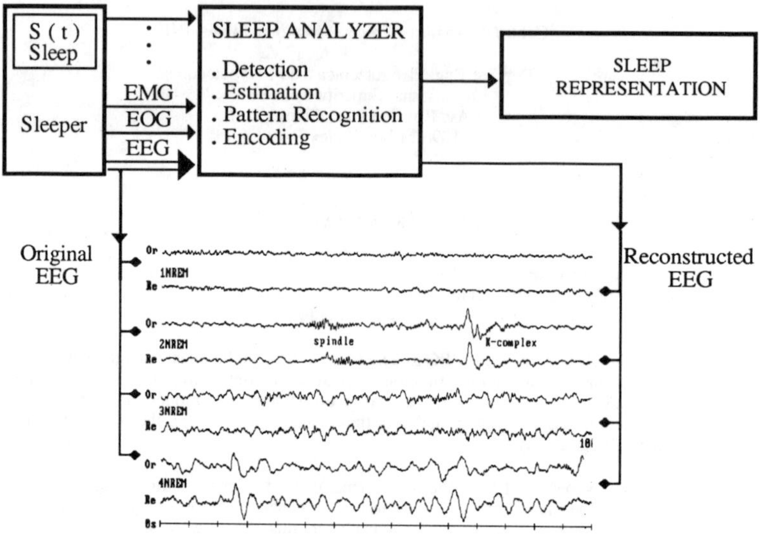

Figure 1: A schematic view of automatic sleep analysis.

The signals normally used as indicators of sleep dynamics are the brain electrical activity given by the electroencephalogram (EEG), the ocular movements, measured by the electrooculogram (EOG), and the muscular tonus, expressed by the electromiogram (EMG).

Being an all-body phenomenon, the assessment of sleep has to take into account other important variables such as temperature, respiration, cardiac activity, vasoconstriction.

The amount of information corresponding to a continuous poligraphic recording along the night is enormous: eight hours are translated into about 1500 pages of paper. In interpreting this writing, the neurophysiologist implements a data reduction procedure based usually on the Rechtschaffen and Kales (R&K) [3] paradigm. The sleeper is considered to stay in one of six stages: Wakefullnes - W; REM sleep (rapid eye movement sleep) and four non REM stages (1, 2, 3 and 4NREM).

By visual inspection, page-by-page, going ahead and back in problematic situation (decision in context), the classifier attributes, according to the R&K rules, one stage to each page (normally corresponding to 20 or 30 seconds). In this way, after 4 to 5 attentive hours, a graphic description of stage evolution, called the hypnogram, is produced. Figure 2 is a typical example.

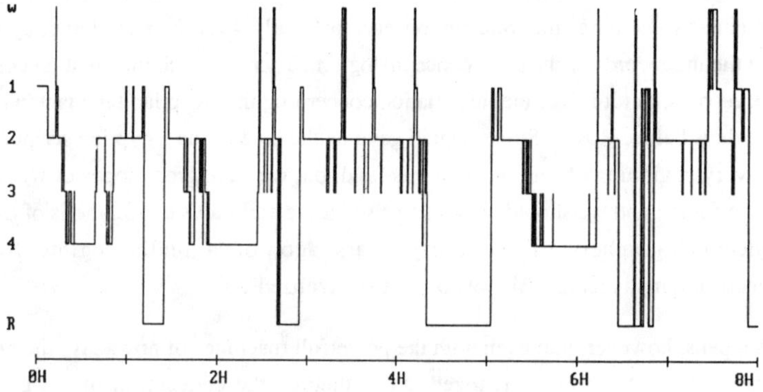

Figure 2: Hypnogram produced according to the R&K criteria

The hypnogram, which in terms of figure 1 is a representation of S(t), shows only the global features of sleep dynamics, that is, its **macrostructure**. However, within each stage, sleep exhibits a characteristic organization whose building blocks are transient events with variable morphologies and random occurence, usually referred to as phasic events; this forms the sleep **microstructure**.

A critical problem in visual sleep analysis is the qualitative and subjective criteria it is based upon, giving rise to the unavoidable inter-classifier disagreement. To those problems, it has to be added the human difficulty in manipulating big information sequences, and the time spent with such a task.

There has been strong and well supported criticism on the weakness of visual analysis and the limitations and contradictions of the R&K rules. Most of the automatic sleep analyzers developed so far tend to mimic the "experienced human classifier". Those systems follow the classical pattern recognition procedure: a) feature extraction from the raw input data; b) supervised classification according the R&K rules. As claimed by Kubiki et al. [4] and recognized by many researchers in the field, automatic sleep analysis should emphasize the computer capabilities and the use of new conceptual tools, irrespective of the rules of R&K; these, should be considered only as established and sometimes necessary convention.

The next sections are mainly concerned with the sleep EEG both in terms of micro and macrostructure.

2. SLEEP EEG MODEL. PHASIC AND TONIC ACTIVITY

In 1956, Wiener [5] wrote: "Brain waves speak a language of their own, but this language is not something that one can observe precisely with the naked eye, by merely looking at the ink records of the electroencephalogram. There is much information contained in this ink records, but it is like the information concerning the Egyptian language which we had in the days before Roseta Stone, which gave us the clue to the Egyptian script (...) We are at the very beginning of our work in this field, but we have great hopes of what it will offer for the future; and we should not be surprised if the ambitious expectations of the early electroencephalographers of the thirty years ago, of a really legible form of electroencephalograph record, will now begin to be realized".

It happens, however, that even with the powerfull machines of nowadays the hopes of Wiener are still far from being realized. Nevertheless, the conception of the EEG as a message to be deciphered remains as a challenge, and a problem to be worked out with the methods and concepts of communication and information theories.

Implicit in this problem is the characterization of the message source, in other words, the brain; this demands the contribution of neurociences and dynamic systems theory (See Lopes da Silva [6]).

It is usuall to consider the EEG as formed by: 1) **tonic activity,** which is a combination of various dominant rhythms (delta, theta, alpha, sigma, etc.); and 2) **phasic activity**, consisting of transient phenomena like spindles, K complexes, vertex sharp waves, sawtooth waves, etc (See Niedermeyer [7]).

A first step towards the interpretation of the EEG message would consist in following the evolution of tonic activity and detecting the relevant phasic events. This can be accomplished in an optimal sense if structured prior knowledge describing the main features of sleep EEG is available.

The model of sleep EEG herein presented has been developed by Rosa [8]; it is essentially a refinement and a generalization of the one considered by Rosa et al. [9], which is based on a blend of neurophysiological data with pragmatic engineering criteria.

Figure 3 is a schematic view of the model corresponding to M basic rhythms and a single channel, i. e., the cortex activity as seen by only one electrode placed on the scalp.

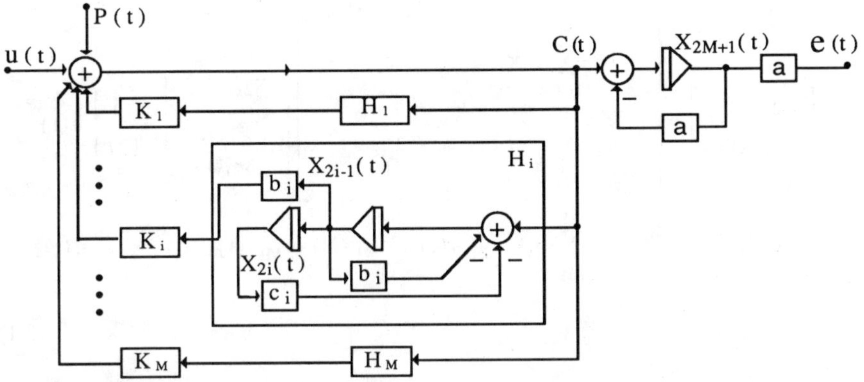

Figure 3: Schematic representation of the stochastic sleep EEG model (single Channel)

Each feedback loop, consisting of a band pass filter, Hi, with variable gain, Ki, models the resonant behavior of the neuronal network correponding to a specific frequency band. The gain takes into account the general effects of inhibitory and excitatory actions within the loop network, and can be interpreted as its level of synchronism. The noise acting on the neuronal network is the sum of billions of neuronal wideband independent actions; accordingly it is a white Gaussian noise. Tonic activity is generated as the response of the system to this noisy input, u(t).

Exogenous and endogenous stimuli, produced by autonomous systems, will be represented by p(t). This input , added to the driving noise u(t), is a pulse whose shape, amplitude and duration depends on the stimulus and, possibly, on the brain state. Phasic events are interpreted as global transient responses of the overall structure. However, due to the pulse durations and the characteristics of the resonant filters, only the delta and theta responses have significant effect, producing vertex waves and K complexes, respectively. Small values of K(delta) give rise to transient responses with small overshoot, the shape being more the one of vertex waves (delta type sharp waves); as K(delta) increases so the usual shape of K complexes appears. As for the theta loops, moderate values of K(theta) give rise to single vertex waves (theta type sharps waves), while high values of K(theta) produce responses interpretable as paroxistic theta bursts.

The low pass filter of figure 3 relates the cortex activity, c(t), with the scalp EEG, e(t).

The loops and the output state variables are given by

$$\begin{bmatrix} \dot{x}_{2i-1}(t) \\ \dot{x}_{2i}(t) \end{bmatrix} = \begin{bmatrix} -b_i+\sum_{i=1}^{M} b_i k_i & -c_i \\ 1 & 0 \end{bmatrix} \begin{bmatrix} x_{2i-1}(t) \\ x_{2i}(t) \end{bmatrix} + \begin{bmatrix} 1 \\ 0 \end{bmatrix} u(t)$$

$$\dot{x}_{2M+1}(t) = \sum_{i=1}^{M} b_i k_i x_{2i-1}(t) - a x_{2M+1}(t) + u(t). \tag{3.1}$$

The state vector $\mathbf{x}(t) = [x_1(t), x_2(t), \cdots, x_{2i-1}(t), x_{2i}(t), \cdots, x_{2M+1}(t)]^T$ is formally described by the stochastic differencial equation

$$d\mathbf{x}(t) = A_c\, \mathbf{x}(t)\, dt + B_c\, dw(t)\,, \tag{3.2}$$

where $\{w(t)\}$ is a scalar Brownian motion with variance parameter q_c ($E[dw(t)] = 0$, $E[dw(t)^2] = q_c\, dt$), and A_c and B_c are matrices easily obtained from equations (3.1).

The solution of (3.2) is the continuous Markov process $\{\mathbf{x}(t)\}$; the scalp EEG is proportional to its last component: $\{e(t)\} = \{a\, x_{2M+1}(t)\}$.

The discrete model equivalent to equation (3.2) is the stochastic difference equation

$$\mathbf{x}_{n+1} = A_d \mathbf{x}_n + B_d\, u_n, \tag{3.3}$$

where $x_n = x(n\Delta)$, $A_d = A_c\Delta + I$, $B_d = B_c\Delta$, and $u_n = [\,w(n+1) - w(n)]/\Delta$ with $E[u_n] = 0$, and $E[u^2{}_n] = q_c/\Delta$.

The sampling interval, Δ, must be sufficiently small so that equations (3.3) and (3.2) describe essentially the same process.

Tonic activity depends on the filter parameters a, b_i, c_i, the feedback gains K_i and the noise variance q_c. Although the spectral characteristics of the dominant rhythms may vary along the night, the values of b_i and c_i have been set constant and adjusted for a sufficiently wide bandwidth, in order to cover the normal frequency ranges. The low-pass filter parameter, a, is also kept constant. Thus, the feedback gains K_i and the noise variance q_c remain, in this model, the main modulators of tonic activity.

A sudden increase in the sigma feedback loop gain, lasting a few seconds, motivates the appearence of sigma spindles (see figure 1); a sudden increase in the delta feedback loop gain explains the behavior shown in figure 4, representing a multichannel EEG recording. A K complex, (KC) - occurring spontaneously or as a response to an auditory stimulus - induces the transition from a low delta activity (typical of 1NREM sleep stage) to a high level delta activity (typical of 4NREM sleep stage); this suggests that the input pulse can trigger a jump in the delta gain, thus giving rise to a notorious state transition, accompanied by the generation of a K complex. Note the special morphology of this K complex, with two clearly delayed peaks: the first is a theta loop pulse response (theta vertex sharp wave); the second corresponds to the impulsive excitation of the delta loop. To describe this morphology and a set of other experimental facts the model has to be modified; this problem is considered in Rosa and Leitão [10].

For practical purposes, and as a result of a series of simulation studies, the inputs p(t) were shaped as square pulses with duration of about 150ns and adaptive amplitude A, such that the signal-to-noise ratio, A^2 / q_c, is of the order of 6 dB.

Based on the model (3.3) and on the preceding assumptions, an optimal monitor (estimator/detector) has been developed. The system parameters K_i and q_c are estimated according to the maximum likelihood criteria; the detector of K complexes and vertex sharp waves implements a combination of Bayesian and χ^2 tests.

Model and monitor have been evaluated by extensive simulations and applied for long periods, in several situations .

By recording the evolution of the estimated parameters and the time localization of the detected pulses, the analyzer builds up a representation of the original signal. Since the estimated parameters remain constant for long periods, only the significant changes have to be stored; the result is a significant data reduction. Thus, besides estimation and detection, the analyzer has also encoding properties.

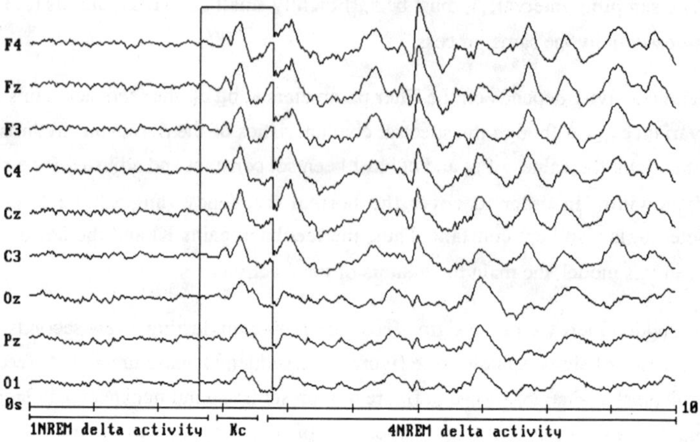

Figure 4 - Poligraphic EEG recording showing the transition of 1REM delta activity to a 4REM delta activity starting with a K complex.

From the resulting compact representation, it is possible to "reconstruct" the original signal by decoding and simulation, as illustrated in figure 1, where short (15 seconds) EEG epochs corresponding to stages 1, 2, 3 and 4NREM are shown. Note that not only the delta tonic activities look alike but also the original and the simulated spindles and K complexes are almost time coincident.

A further elaboration on this subject leads to the representation of the input signals as a sequence of symbols. Problems of algorithmic complexity, in the sense given by Rissanen [11], Kolmogorov [12], Solomonoff [13], Chaitin [14], arise in this context.

In the past few years there has been a series of studies showing that the EEG signals may exhibit cahotic behavior , reflecting the existence of strange attractors governing the evolution of neuronal populations, at least in certain conditions. The papers of Babloyantz et al. [15] and Röschke and Aldenhoff [16], among others, contain results concerning sleep EEG. To sufficiently long EEG epochs belonging to different sleep stages, various algorithms have been applied to determine the attractor's dimensions. The values obtained seem to have a good discriminating power. However, we would like to point out that the EEG is only short term stationary and this fact may not be compatible with the long epochs needed in the above mentioned analysis.

3. MACROSTRUCTURE ANALYSIS

The temporal organization of sleep as given by the hypnogram, i. e., the sleep macrostructure, has been modelled by some authors in terms of Markov chains (see Kemp [17] and the references therein).

In the work reported by Fred and Leitão [18] the perspective is to look at the hypnogram as expression of a language whose structure is analyzed and modelled in terms of stochastic grammars.

The study consisted in comparing these two populations: 17 normal subjects and 10 psychiatric patients (dysthymic disorders). Patients had subtle and borderline abnormalities which concerned mainly the amount of deep sleep (slow wave sleep). Visual stage classification (R&K) was performed on standard polysomnography recordings (14 channels).

Figure 4 shows a diagram of the experimental protocol used.

- For each population and taking as symbols the usual hypnogram stages, a stochastic regular grammar was inferred using the Crespi - Reghizzi (See Fu et al. [19]) method without prior structure information.
- In order to evaluate the discriminating power of the representations/grammars used, each sample hypnogram was later classified. An a priori probability was assigned to each class and the a posteriori probability was determined using Bayes law.

The results obtained were analyzed under two main aspects: structural modelling and discriminating power.

From the structural point of view, the context-free grammars used are a natural way of representation , being able to describe the tendencies of sleep cyclicity. The repetitions (cycles) are modelled as recursive structures of rules, which allow the extension of the language beyond the training set.

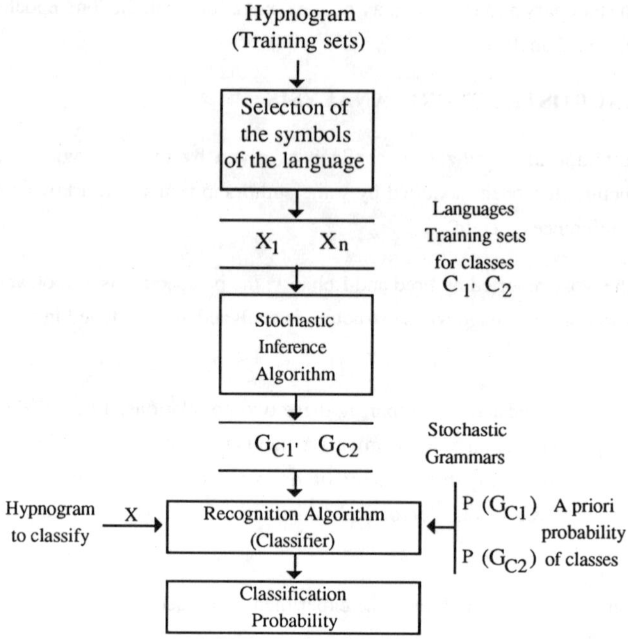

Figure 5: Modelling/Recognition diagram

Results show clear differences in the grammatical complexity associated with each population, as consistently measured by the number of rules obtained. Interpretation of these results leads to relating sleep disturbance of psychiatric patients to a decrease in organization complexity.

4. EEG MICROSTRUCTURE

The work summarized by Largo et al.[20] reflects the perspectives and the first results of an ongoing research project on the relations between stage transitions and microstructure.

The main concern are the periods of 2NREM before transition to REM, where the mobility and the appearance of alternating patterns are very clear. Figure 5 shows two contiguous epochs of EEG belonging to those two stages.

In animal sleep studies the occurrence of REM is preceded by muscular atonia and PGO (ponto-geniculo-occipital) spikes. In man, no clear-cut features predicting the transition to REM sleep have so far been identified, but references to increased densities of phasic events, like K complexes, are available in the literature.

Based on empirical evidence, we started predicting REM sleep. Predictions were based on visual inspection and afterwards assessed as "true" or "false"; they usually occurred during the 10 minutes preceding a REM episode.

In order to evaluate the empirically motivated conjectures, sets of single channel EEG recordings were automatically analyzed in 10 normal subjects. Some specific features were considered: K complexes, sawtooth-like bursts, EEG desynchronization.

EEG segmentation has been performed by associating the K complexes detector descibed by Rosa [8] with adaptive parametric segmentation (auto-regressive modelling) . Results of automatic clustering, have been controlled by a human classifier observing the segments of the EEG and their spectra.

Using these methods, the following results could be observed in epochs preceding REM: increased K density; higher percentage of sawtooth-like activity; longer segments of EEG desynchronization. These are preliminary results, demanding the use of more sophisticated syntactic methods and string analysis as the ones considered in the previous section. Thus, the description of the EEG as a sequence of symbols is, in this study, particularly important.

This perspective is not new. In the hierarchical approach of EEG proposed by Sanderson et al.[21] , the signal is considered as the output of a composite source model. The sequences of subsources are regarded as character strings and treated using Markov chains or syntactic techniques. Although not going into details, the paper advocates the usefulnes of such an approach in brain sleep research. The predictive ability of the proposed models is well-suited to monitoring applications where long-term as well as short-term change detection are an essential requirements.

So far, in our work, the segmentation mentioned above has been limited to a single channel EEG. A further step will be the adaptive segmentation of the multichannel EEG. For a big number of electrodes the problem is equivalent to the study of sudden modifications in scalp field topography. This has to do with the perspective of Lehmann [22]: stable field landscapes correspond to brain microstates; these microstates are conceived to be superimposed on the gross brain states.

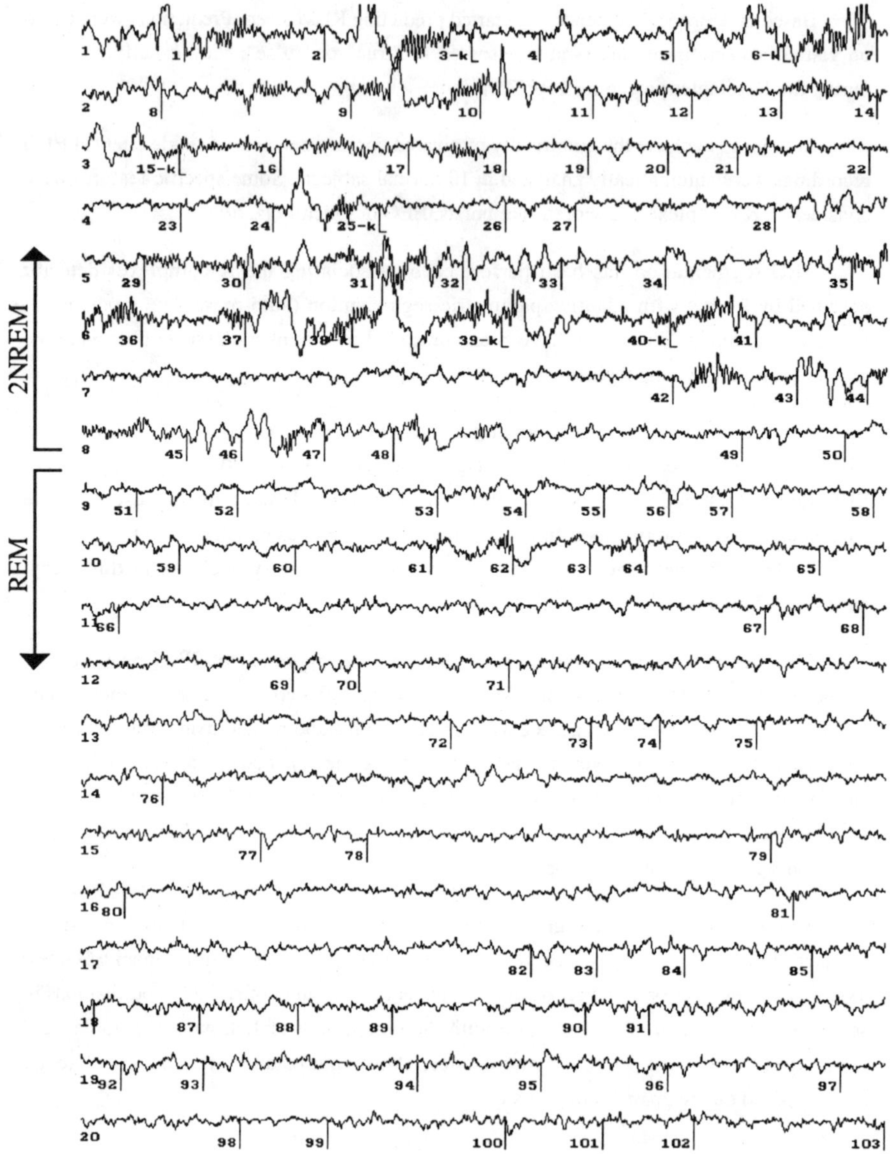

Figure 6: Single channel recording showing 2NREM preceding REM.

5. ACKNOWLEDGEMENTS

The present work reflects knowledge and experience acquired along years of cooperation with Dr. Teresa Paiva, Head of the Laboratory of Electroencephalography, Centro de Estudos Egas Moniz, Lisboa. We acknowledge her contribution to this paper.

We also acknowledge the contributions from our coworkers Rogério Largo and Ana Fred.

6. REFERENCES

1. W. B. Webb, "Sleep", in the *Encyclopaedia of Ignorance*, Pergamon Press, 1977.

2. Nicolis and Prigonine, *Exploring Complexity*, Freeman, 1989.

3. A. Rechtschaffen and A. Kales, *A Manual of Standardized Terminology, Techniques and Scoring System for Sleep Stages of Human Subjects*, U. S. Government Printing Office, Washington DC, 1968.

4. St. Kubiki, W. M. Herrmann, L. Höller, "Critical Comments on the Rules by Rechtschaffen and Kales Concerning the Visual Evaluation of EEG Sleep Records", in *Methods of Sleep Research*, Edited by St. Kubicki and W. M. Herrmann, Gustav Fischer Verlag, 1985.

5. N. Wiener, *I am a Mathematician*, The MIT Press, 1964.

6. F. H. Lopes da Silva, "Dynamics of EEGs as Signals of Neuronal Populations: Models and Theoretical Considerations", in *Electroencephalografy - Basic Principles, Clinical Applicaations and Related Fields*, E. Niedermeyer and F. H. Lopes da Silva (Eds). Urban & Schwartzenberg, pp: 15-28, 1987

7. E. Niedermeyer, "Sleep and EEG", in *Electroencephalografy - Basic Principles, Clinical Applicaations and Related Fields*, E. Niedermeyer and F. H. Lopes da Silva (Eds). Urban & Schwartzenberg, pp: 15-28, 1987

8. A. C. Rosa, *Detecção Óptima de Complexos K com Base num Modelo Estocástico do EEG do Sono*, Doctoral Thesis, Instituto Superior Técnico, 1989.

9. A. C. Rosa, B. Kemp, T. Paiva, F. H. Lopes da Silva and H. A. C. Kamphuisen, "Model - Based Detector of Vertex Waves and K - Complexes in Sleep Electroencephalogram", *Electroenceph. Clin. Neurophysiol.*, 78: 71-79, Jannuary, 1991.

10. A. C. da Rosa and J. M. N. Leitão, "A Stochastic Model of Sleep EEG: Tonic and Phasic Activity", submitted for publication.

11. J. Rissanen, *Stochastic Complexity in Statistical Inquiry*, World Scientific, New Jersey, 1989.

12. A. N. Kolmogorov, "Logical Basis for Information Theory and Probability Theory", *IEEE Trans. Inform. Theory*, Vol. IT-14, nº 5, Sept. 1968.

13. R. G. Solomonoff, "Complexity - Based Induction Systems: Comparisons and Convergence Theorems", *IEEE Trans. Inform. Theory*, Vol. IT-24, nº 4, July 1978.

14. G. J. Chaitin, "A Theory of Program Size Formally Identical to Information Theory", *I. Ass. Comput. Mach.*, Vol. 22, pp. 329-340, 1975.

15. A. Babloyantz, J. M. Salazar and C. Nicolis, "Evidence of Cahotic Dynamics of Brain Activity During the Sleep Cycle", *Phys. Letters*, 111 A: 152-156, 1985.

16. J. Röschke and J. Aldenhoff, "The Dimensionality of Human's Electroencephalogram During Sleep", *Biol. Cybern.*, **64**, 307-313 (1991).

17. B. Kemp, *Model-Based Monitoring of Human Sleep Stages*, PhD Thesis, University of Twente, 1987.

18. A. Fred and J. M. N. Leitão, "Use of Stochastic Grammars for Hypnogram Analysis", in *Proc. of the 11th IAPR International Conference on Pattern Recognition,* 1992.

19. K-S. Fu, L. Taylor and T. L. Booth, "Grammatical Inference: Introduction and Survey", Parts I and II, *IEEE Transactions on Patterns Analysis and Machine Intelligence*, PAMI-8: 343-359, May 1986.

20. R. Largo, J. M. N. Leitão, A. C. Rosa, T. Paiva, "Sleep EEG Patterns Preceding REM", *Sleep Research*, 20-A, 39, 1991 (abstract).

21. A. C. Sanderson, J. Segen and E. Rickey, "Hierarchical Modelling of EEG Signals", *IEEE Transactions on Pattern Analysis and Machine Intelligence*, PAMI-2, nº. 5, September 1975.

22. D. Lehmann, "Microstates of the Brain in EEP and ERP Mapping Studies", in *Brain Dynamics*, Eds. E. Bazar and T. H. Bullok, Springer - Verlag , 1989.

PREDICTION OF TIME SERIES

J. Santana

Dept. de Engenharia Electrotécnica e Computadores, Instituto Superior Técnico, Av. Rovisco Pais 1, 1096 LIsboa Codex, Portugal

R. Vilela Mendes

Centro de Física da Matéria Condensada, Av. Gama Pinto 2, 1699 Lisboa Codex, Portugal and TH-Division, CERN, Genève 23, CH-1211, Suisse

Prediction of time series is a problem one encounters in many fields of science and technology. We survey the methods that have been proposed to deal with this problem and then describe some work on predictors based on the adaptative filtering approach. The main design requirement of the predictors that were developed, is that they should provide a fast on-line adaptation to changes in the nature of the process that generates the time series.

1. Introduction

In the study of natural phenomena, social and economic events or human behavior, one of the main objectives of a scientific approach is to achieve some sort of forecasting capability, i.e. given the knowledge about the past behavior of a system to make meaningful predictions about its future.

In some fairly stable situations (satellite orbits for example) we are interested in the long-term predictability of the system. However, many complex systems display the phenomenon of sensitive dependence to initial conditions (exponential divergence of nearby orbits), and this makes long-term predictability a hopeless task and a useless concept. Some kind of predictability is however mantained in the short run and this is usually sufficient if, for example, on deals with adaptative systems in interaction with an external environment. Incorporating the new information as it comes along at each new time step, short-term predictability is all that is actually needed for the system to react adequately to the environment. Short term prediction may be modelled by the problem of, given a certain number of elements of a time series of scalar quantities,

$$x_{t-1}, x_{t-2}, \dots x_{t-k+1}, x_{t-k}$$

to guess the next elements x_t in the series. More complicated situations where x_t is a multicomponent object or prediction of a finite (> 1) number of elements in the future is required, may be reduced to this case. Therefore, for the sake of simplicity, we will restrict ourselves to the scalar one-step ahead prediction problem.

The plan of the paper is as follows: First we survey some of the approaches that have been used to deal with the problem of prediction of time series. We comment on their range of applicability, their merits and limitations. Finally we conclude by listing a few of the requirements that seem to be desirable in a predictor. In sect. 3 we report on some work on predictors that develop one of the approaches surveyed in sect.2. We do not, of course, claim to have achieved all the requirements listed at the end of sect. 2 although these requirements were certainly a source of inspiration.

The different approaches discussed in Sect. 2 are, to our knowledge, representative of the main attempts to deal with this problem, although the references we list for each one of them are by no means supposed to be complete. They only represent the papers we happen to know best or those that better represent each one of our own comments in the survey. Our division of time series predictiors into classes emphasizes more the different points of view than the strict inequivalence of the methods. In many cases they may be shown to be reducible to one another. For example, in the sense that in most methods x_t is constructed directly from the past values of the time series, they might all be called autoregressive. Also, the connectionist approach may be considered as a particular case of the dynamical systems approach, with the representation step implemented by a neural network architecture, etc.

2. Survey of time series prediction techniques

2.1 Autoregressive models

The simplest one is the linear autoregressive model (LAR)

$$x_t = c_0 + c_1 x_{t-1} + \ldots + c_k x_{t-k} + \varepsilon_t \tag{2.1}$$

where $(c_0 \ldots c_k)$ are constants chosen by fitting a set of data points from the past data and ε_t, supposed to represent the part of the data that cannot be fitted by the model, is interpreted as a random variable with zero conditional expectation

$$E\left(\varepsilon_t \mid x_{t-1} \ldots x_{t-k}\right) = 0$$

A linear regressive model was used as early as 1927 by Yule[1] to study the sunspot number time series.

The LAR is somewhat limited in that the dynamical function $x_t = f(x_{t-1}, \ldots. x_{t-k})$ is approximated by a single hyperplane. There is also no clear control over the optimum number k of past points to be kept in the regression. This is an important issue because overfitting may overemphasize inessential (noisy) features of the data and spoil the predictive power of the method. Despite its simplicity the LAR has enjoyed, for the one-step ahead problem, a fair amount of sucess in many fields[2].

A natural generalization of the simplest autoregressive model consist in considering state-dependent coeficients[4,5,6]. Bilinear, exponential, threshold autoregressive and local linear models have been constructed (see Ref. 3). Another possibility consists in using higher order polynomials for the autoregressive model[7]. Most popular is probably the threshold autoregressive model[5]. It is based on the idea of piecewise linearization (in state space). Eq.(2.1) is replaced by

$$x_t = c_0^{(j)} + c_1^{(j)} x_{t-1} + \ldots + c_k^{(j)} x_{t-k} + \varepsilon_k^{(j)} \qquad (2.2)$$

The real line is supposed to be divided into p intervals and the value of $j \in (1,\ldots,p)$ is determined by the interval in which one of the x_{t-d} $(1 \leq d \leq k)$ falls. It means that the deterministic part of the process is approximated by p hyperplanes, the choice of the hyperplane depending at each time on the interval where x_{t-d} falls (i.e. the "threshold" above which this variable lies).

2.2 The dynamical systems approach

The idea underlying this approach is that, if the process generating the time series is effectively low dimensional, then it is possible to reconstruct the dynamical law from the time series. This approach is sucessful in many cases, because even very complicated dynamics sometimes live on low-dimensional attractors. If however the time series is the manifestation of an effectively high dimensional process, then this method is not very useful because the amont of data points needed for the reconstruction becomes prohibitive. Another limitation of the method is that attractor reconstruction implicitely assumes that the process is stationary.

The reconstrution is made using delayed coordinates and creating a state vector[8,9]

$$X_t = (x_t, x_{t-1} \ldots x_{t-d})$$

If D is the dimension of the manifold that contains the attractor the minimal embedding dimension d, needed to fully describe the deterministic dynamics, lies in the range $D \leq d \leq 2D+1$. The embedding dimension is often chosen by trial and error or more systematically from mutual information[10] or principal value decomposition[11].

Once the state space representation of the time series, in a space of dimension d, is obtained, the next step is to fit a dynamical model f to the data.

$$X_{t+1} = f(X_t) \qquad (2.3)$$

This may be done in a global or a local way. Global means that a function f defined over all the state space is fitted to the past data. The global function might be a polynomial[12,13] a ratio of two polynomials[14], a superposition of radial basis functions[15] or a weighed superposition of local maps[16]. In the local approach[17-19] the state space is divided into several (overlapping) neighborhoods and a different function is fitted to the data in each neighborhood. This method allows for the use of simpler (computer time efficient) fitting functions which, because they are only required to hold locally, may still describe the complex nonlinear features of the data.

Whenever the dynamics that generates the time series is low dimensional, stationary and highly non-linear, the method just described yields performances unmatched by any other method. By estimating the Lyapunov exponents of the dynamics the method also provides some control of the prediction error. With an efficient (decision tree) algorithm for the choice of neighborhoods and the use of simple local functions, such that the parameters can be obtained by solving a linear problem, the computer time requirements can be kept at a modest level. For (unknown) dynamics that is intrinsically high dimensional or non-stationary, the need for high dimensions in the state space or the fitting of new dynamical models to account for the changing nature of the process, make the method computationally intractable. Even if, with sufficient computer power, high dimensions and nonstationarity can, in principle, be dealt with, the method would not seem appropriate for an on-line adaptative predictor.

2.3 The connectionist approach

In this approach one considers a neural network, to the input nodes the values $x_{t-1}...x_{t-k}$ are fed and the output should represent x_t[20-24]. After being trained on a sufficiently large data set, the network should represent the function f of Eq(2.3) and, in this sense, the network approach may be considered as a particular instance of the dynamical model implementation step discussed in 2.2. The nonlinear nature of the network is located in the activation functions of the hidden units. Both sigmoid and Gaussian and radial basis functions have been used. A critical issue is the size of the network. Too small a network will not be able to represent complex functions, and too large a network will have too many free parameters leading to overfitting. By emphasizing inessential (or noisy) features of the data, too large a network leads to poor generalization. Several solutions have been proposed to deal with this problem: stopping the training at an ealier stage, trial and error clipping of the links or, more systematically[22,23], using a back-propagation cost function.

$$\sum_{k\in T} (\text{target}_k\text{-output}_k) + \lambda \sum_{ij} \frac{w_{ij}^2}{1 + w_{ij}^2}$$

which favours reduced connection weights w_{ij}.

A basic drawback of the network approach is the growth of the training time for large data sets. This may be somewhat improved by using nonlinear least square methods[21] instead of back-propagation but, for example, on-line training and adaptation to a time series seems out of question.

Instead of simply predicting the next value in the time series, it is sometimes interesting to estimate also the probability of occurrence of the other possible values. Neural networks were also used in this context to forecast probabilities of forthcoming symbols in a symbol sequence[25] .

2.4 Prediction of band-limited functions

If the time series is generated by a band limited process then, when the sampling rate is larger than the Nyquist rate, the x_t's may be predicted from the past samples. The classical result is Shannon's sampling theorem [26,27] that states that a function for which the Fourier transform vanishes outside the band [-W, W] can be exactly reconstructed from its values at discrete times separated by $T=\frac{\pi}{W}$ (T^{-1} is called the Nyquist rate)

$$x(t) = \sum_{-\infty}^{\infty} x(t_n) \frac{\sin W(t-t_n)}{W(t-t_n)} \tag{2.4}$$

Notice however that Eq. (2.4) is not causal, in the sense that the reconstruction of the signal at time t requires the knowledge of all its samples in the past and the future. Clearly Eq.(2.4) cannot be used as a predictor of time series. It has however been shown by several authors[28-34] that, as long as the sampling rate is larger than the Nyquist rate, then the signal can be reconstructed from its past samples. The idea is fairly simple. Let the sampling interval be normalized to 1. Then the band is $[-\pi, \pi]$. Consider a finite impulse response (FIR) high-pass filter h with stop band edge $\omega_M < \pi$. Then, if the process that generates the time series has a band in the interval $[-\omega_M, \omega_M]$, when the signal is filtered the output becomes negligible

$$\sum_{k=0}^{L-1} h(k)\, x_{n-k} \approx 0 \tag{2.5}$$

If $h(0) \neq 0$, (2.5) can be inverted to obtain a predictor

$$x_n \approx -\frac{1}{h(0)}\sum_{k=1}^{L-1} h(k)\, x_{n-k} \tag{2.6}$$

Therefore, in the band-limited hypothesis, the coefficients of the high-pass filter are the coefficients of an exact linear autoregressive predictor. The main shortcoming of the method, in the simple form described above, is the need to have a good estimate of the band of the process that generates the time series. In Sect. 3 we test a scheme where this restriction is relaxed by letting the system adjust itself to the optimal band. For processes which effectively cover a band larger than π (in normalized units), there is a certain amount of information that cannot be extracted from the past samples, and the way to check that this is really happening is to evaluate the right hand side of (2.5) for real data and with $\omega_M \approx \pi$.

2.5 Probabilistic approaches

Here we consider $\{x_t\}$ to be a stochastic process living in a L^2 space which we denote by $\mathcal{H}$. I.e. $\{x_t\}$ is a square integrable process in the sense that $E[x_t^2] < \infty$, E denoting the expectation value. Denote by y_t the process of the past samples $y_t = (x_{t-1}, x_{t-2} \ldots)$. Then the prediction problem consists in obtaining the best estimate $\bar{x}_t$ for x_t given y_t

$$\bar{x}_t = \text{est}\,(x_t | y_t) \doteq h(y_t) \tag{2.7}$$

Let $\mathcal{H}(y)$ be the subspace of the filtered observations

$$\mathcal{H}(y) = \{\ g(y) \mid g(y) \in \mathcal{H}\ \} \tag{2.8}$$

I.e. $\mathcal{H}(y)$ is the subspace generated by the action of all possible transformations (filters) on y, subject only to the condition that g(y) is still a square integrable process. The difference between the actual process x and the estimate $\bar{x}$ is called the innovation. To minimize (in the squared mean) the innovation, the estimate $\bar{x}$ should be the projection of x on $\mathcal{H}(y)$, i. e.

$$x - \bar{x} \perp \mathcal{H}(y) \tag{2.9}$$

This implies that the ideal filter which provides the best estimate coincides with the conditional expectation of x given y.

$$h(y) = E[\ x \mid y\ ] \tag{2.10}$$

To actually compute the filter, additional restrictions are usually put on $\mathcal{H}(y)$ which may or may not introduce new innovation errors. For example for Gaussian processes, linearity of the filter is harmless. Several algorithms have been developed for the construction of the filters (see Ref.[32], in particular the lectures by B. Picinbono and J.M.F. Moura).

Other approaches for stationary process are based on the ergodic theorem. Ornstein[36] analysed an ergodic process, with values zero and one, and developed a guessing scheme for the probability of the next output given a finite segment of the past, or equivalently, for the probability of the previous output given a finite segment of the future. The guessing scheme analyses the repetition of the finite segment that starts at time zero and the frequency of their sucessor events. For an ergodic process, the scheme is shown to converge to the probability given the complete past (or the future). This is easily understood because the ergodic theorem implies that the probability of any finite string is equal to the frequency with which it will occur in the past (or the future).

We end up this section by listing a few of the requirements desirable in a predictor of time series, mainly if it is intended to be used in real time and on data of unknown nature.

1. The predictor should not only make a guess, but also assign to it a probability or some sort of estimate of the reliability of the guess.

2. On-line adaptation should be incorporated as a feature so that, if the process that generates the time series changes, the predictor is able to adapt itself to the changing nature of the process.

3. The predictor should make use of all the information available and the guess should be maximal. Although this motion is not always easy to quantize, it means essentially that no better guess (with the same probability) might ever be done with the same data.

4. The predictor guess should be robust. I.e. it should not overemphasize inessential details of the data. This requirement is extremely inportant for noisy processes.

3. On-line adaptative filter predictors

Here we describe some work on predictor systems which elaborate on the filtering techniques described at the end of the previous section. The idea of using adaptative filters as predictors goes back to Gabor, Wilby and Woodcok[7]. In the systems that we explore, some adaptation is indeed provided in each filter, however the basic idea differs from the one in Ref.7, in that we consider a large collection of filters, exploring a wide range of transfer functions, with a selection unit that at each time chooses the most appropriate filter for the prediction. With digital techniques and FIR filters, collections as large as one desires may easily be implemented.

The basic aim was the development of predictors satisfying the requirement of real-time on-line adaptation to process changes Two systems were developed. The first attempts, at each time step, to search for the optimal bandwidth in the predictor filter, whereas the second tries to select the filtering most appropriate for each subdomain of the input variables. The main characteristics are therefore the locality in time, for the first one, and locality in phase space, for the second one

3.1 Band-searching predictor

The general block diagram of this system is sketched in Fig.1

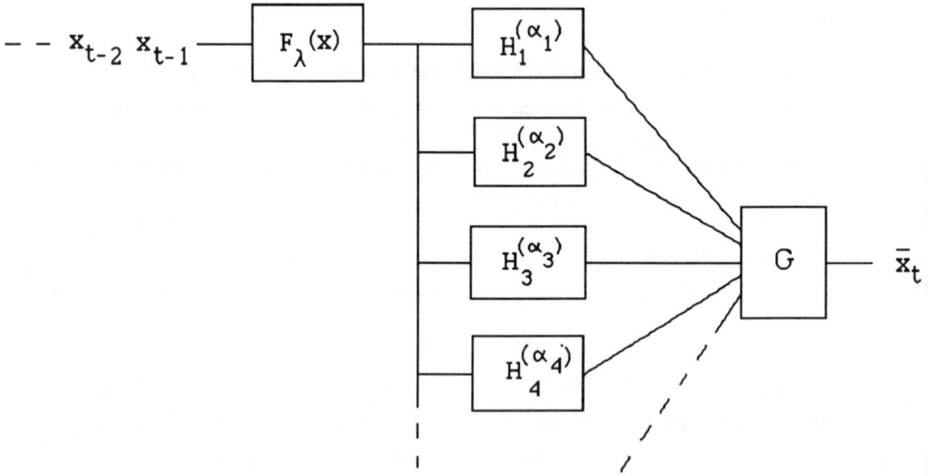

Fig. 1 - Band-searching adaptative filter predictor

$F_{\lambda(x)}$ is a linear or nonlinear invertible function depending on a set λ of adjustable parameters and $\{H_i^{(\alpha_i)}\}$ is a set of FIR filters which explore a large range of frequency bands. As a rule the parameters α_i are intended to adjust the scale of the filters band, not to change their qualitative features. The filters may be nonlinear. The only requirement is that they have to be invertible on the last input, i. e. if $H_i(x_{t-1},x_{t-2}, \dots)=c_i$ it exists a unique function H_i' such that

$$x_{t-1} = H_i'(c_i, x_{t-2}, x_{t-3}, \dots) \tag{3.1}$$

The selecting unit G inspects the outputs of the filters, selects the one with the smaller output and then uses its parameters to predict the next output . Let

$$y_t = (x_{t-1}, x_{t-2}, \dots) \tag{3.2}$$

Then G selects the filter H_{i_m} that has the minimal output c_{i_m} at time t - 1 and the prediction is

$$\bar{x}_t = F_\lambda^{-1}\left(H'_{i_m}\left(c_i, x_{t-1}, x_{t-2}, \dots\right)\right) \tag{3.3}$$

In the simplest case, when F_λ=x and the filters are linear, i. e.

$$H_i(y) = a_1 x_{t-1} + a_2 x_{t-2} + \dots = c_i \tag{3.4}$$

the filter with the smallest c_i at time t-1 is selected and

$$\bar{x}_t = \frac{1}{a_1}\{c_i - a_1 x_{t-1} + a_2 x_{t-2} - \cdots\} \tag{3.5}$$

In this case the system is an automatic state-dependent autoregressive model, where the state-dependence can be considered local either in time or in state space. However, contrary to the threshold autoregressive model where one single variable selects the hyperplane to be used, here the selection is made by a cooperative effect of all variables. Even with linear filters, because of the local piecewise nature of the selection, the system may be able to deal adequately with intrinsically nonlinear effects. For non-linear filters the main limitation is that the nonlinear transfer function

$$y_i = h_0 + \sum_{j=0}^{k} h_1(j)\, x_{i-j} + \sum_{j=0}^{k}\sum_{r=0}^{k} h_2(j,r)\, x_{i-j}\, x_{i-r} \tag{3.6}$$

should be chosen in such a way that the equation $y_i=c_i$ may be inverted to obtain x_i as a function of $x_{i-1} \dots x_{i-k}$. To pass the input through an (invertible) nonlinear function and use linear filters is equivalent to use nonlinear filters with diagonal kernel[37]. The role of the nonlinear function is to adapt the sequence to the action of the linear filters. The logical approach should be to start with some set of parameters λ in the nonlinear function and then to change λ to minimize the average minimum c_i.

The appropriate choice of the collection of filters is the essential point in this approach. One choice we found to be reasonable is a collection of soft stop-band filters H_i which become sharper as i increases (see Fig. 2)

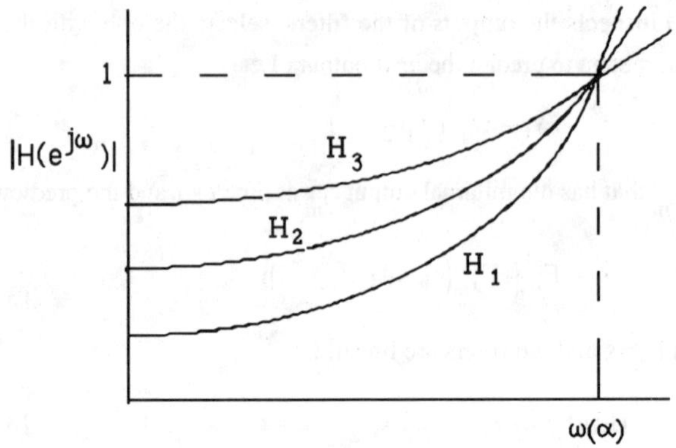

Fig. 2 Transfer function amplitude for a collection of soft stop-band filters

The adjustable parameter α defines the band-width. The simplest choice has a sequence of z-transformed transfer functions

$$H_i(z) = \left(1 - \frac{\alpha}{z}\right)^i \tag{3.7}$$

This choice has the appealling feature that the FIR sequence starts with the smallest possible number of past values which increases systematically with i. Sucessive filters H_i suppress more and more frequencies in the range $[0, \omega(\alpha)]$ and amplify each time more the oscillations with frequencies $\omega > \omega(\alpha)$. Therefore one expects that in the sequence $H_1, H_2, \ldots$ the outputs will decrease up to a point, increasing afterwards due to the amplification of small components (and noise) ouside the relevant band. If, conversely, the outputs c_i of the filters grow monotously with i, starting already at $i = 1$, this means that there are important components of the signal above $\omega(\alpha)$ and α should be adjusted accordingly.

The adaptative strategy that we have used, in a pratical implementation, is to have a collection $H_i^{(\alpha_j)}$ of filters for a discrete $\{\alpha_j\}$ set. One then obtains a matrix $M_i^{(\alpha_j)}$ of filter outputs. To suppress noise fluctuations $M_i^{(\alpha_j)}$ is fitted to a smooth M(x,y) function, the minimum of M(x,y) is obtained and the filter used for the prediction is the one with labels $((\alpha_j),i)$ closer to the coordinates of the minimum of the smoothing function.

The feature of on-line adaptation to changing processes is intrinsic to the scheme. With a sequence H_i of FIR filters which react to an increasing number of past values, robustness is implicit in the limitation to the filters with the minimum c_i. I. e. the order of the regression is stopped before it starts to amplify the noise or small details of the data. On the other hand an estimate of the reliability of the guess $\overline{x}_t$ is provided by the amplitude of c_i minimum.

In Figs.3a-d we show the results of the application of the algorithm to several time series. Figs.3a and 3b compare the inflation in Portugal in the period January 1984 - November 1990 (solid line) with the output of the predictor (crosses). In Fig.3b the data is filtered before being fed to the predictor. Figs.3c and 3d present the results for the sunspot number time series.

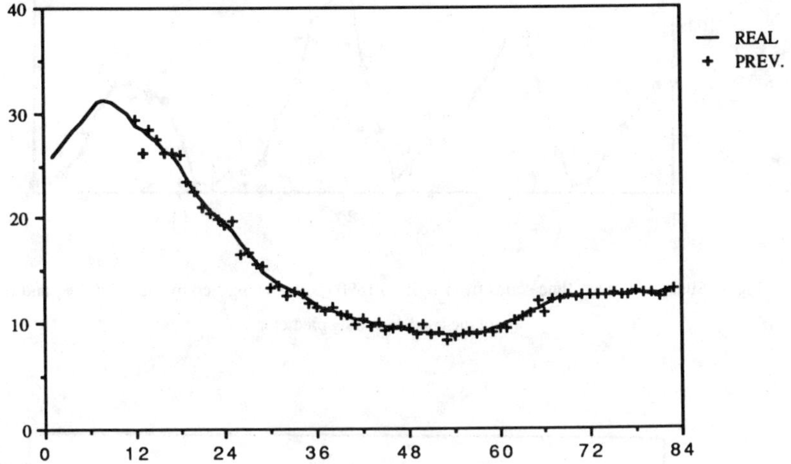

Fig.3a Inflation in Portugal from January 1984 to November 1990 (solid line) versus the output of the predictor (crosses)

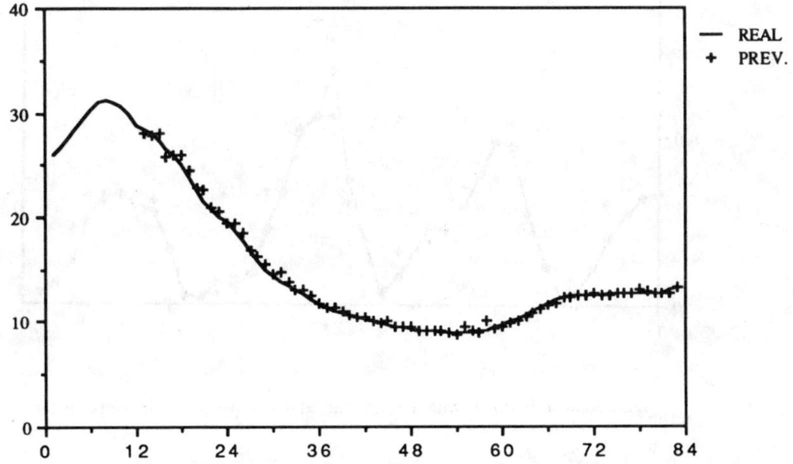

Fig.3b Same as in Fig.3a with the data filtered before being used in the predictor

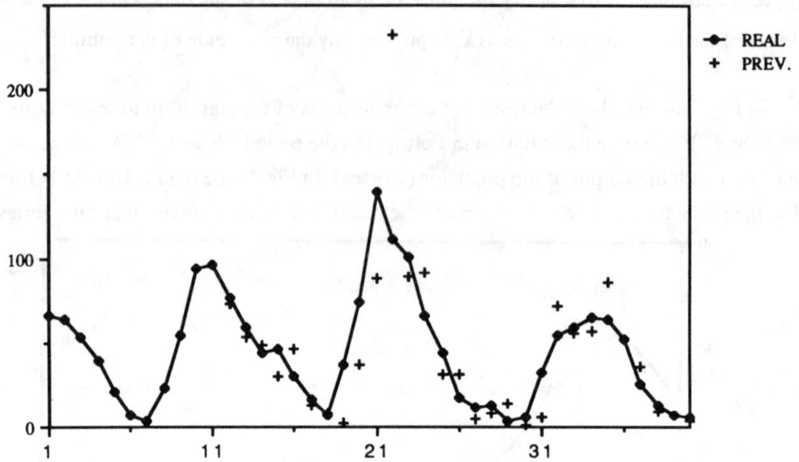

Fig.3c Sunspot number time series from 1850 to 1890 (points connected by the solid line) and the output of the band-searching predictor

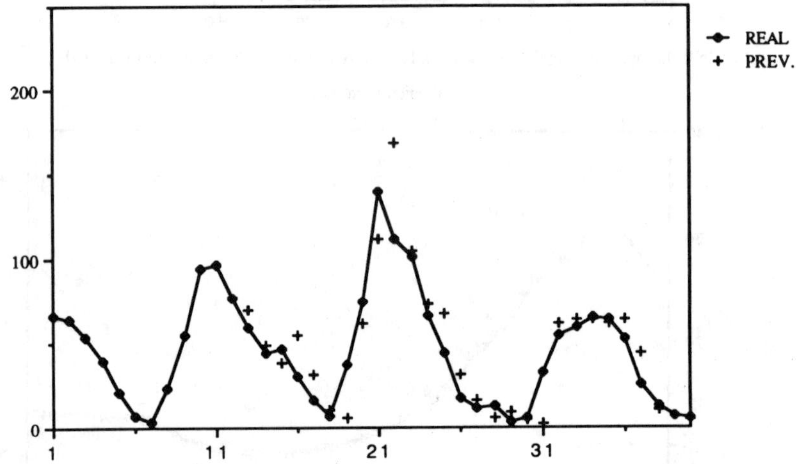

Fig.3d Same as in Fig.3c with the data filtered before being used in the predictor

3.2 Phase-space local adaptative filter predictor

The construction of a predictor of this type may be summarized in the following steps:

1. Make a partition of phase space into k parts.

2. Generate a tree of FIR adaptative filters labelled by the partition.

3. Use the best local agreement for the next guess.

Fig.4 shows the corresponding diagram for the case k=2, the phase space parts being labelled by 0 and 1.

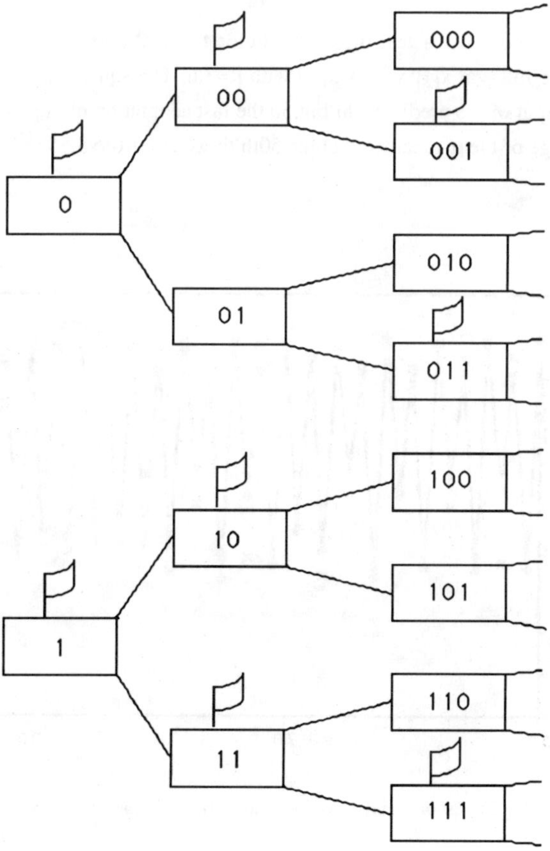

Fig.4 Filter diagram of a phase-space local predictor when the partition number k is 2

From left to right, the columns contain FIR filters of sucessively higher orders. For example the filter labelled by [010] is a order-3 filter and is associated to phase space points such that $x_{t-1} \in 0$, $x_{t-2} \in 1$ and $x_{t-3} \in 0$. In the algorithm there is a system of moving flags to signal the best past predictions (see below).

Let the maximal FIR order be p. At each time step the set $(x_{t-1} \ldots x_{t-p})$, where $(x_{t-1} \in i_1 \ldots x_{t-p} \in i_p)$, defines a branching line going from the filter $[i_1 \ldots i_p]$ down to $[i_1]$. The higher order filter $[\alpha]$, in that line, that has a flag is used to predictor $\overline{x}_t$. Once the actual x_t is known the filter $[\beta]$, in the line, that would have led to the smallest $|x_t - \overline{x}_t|$ receives a flag and $[\alpha]$ loses its flag if $\alpha \neq \beta$. Then all filters in the line are adapted using the value od x_t. The predictor therefore contains the capacity of local adaptation both in time and in phase-space.

In Fig.5a we show a test of the predictor for k=2 and a time series generated by a mapping of the interval $x_t = \mu x_{t-1}(1-x_{t-1})$ with $\mu=3.6$. The squares are the data points and the crosses the output of the predictor. In Fig.5b the fast adaptation of the predictor is exhibited by a sudden change of the μ-parameter, at the 50th time step, from 3.55 to 3.4.

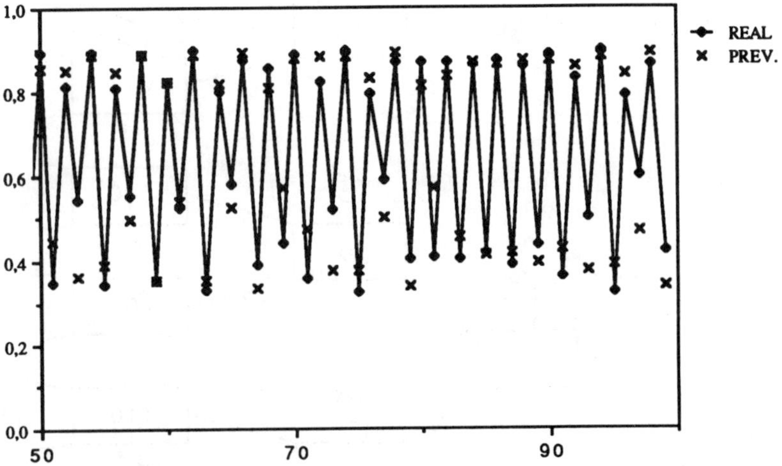

Fig.5a Test of the predictor for a time series generated by $x_t = \mu x_{t-1}(1-x_{t-1})$; $\mu=3.6$

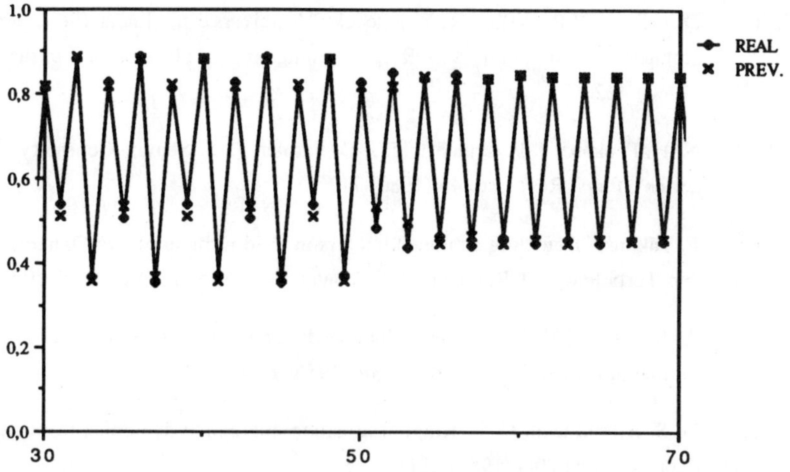

Fig.5b Adaptation to a sudden change of parameter (μ=3.55 to 3.4 at step 50)

References

[1] G.U. Yule; "On the method of investigating periodicities in distributed series with special reference to Wolfer's sunspot numbers", Phil. Trans. Roy. Soc. London A 226 (1927) 267-298.

[2] G. E. P. Box, G. M. Jenkins; "Time Series Analysis, Forecasting and Control"; Holden-Day, San Francisco, 1970.

[3] M. B. Priestley; "Non-linear and Non-stationary Time Series Analysis"; Academic Press, 1988.

[4] M. B. Priestley; "State dependent models: a general approach to non-linear time series analysis; Journal of Time Series Analysis" 1 (1980) 47-71.

[5] H. Tong, K. S. Lim; "Threshold autoregression, limit cycles and cyclical data", J. R. Statist. Soc. B42 (1980) 245-292.

[6] V. Haggan, S. M. Heravi, M. B. Priestley; "A study of the application of state-dependent models in non-linear time series analysis", Journal of Time Series Analysis 5 (1984) 69-102.

[7] D. Gabor, W.P. Wilby, R. Woodcock; "A universal nonlinear filter, predictor and simulator which optimizes itself by a learning process"; Proceedings of the IEE, 108 B (1960) 422.

[8] N. H. Packard, J. P. Crutchfield, J. D. Farmer, R. S. Shaw; "Geometry from a time series", Phys. Rev. Lett. 45 (1980) 712.

[9] F. Takens; "Detecting strange attractors in fluid turbulence", in Dynamical Systems and Turbulence, D. Rand and L. S. Young, Eds., Springer-Verlag, Berlin, 1981.

[10] A. M. Fraser, H. L. Swinney; "Independent coordinates for strange attractors from mutual information" Phys. Rev. 33A (1986) 1134.

[11] D. S. Broomhead, G. P. King; "Extracting qualitative dynamics from experimental data", Physica 20D (1987) 217.

[12] J. Cremers, A. Hübler; "Construction of differential equations from experimental data", Z. Naturforsch. 42a (1987) 797.

[13] J. P. Crutchfield, B. S. McNamara, "Equations of motion from a data series", Complex Systems 1 (1987) 417.

[14] B. J. Bayly, I. Goldhirsch, S.A. Orszag; "Independent degrees of freedom of dynamical systems" Princeton Univ. preprint, 1986.

[15] M. Casdagli; "Nonlinear prediction of chaotic time series"; Physica D35 (1989) 335.

[16] K. Stokbro, D. K. Umberger, J. A. Hertez; "Exploiting neurons with localized receptive fields to learn chaos", Complex Systems 4(1990) 603.

[17] J. Doyne Farmer, J. J. Sidorowich; "Predicting chaotic time series", Phys. Rev. Lett. 59 (1987) 845.

[18] J. Doyne Farmer, J. J. Sidorowich; "Exploiting chaos to predict the future and reduce noise", in Evolution, Learning and Cognition. Y. C. Lee, Ed. World Scientific Press, Singapore, 1988.

[19] M. Casdagli, S. Eubank, J. Doyne Farmer, J. Gibson; "State space reconstruction in the presence of noise", Physica D51 (1991) 52.

[20] A. S. Lapedes, R. Farber; "Nonlinear signal processing using neural networks: prediction and system modeling", Los Alamos report LA-UR-87-2662.

[21] H. White; "Economic prediction using neural networks: The case of IBM daily stock returns" IEEE Int. Conf. on Neural Networks, Page II-451, San Diego, 1988.

[22] A. S. Weigend, B. A. Huberman, D.E. Rumelhart; "Predicting the future: A connectionist approach"; International Journal of Neural Systems 1 (1990) 193.

[23] A. S. Weigend, D. E. Rumelhart, B. A. Huberman; "Generalization by weight-elimination with application to forecasting", in "Advances in Neural Information Processing 3", R. P. Lippmann, J. Moody and D. S. Touretzky, San Mateo CA, Morgan Kaufmann, 1991.

[24] D. Lowe, A. R. Webb; "Time series prediction by adaptative networks: a dynamical systems perspective", IEE Proceedings F 138 (1991) 17.

[25] A. Bergman, P. Grassberger, T.P. Meyer; "Forecasting probabilities with neural networks", Wuppertal prepint WU B 91-8.

[26] C. E. Shannon; "Communication in the presence of noise", Proc. IRE 37 (1949) 10.

[27] For the history of the sampling theorem and related results see J. R. Higgins, "Five short stories about the cardinal series", Bull. Am. Math. Soc. 12 (1985) 45.

[28] L. A. Wainstein, V. D. Zubakov: "Extraction of Signals from Noise", Prentice-Hall, Englewood Cliffs NJ, 1962.

[29] F. Bently, "Error free recovery of signals from irregular samples" SIAM Rev. 8 (1966) 322.

[30] J. L. Brown, Jr.; "Uniform linear prediction of band limited processes from samples", IEEE Trans. Inform. Theory IT-18 (1962) 662.

[31] T. T. Fjallbrandt; "Interpolation and extrapolation in nonuniform sampling sequences with average sampling rate below the Nyquist rate"; Electron. Lett. 11 (1975) 264.

[32] A. Papoulis; "A note on the predictability of band-limited processes"; Proc. IEEE 73 (1985) 1332.

[33] J. L. Brown, Jr.; "On the prediction of a band-limited signal from past samples" Proc. IEEE 74 (1986) 1596.

[34] P. P. Vaidyanathan; "On predicting a band-limited signal based on past sample values", Proc. IEEE 75 (1987) 1125.

[35] J.-L. Lacoume, T. S. Durrani, R. Stora, Eds.; "Signal processing", Les Houches Session XLV 1985, North-Holland, Amsterdam 1987.

[36] D. S. Ornstein; "Guessing the next output of a stationary process", Israel Journal of Math. 30 (1978) 292.

[37] See I. Pitas, A. N. Venetsanopoulos; "Nonlinear Digital Filters. Principles and Applications", Kluwer Acad. Pub. Dordrecht, 1990.

Production Lines

STOCHASTIC DISPERSIVE TRANSPORT. AN EXCURSION FROM STATISTICAL PHYSICS TO AUTOMATED PRODUCTION LINE DESIGN

M.-O. HONGLER

Institut de Microtechnique

Ecole polytechnique fédérale de Lausanne

CH-1015 Lausanne, Switzerland

ABSTRACT

The sediment transport dynamics and the population level of a buffer in automated production line systems can be described by the same class of stochastic differential equations. The ubiquous noise is generated by continous time Markov chains. The probability densities which describe the dynamics are governed by high order hyperbolic systems of partial differential equations. While this hyperbolic nature clearly exhibits a non-diffusive character of the processes ; (diffusion would imply a parabolic evolution of the probability densities), we nevertheless can use a central limit theorem which holds for the large times regimes. This enables analytical estimations of the time evolution of the moments of these processes. A particular emphasis is devoted to non-Markovian, dichotomous alternating renewal processes which enter directly into the description of the applications presented.

1. Introduction.

The purely deterministic description of numerous transport phenomena is very often incomplete in obtaining a reliable prediction of the expected flows and other macro- and mesoscopic quantities of interest. Indeed, stochastic perturbations always corrupt the mean values of the external control parameters governing the system and hence, very often can strongly affect the expected behaviour of the relevant observables. Therefore, one should definitely include into the modelization the influence of noise sources. The estimation of the statistical properties of stochastic flows of matter and/or informations, is a common need in both science and technology. One of the pioneers in this class of mathematical modelling problems is Sydney Goldstein who is reported [1] to have said that "to be a good applied mathematician, you have to be one third physicist, one third engineer and one third mathematician ". This exact cocktail of expertises are those needed to bridge the gaps existing between certain aspects of physics and technology. In the spirit of Goldstein's guiding lines, let us, more specifically, introduce the class of problems to be discussed by considering separately the three different points of view.

1.1 Stochastic transport of particles.- A topic in statistical physics

The theoretical discussion of the stochastic transport of particles and sediment were first considered in the beginning of the century in connection with the analysis of the rate of movements of underground waters. In these first investigations, salt was used as a tracer in ground water flow, the objective being to study and modelize the stochastic motion of the tracer particles. Such transport problems remain in actuality. In particular, the increasing rate of pollution of ground water sources by chemical polluants makes it clear that a detailed description of the dynamics involved is important, if not imperative. From the physicist's point of view, the transport of sediment is one of the topics belonging to the off-equilibrium statistical mechanics. In its simplest form, the model of the transport of sediment is described by the one-dimensional motion of a particle subject to a finite number of discrete velocities which are switched at randomly selected times [2, 3, 4]. A closely related situation is encountered in the problem of chromatography [5]. The basis for the stochastic model is that the chromatographic process is viewed as a series of sorptions and desorptions as a particle travels along the chromatographic column, such that the probability of changing phase during a short time interval is essentially proportinal to the time interval, the proportionality factor being the transition rate constant.

1.2 Selection of the capacity of buffers in automated production line.- A production engineer problem [6-14]

The need to increase the quality and the competitivity in the massive production of commonly diffused products (i. e. television sets, camera, computers etc...) has urged the production engineer community to develop large and complex automated production lines. In their simplest topology, these systems can be considered as a succession of working zones (i.e. machines) separated by storage zones (i.e. buffers). A flow of parts enters into the first working area and then proceeds downstream through all buffers and successive working zones until at the output of the last machine is the completed product ready to sell. In actual industrial contexts, the machines are always prone to failure. The failures of these machines interrupt, at random instants, the flow of parts. The breakdowns propagate up and down the line and can thereby considerably lower the efficiency of the system, especially for lines composed of a large number of machines. To partly prevent this loss of productivity, the line is equipped with buffers located between the machines. These buffers lower the effects of the interaction between the machines. Hence, they enable to partly compensate for the local fluctuations in the flow of parts. When judiciously selected, the buffers can significantly improve the efficiency of the production line. Obviously, the capacity of the buffers is finite and storage places are often very expensive. Consequently, one of the major issues one should be concerned of in automated line design is to optimally choose the size and location of the buffers in order to match a required production "throughput" (i.e, flow of parts). The optimal selection of the capacity of the buffers can be estimated by adequate knowledge of the dynamics of the content of the storage areas. Typically , the filling level of the buffers is described by a stochastic processs. Indeed, the input and the output flows in the storage are random processes which are ruled by the time between failures and the time needed for the repairs.

1.3 The probabilistic analysis of hyperbolic systems of partial differential equations.-The mathematician's framework.

The probabilistic analysis of hyperbolic systems of differential equations was pioneered by S. Goldstein.[15] and M Kac[16] This subject arises in the probabilistic characterization of the solutions of stochastic differential equations [17-21]. The noise present in the dynamics is generated by continuous time Markov chains. The simplest case is the well known and commonly encouterd telegraphic noise or dichotomous noise. In these systems, the transition probability densities have compact supports. The sizes of the support evolve with the characteristics of the Chapman - Kolmogorov equations which are hyperbolic in these cases. This behaviour has to be opposed to what is observed for diffusion processes, (i.e. the use of white noise in the stochastic differential equations) where the transition probability densities obey to parabolic equations known as the Fokker-Planck equation to the physics community. In asymptotic time regimes, it is howewer possible to establish a central limit theorem [19] which is particularly useful for most applications.

2. The fundamental equation of the motion.

2.1. Probabilistic analysis of systems of hyperbolic partial differential equations

Consider a particle which travels in one dimension at one of the possible velocities $v_1.v_2,...v_n$ making transitions from one velocity to another according to a Markov chain with infinitesimal generator is a n x n matrix $(Q)_{ij} = q_{ij}$, i, j = 1, 2, ... n.The transition probability densities P_i (x, t) to find the particle at position x at time t and velocity v_i, satisfy an hyperbolic system of differential equations [17-19]:

$$\frac{\partial}{\partial t} P_i(x, t) = v_i \frac{\partial}{\partial x} P_i(x, t) + \sum_{j=1}^{n} q_{ij} P_j(x, t) ; \qquad i = 1,2 ...n \tag{2.1}$$

The marginal probability density $\Sigma(x, t) = \Sigma_n P_i$ (x, t) which represents the probability of finding the particle at position x at time t whatever the state of the velocity is , can be shown to satisfy a nth order hyperbolic differential equation which has the form [19] :

$$A\left(\frac{\partial}{\partial t}, \frac{\partial}{\partial x}\right) \Sigma(x, t) = 0 \tag{2.2}$$

with the initial conditions :

$$\frac{\partial^k}{\partial t^k} \Sigma(x, t=0) = 0 \qquad k = 1, 2, ..., n-1 \tag{2.3}$$

The initial conditions Eq. 2.3 ensure that the normalization condition is fulfilled, (see the discussion in [22]).

The nth order differential operator $A(\xi, \gamma)$ is expressible as such :

$$A(\xi, \gamma) = \mathrm{Det}\left(Q - \xi V - \gamma Id\right) \tag{2.4}$$

where $(V)_{ij} = v_i \delta_{ij}$ and Id is the n x n identity matrix.

2.2. A special class of processes. The alternating renewal processes [23].

As an illustration, which we will use later, let us consider a particle which equation of the motion is given by a stochastic differential equation :

$$\frac{d}{dt} x(t) = v(t) \qquad v(t) \in \{v_0, v_1\} \text{ with } v_0 \leq 0 \text{ and } v_1 \geq 0 \tag{2.5}$$

where the noise source v(t) is a dichotomous process. The sojourn time in the states v_o and v_1 are respectively given by the probability distributions $F_\lambda(t)$, $F_\mu(t)$ with densities $f_\lambda(t)$, $f_\mu(t)$. The densities $f_\lambda(t)$ and $f_\mu(t)$ will be here chosen to be of the generalized Erlang type. They can be easely expressed in terms of their Laplace transforms :

$$G_\lambda(u) = \int_0^\infty \exp(-ut) f_\lambda(t)\, dt = \prod_{k=1}^{M} \left[\frac{\lambda_k}{\lambda_k + u}\right] \tag{2.6}$$

and

$$G_\mu(u) = \int_0^\infty \exp(-ut) f_\mu(t)\, dt = \prod_{k=1}^{N} \left[\frac{\mu_k}{\mu_k + u}\right]. \tag{2.7}$$

From Eqs 2.6 and 2.7 , we can directly calculate the mean sojourn times τ_λ and τ_μ in the states v_o respectively v_1. We have therefore :

$$\tau_\lambda = \frac{\partial}{\partial u} G_\lambda(u = 0) = \sum_{k=1}^{M} \frac{1}{\lambda_k} \tag{2.8}$$

and

$$\tau_\mu = \frac{\partial}{\partial u} G_\mu(u = 0) = \sum_{k=1}^{N} \frac{1}{\mu_k} \tag{2.9}$$

Similarly, the coefficients of variation CV_λ and CV_μ read :

$$(CV_\lambda)^2 = \frac{1}{(\tau_\lambda)^2}\left[\frac{\partial^2}{\partial u^2} G_\lambda(u = 0) - (\tau_\lambda)^2\right] = \left(\frac{1}{\tau_\lambda}\right)^2 \sum_{k=1}^{M} \left(\frac{1}{\lambda_k}\right)^2 \tag{2.10}$$

and

$$(CV_\mu)^2 = \frac{1}{(\tau_\mu)^2}\left[\frac{\partial^2}{\partial u^2} G_\mu(u = 0) - (\tau_\mu)^2\right] = \left(\frac{1}{\tau_\mu}\right)^2 \sum_{k=1}^{N} \left(\frac{1}{\mu_k}\right)^2 \tag{2.11}$$

The generalized Erlang distribution Eq. 2.6 is obtained from continuous time Markov chain with M stages, each stage having a λ_k-exponential sojourn time (k = 1, 2, ...M) and similarly for the μ process Eq. 2.7. Schematically, this situation can be described as shown in Fig. 2.1.

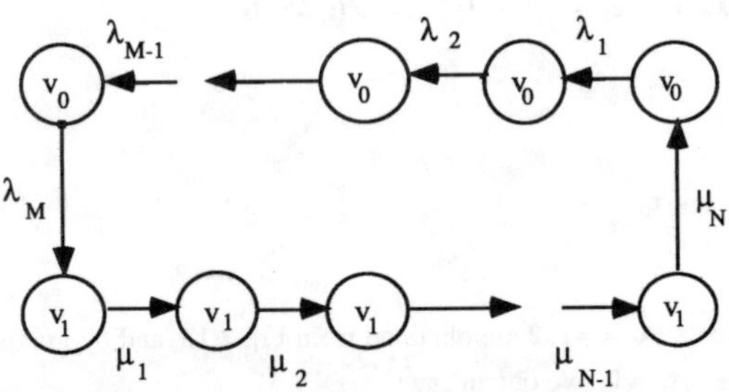

Fig 2.1 Representation of a two stage v_0 and v_1 alternating renewal process. Between each stage, the transition is exponetially distributed with parameters λ_j =1, 2, ... M for the stages associated with v_0 and μ_j j = 1, 2, ... N for those describing v_1.

For this process, the infinitesimal matrix generator Q has the form :

$$Q = \begin{pmatrix} Q_{11} & Q_{12} \\ Q_{21} & Q_{22} \end{pmatrix} \tag{2.12}$$

with :

$$(Q_{11})_{ij} = -\lambda_i \delta_{ij} + \lambda_j \delta_{i\,j+1} \qquad i = 1, 2, \ldots M \; ; \; j = 1, 2, \ldots M-1 \tag{2.12a}$$

$$(Q_{12})_{ij} = \mu_N \delta_{1i} \delta_{Nj} \quad i = 1, 2, \ldots M \; ; \; j = 1, 2, \ldots N \tag{2,12b}$$

$$(Q_{21})_{ij} = \lambda_M \delta_{1i} \delta_{Mj} \quad i = 1, 2, \ldots N \; ; \; j = 1, 2, \ldots M \tag{2.12c}$$

$$(Q_{22})_{ij} = -\mu_i \delta_{ij} + \mu_j \delta_{i\,j+1} \qquad i = 1, 2, \ldots N \; ; \; j = 1, 2, \ldots N-1 \tag{2,12d}$$

Introducing Eqs 2.12a-d into Eq 2.4, it is straightforward to establish :

$$A(\xi, \gamma) = \prod_{k=1}^{M} \left(-\lambda_k - \xi v_0 - \gamma\right) \prod_{k=1}^{N} \left(-\mu_k - \xi v_1 - \gamma\right) + (-1)^{N+M-1} \prod_{k=1}^{M} \lambda_k \prod_{k=1}^{N} \mu_k \tag{2.13}$$

For the representations $\xi = \frac{\partial}{\partial x}$ and $\gamma = \frac{\partial}{\partial t}$, the operator $A(\xi, \gamma)$ given by Eq. 2.13 is therefore a $(N+M)^{th}$ order differential operator.

In large time regimes, the dynamics is governed by a central limit theorem [19]. In this case, the central moments $< (x(t) - < x(t) >)^m >$ obey a diffusion equation of the form :

$$\int_{-\infty}^{\infty} \left\{ \omega_1 \frac{\partial}{\partial t} \Sigma(x, t) + \omega_2 \frac{\partial^2}{\partial x^2} \Sigma(x, t) \right\} (x - \langle x(t) \rangle)^m \, dx \tag{2.14}$$

where :

$$\langle x(t) \rangle \cong \frac{v_0 \tau_\lambda + v_1 \tau_\mu}{\tau_\lambda + \tau_\mu} \, t \tag{2.15}$$

The coefficients ω_k $k = 1, 2$ are obtained from Eq. 2.13 and by imposing the property : $<v> = v_0\tau_\lambda + v_1\tau_\mu = 0$. We obtain :

$$\omega_1 = \left(\prod_{k=1}^{M} \lambda_k \right) \left(\prod_{k=1}^{N} \mu_k \right) [\tau_\lambda + \tau_\mu] \tag{2.16}$$

$$\omega_2 = \left(\prod_{k=1}^{M} \lambda_k \right) \left(\prod_{k=1}^{N} \mu_k \right) \left[\frac{v_0^2}{2} (\tau_\lambda)^2 \, CV_\lambda^2 + \frac{v_1^2}{2} (\tau_\mu)^2 \, CV_\mu^2 \right] \tag{2.17}$$

Therefore, we can write :

$$\langle (x(t) - \langle x(t) \rangle)^2 \rangle = \frac{v_0^2 \, \tau_\lambda^2 \, CV_\lambda^2 + v_1^2 \, \tau_\mu^2 \, CV_\mu^2}{\tau_\lambda + \tau_\mu} \, t \tag{2.18}$$

2.3 Transition probability density for a non-markovian alternating process [23].

For the process Eq. 2.5, it is also possible to derive the exact transition probability density $\Sigma(x, t)$. The successive sojourn times of $v(t)$ in the states $\{v_0\}$ and $\{v_1\}$ are respectively denoted by $\{\xi_i\}$ and $\{\eta_i\}$ $i = 1, 2 \ldots$ .The selection of the switching times obeys an alternating renewal process characterized by $F_\lambda(t)$ and $F_\mu(t)$. Hence for the times t^* which coincide with one of the renewal times, if $v(t = 0) = \{v_0\}$, we either have :

$$t^* = \xi_1+\eta_1+\xi_1+\eta_2+.......+ \xi_k \quad \text{or} \quad t^* = \xi_1+\eta_1+\xi_1+\eta_2+.......+ \eta_k \tag{2.19}$$

The $\{\xi_i\}$'s are $F_\lambda(t)$ - distributed and the $\{\eta_i\}$'s are $F_\mu(t)$ - distributed. By $N_\xi(t)$ and $N_\eta(t)$, we respectively denote the renewals processes generated by the $\{\xi_i\}$'s and the $\{\eta_i\}$'s. In other words, $N_\xi(t)$ and $N_\eta(t)$ count the number of occurences of ξ and η up to the time t.

We have represented a typical realization of this process in Fig. 2.2.

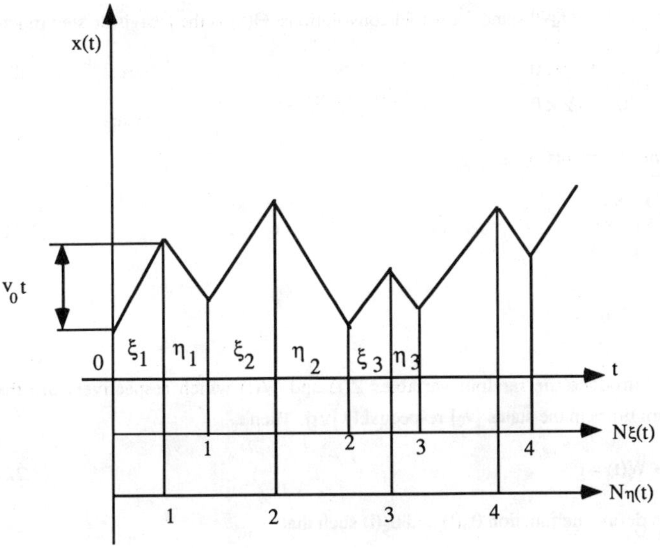

Fig. 2.2 A realization of an alternating renewal process.

Assume that the alternating renewal process is in equilibrium [23]. Then the initial conditions is a ponderated superposition of the solution obtained for $v(t=0)=v_0$ or for $v(t=0)=v_1$. The respective ponderations are chosen as p_0 and p_1 which respectively denote the average time spent in the states $\{v_0\}$, $\{v_1\}$ when time t tends to infinity.

With these assumptions, we can deduce the following proposition :

Proposition.

$$\Sigma(x,t) = \frac{p_0}{v_1 - v_0} \Pr(N_\eta(t) = 0)\, \delta(\chi_0) + \frac{p_1}{v_1 - v_0} \Pr(N_\xi(t) = 0)\, \delta(\chi_2) +$$

$$+ \left\{ \frac{p_0}{v_1 - v_0} \sum_{n=1}^{\infty} f_\mu^{(n)}(\chi_0) \Pr(N_\xi(\chi_1) = n) + \frac{p_0}{v_1 - v_0} \sum_{n=1}^{\infty} f_\lambda^{(n)}(\chi_1) \Pr(N_\eta(\chi_0) = n-1) \right\} \Theta(\chi_0) +$$

$$+ \left\{ \frac{p_1}{v_1 - v_0} \sum_{n=1}^{\infty} f_\mu^{(n)}(\chi_0) \Pr(N_\xi(\chi_1) = n-1) + \frac{p_1}{v_1 - v_0} \sum_{n=1}^{\infty} f^{(n)}(\chi_1) \Pr(N_\eta(\chi_0) = n) \right\} \Theta(\chi_1) \tag{2.20}$$

where $f_\lambda^{(n)}$ and $f_\mu^{(n)}$ stand for n-fold convolutions, $\Theta(z)$ is the Heaviside step function,

$$\Theta(z) = \begin{cases} 1 & \text{if } z \geq 0 \\ 0 & \text{if } z < 0 \end{cases}$$

and the arguments χ_0 and χ_1 are :

$$\chi_0 = \frac{x - v_0 t}{v_1 - v_0} \quad \text{and} \quad \chi_1 = \frac{v_1 t - x}{v_1 - v_0}. \tag{2.21}$$

Proof.

Introduce the random variables Z(t) and W(t) which respectively are the total sojourn times in the states $\{v_0\}$ respectively $\{v_1\}$. Then :

$$Z(t) + W(t) = t. \tag{2.22}$$

Let us define the function $\theta_\eta(t)$ and $\theta_\xi(t)$ such that:

$$\theta_\eta(t) = \begin{cases} \eta_1 + \eta_2 + \eta_3 + \ldots + \eta_{N_\xi(t)} & \text{if } N_\xi(t) > 0 \\ 0 & \text{if } N_\xi(t) = 0. \end{cases} \tag{2.23a}$$

and

$$\theta_\xi(t) = \begin{cases} \xi_1 + \xi_2 + \xi_3 + \dots + \xi_{N_\eta(t)} & \text{if } N_\eta(t) > 0 \\ 0 & \text{if } N_\eta(t) = 0. \end{cases} \tag{2.23b}$$

It can be deduced that [24]:

$$\Pr\left(W(t) \le x\right) = \Pr\left(\theta_\eta(t-x) \le x\right) \qquad \text{if } v(t=0) = v_0 \tag{2.24a}$$

$$\Pr\left(Z(t) \le x\right) = \Pr\left(\theta_\xi(t-x) \le x\right), \qquad \text{if } v(t=0) = v_1, \tag{2.24b}$$

where $\Pr(Y(t) \le x)$ is the (cumulative) distribution function of $Y(t)$.

Since :

$$X(t) = v_0 Z(t) + v_1 W(t), \tag{2.25}$$

we can either write :

$$X(t) = v_0\, t + (v_1 - v_0) W(t) \tag{2.26a}$$

or

$$X(t) = v_1 t - (v_1 - v_0)\, Z(t) \;. \tag{2.26b}$$

From Eqs. 2.26a, 2.26b, we can write:

$$\Pi_0 = \Pr\left(X(t) \le x\right) = \Pr\left(v_0\, t + (v_1 - v_0) W(t) \le x\right) = \Pr\left(W(t) \le \chi_0\right) \Theta\,(\chi_0) \tag{2.27a}$$

and

$$\Pi_1 = \Pr\left(X(t) \le x\right) = \Pr\left(v_1\, t - (v_1 - v_0)\, Z(t) \le x\right) = \Pr\left(Z(t) \ge \chi_1\right) \Theta\,(\chi_1) =$$

$$= [1 - \Pr\,(Z(t) < \chi_1)]\; \Theta\,(\chi_1) \tag{2.27b}$$

where the functions χ_0 and χ_1 are defined in Eq. 2.21.

From Eqs. 2.25a, 2.25b, 2.26a and 2.26b we have:

$$\Pr\,(W(t) \le x) = \Pr\,(\theta_\eta(t - x) \le x) = \sum_{n=1}^{\infty} F_\mu^{(n)}(x) \left[\Pr\left(N_\xi(t\text{-}x)\right) = n\right]$$

$$= \sum_{n=1}^{\infty} F_\mu^{(n)}(x) \left\{F_\lambda^{(n)}(t\text{-}x) - F_\lambda^{(n+1)}(t\text{-}x)\right\} \quad \text{for } x > 0 \tag{2.28a}$$

and

$$\Pr(W(t)=0) = \Pr\left\{N_\eta(t)=0\right\} \quad \text{if } x = 0 . \tag{2.28b}$$

Similarly

$$\Pr(Z(t) \le x) = \Pr(\theta_\xi(t - x) \le x) = \sum_{n=1}^{\infty} F_\lambda^{(n)}(x) \Pr\left(N_\eta(t-x) = n\right)$$

$$= \sum_{n=1}^{\infty} F_\lambda^{(n)}(x) \left\{F_\mu^{(n)}(t-x) - F_\mu^{(n+1)}(t-x)\right\} \quad \text{if } x > 0 \tag{2.29a}$$

and

$$\Pr(Z(t)=0) = \Pr\left\{N_\xi(t)=0\right\} \quad \text{if } x = 0 \tag{2.29b}$$

In Eqs. 2.28a, b and 2.29a, b, the notation $F_\lambda{}^{(n)}(x)$ and $F_\mu{}^{(n)}(x)$ stands for the n-fold convolution of $F_\lambda(x)$ and $F_\mu(x)$.

Due to the equilibrium assumption, the process $v(t = 0)$ is either in the state $\{v_0\}$ or $\{v_1\}$ with the respective probabilities p_0 or p_1. Accordingly, we use the result Eq. 2.27a when $v(t = 0) = v_0$ and the result Eq.2.27b when $v(t = 0) = v_1$.

Hence, summing up the expressions Eqs. 2.27a and 2.27b, we derive the transition probability density of the process in the form :

$$\Sigma(x,t) = \frac{\partial}{\partial x} \Pr(X(t) \le x) = \frac{\partial}{\partial x}\left[p_0\, \Pi_0 + p_1\, \Pi_1\right] . \tag{2.30)}$$

After calculation, we obtain :

$$\Sigma(x,t) = \frac{p_0}{v_1 - v_0} \Pr\left(N_\eta(t)=0\right) \delta(\chi_0) + \frac{p_1}{v_1 - v_0} \Pr\left(N_\xi(t)=0\right) \delta(\chi_1) +$$

$$+\left\{ \frac{p_0}{v_1 - v_0} \sum_{n=1}^{\infty} f_\mu^{(n)}(\chi_0) \Pr\left(N_\xi(\chi_1)=n\right) - \frac{p_0}{v_1 - v_0} \sum_{n=1}^{\infty} F_\mu^{(n)}(\chi_0)\left[f_\lambda^{(n)}(\chi_1) - f_\lambda^{(n+1)}(\chi_1)\right] \right\}\Theta(\chi_0)$$

$$+\left\{ \frac{p_1}{v_1 - v_0} \sum_{n=1}^{\infty} f_\lambda^{(n)}(\chi_1) \Pr\left(N_\eta(\chi_0)=n\right) - \frac{p_1}{v_1 - v_0} \sum_{n=1}^{\infty} F_\lambda^{(n)}(\chi_1)\left[f_\mu^{(n)}(\chi_0) - f_\mu^{(n+1)}(\chi_0)\right] \right\}\Theta(\chi_1) \tag{2.31}$$

By finally observing that we have :

$$-\sum_{n=1}^{\infty} F_{\mu}^{(n)}(\chi_0)\left\{f_{\lambda}^{(n)}(\chi_1) - f_{\lambda}^{(n+1)}(\chi_1)\right\} = \sum_{n=1}^{\infty} f_{\lambda}^{(n)}(\chi_1)\,\Pr\left(N_{\eta}(\chi_0) = n - 1\right) \tag{2.32a}$$

and :

$$-\sum_{n=1}^{\infty} F_{\lambda}^{(n)}(\chi_1)\left\{f_{\mu}^{(n)}(\chi_0) - f_{\mu}^{(n+1)}(\chi_0)\right\} = \sum_{n=1}^{\infty} f_{\mu}^{(n)}(\chi_0)\,\Pr\left(N_{\xi}(\chi_1) = n - 1\right) \tag{2.32b}$$

the proposition follows. QED.

2.4. Discussion of a special case.

To elaborate on the previous section, let us discuss the simplest case belonging to that class of models. This is obviously the Markovian alternating process.

In this case we have :

$$F_{\lambda}(x) = 1 - e^{-\lambda x} \quad \text{and} \quad F_{\mu}(x) = 1 - e^{-\mu x} \tag{2.33}$$

The convolution of the probability densities have the form :

$$f_{\lambda}(x) = \lambda \exp(-\lambda x) \;\Rightarrow\; f_{\lambda}^{(n)}(x) = \mathrm{Er}\left(n, \lambda; x\right) \tag{2.34}$$

$$f_{\mu}(x) = \mu \exp(-\mu x) \;\Rightarrow\; f_{\mu}^{(n)}(x) = \mathrm{Er}\left(n, \mu; x\right) \tag{2.35}$$

where $\mathrm{Er}(n, \mu ; x)$ stands for the Erlang density of order :

$$\mathrm{Er}\left(n, \mu; x\right) = \mu \exp(-\mu x) \frac{(\mu x)^{n-1}}{(n-1)!} \,. \tag{2.36}$$

The other relevant probabilities are :

$$\Pr\left(N_{\xi}(t) = n\right) = \exp(-\lambda t) \frac{(\lambda t)^n}{n!} \tag{2.37a}$$

$$\Pr\left(N_{\eta}(t) = n\right) = \exp(-\mu t) \frac{(\mu t)^n}{n!} \tag{2.37b}$$

In this case Eq. 2.20 takes the form :

$$\Sigma(x,t) = \frac{p_0}{v_1 - v_0}\,\delta(\chi_0)\, e^{-\mu t} + \frac{p_1}{v_1 - v_0}\,\delta(\chi_1)\, e^{-\lambda t} +$$

$$+ \exp -\left(\lambda\chi_1 + \mu\chi_0\right)\left\{\frac{p_0}{v_1 - v_0} S_1\, \Theta(\chi_0) + \frac{p_1}{v_1 - v_0} S_2\, \Theta(\chi_1)\right\} \tag{2.38}$$

with :

$$S_1 = \sum_{n=1}^{\infty} \mu \frac{(\mu\chi_0)^{n-1}(\lambda\chi_1)^{n}}{(n-1)!\,(n)!} + \lambda \frac{(\mu\chi_0)^{n-1}(\lambda\chi_1)^{n-1}}{(n-1)!\,(n-1)!} \tag{2.39a}$$

$$S_2 = \sum_{n=1}^{\infty} \mu \frac{(\mu\chi_0)^{n-1}(\lambda\chi_1)^{n-1}}{(n-1)!\,(n-1)!} + \lambda \frac{(\mu\chi_0)^{n}(\lambda\chi_1)^{n-1}}{(n-1)!\,(n)!} \tag{2.39b}$$

The series expansions of the Bessel functions [25] enable us to write Eq. 2.38 in the final form :

$$\Sigma(x,t) = \frac{p_0}{v_1 - v_0}\,\delta(\chi_0)\,e^{-\mu t} + \frac{p_1}{v_1 - v_0}\,\delta(\chi_1)\,e^{-\lambda t} + \\ + \frac{1}{v_1 - v_0}\exp - \left\{\frac{(\mu-\lambda)\,x + (\lambda v_1 - \mu v_0)\,t}{v_1 - v_0}\right\}\{BES\} \tag{2.40}$$

where BES denotes :

$$BES = I_0(\Gamma(x,t))\left[p_0\,\lambda\,\Theta(\chi_0) + p_1\,\mu\,\Theta(\chi_1)\right] + \\ + \frac{2\,\lambda\mu}{\Gamma(x,t)}\,I_1(\Gamma(x,t))\left[p_0\,\chi_1\,\Theta(\chi_0) + p_1\,\chi_0\,\Theta(\chi_1)\right] \tag{2.41a}$$

with I_0 and I_1 are Bessel functions and $\Gamma(x, t)$ stands for :

$$\Gamma(x,t) = \sqrt{4\,\lambda\,\mu\,\frac{\{-\,v_0 v_1\,t^2 + (v_0 + v_1)\,x\,t - x^2\}}{(v_1 - v_0)^2}} \tag{2.41b}$$

When $v_0 = -\,v_1$ and $\lambda = \mu$, the solution Eq. (2.39) has yet been derived in Ref.(10) directly using the associated Chapman-Kolmogorov equation.

3.Stochastic analysis of the transport of sediments.

3.1. Modelization of longitudinal dispersion in a porous medium [2-5].

Consider a steady uniform flow through a column packed with a porous material and assume that at time $t = 0$ a set of tagged particle are simultaneously injected into the flow. We assume that the particles themselves do not affect the motion of the fluid. Immediately after the release, the particles are dispersed. Each particle undregoes a random walk through the voids and pores of the porous structure. Owing to dead zones present between grains of the porous material and/or to the vorticity zones in the flow, the particles may be trapped and then later released into the current. In other words, a particle flowing through a porous medium progresses through the structure in a serie of steps of random lengths with rest periods of random duration between two consecutives steps. This is a nice illustration of where the dynamics can be efficiently represented by a stochastic process of the type discussed in the last section. In the simplest modelization, the position x(t) of a tagged particle is governed by the stochastic differential equation such as Eq.2.5 with $v_0 = 0$. Implicitly, we approximate the lengths of the random steps by assuming that the particle moves with a constant velocity during a random length of time.

$$\sigma^2 = \langle x(t)^2 \rangle - \langle x(t) \rangle^2 = \frac{v_1^2 \tau_\mu^2 CV_\mu^2}{\tau_\lambda + \tau_\mu} t \tag{3.1}$$

with :

$$\langle x(t) \rangle \cong \frac{v_1 \tau_\mu}{\tau_\lambda + \tau_\mu} t \tag{3.2}$$

When the coefficient of variation is smaller than one, the sojourn time in the state $\{v_1\}$ is distributed according to Eq. 2.7 with M >1. Therefore the process is non-Markovian and we observe that the presence of a "memory" reduces the dispersion of the process. It should be noted that $CV_\lambda = 1$ and $CV_\mu = 1$ correspond to exponentially distributed sojourn times in the states $\{v_0\}$and $\{v_1\}$. Also that the dispersion Eq. 3.1 depends only on τ_λ and not on the coefficient CV_λ. This is in fact not surprising if one observes that when $v_0 = 0$ the variance is not modified by the time evolution.

3.2 The approximative dispersive model of Todorovic [4]

It is interesting to exhibit how Todorovic's solution can be obtained from our more general discussion. Let us consider the dynamics given by Eq. 2.5 with velocities $v_0 = 0$ and $v' = \rho v_1$ with ρ a positive scaling parameter and $\tau' = \tau_\mu/\rho$. Here we deal with the Markovian case and therefore both $F_\lambda(t)$ and $F_\mu(t)$ are the exponential probability densities. In the limit ρ tends to infinity, the system admits an infinte velocity v' but it stays infinitely shortly in the state v' such that the product $v'\tau'$ remains finite.A typical realization of this process is represented in Fig 3.2.

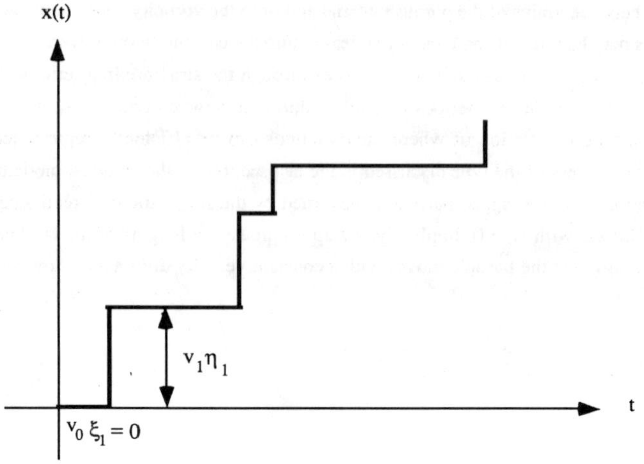

Fig. 3.1. A typical realization of the Todorovic process.

This model accurately represents the dynamics discussed by Todorovic. In this case Eq. 2.2 is :

$$\frac{\partial^2}{\partial t^2}\Sigma(x,t) + (\lambda + \rho\mu)\frac{\partial}{\partial t}\Sigma(x,t) + \rho\, v'\frac{\partial^2}{\partial x \partial t}\Sigma(x,t) + (\lambda\,\rho\, v')\frac{\partial}{\partial x}\Sigma(x,t) = 0 \tag{3.3}$$

In the limit $\rho \to \infty$, Eq. 3.3 we obtain :

$$\mu\frac{\partial}{\partial t}\Sigma(x,t) + v'\frac{\partial^2}{\partial x \partial t}\Sigma(x,t) + \lambda\, v'\frac{\partial}{\partial x}\Sigma(x,t) = 0 \tag{3.4}$$

which has been given in [4].

Moreover, in this limit, we have $p_0 = 1$ and $p_1 = 0$ and Eq. 2.44 reduces to :

$$\Sigma(x,t) = 2\,\delta(x)\exp - (\mu t) + \frac{\lambda}{v'}\left[\exp - \left\{\mu\, t + \frac{\lambda}{v'}x\right\}\right]\left(\frac{\mu t}{\lambda\,(v')^{-1}\,x}\right)^{\frac{1}{2}} I_1\left(2\sqrt{\lambda\,\mu\,(v')^{-1}\,x\,t}\right) \tag{3.5}$$

which has also been calculated in [4]. Finally for asymptotically large times, the mean and the variance are given by Eqs. 3.1 and 3.2 with $CV_\mu = 1$.

4. Transcient filling of a buffer in a simple production line.

4. 1. The importance of buffers in a simple production lines and its population dynamics.

During the last forty years, a considerable effort has been devoted to the study of the stochastic flow of parts in complex production devices. At the conception level of a new line, it is indeed crucial to be able to estimate, prior to simulations, the expected throughput of a production line. It would be pointless and somehow perilous to try to quote here an extensive list of papers devoted to this important topic. However, it should be noted that our description and modelization has been pioneered by [7] and further considered by [8-14]. In these studies, the time variable is continuous and it is assumed that the population of the buffer can also be described by a continuous variable. While extensive calculations are devoted to the extraction of invariant probability measures which adequately describe the stationary regimes of production, little attention is focused on the transcient phenomena. In particular, the time evolution of the buffers population has not been studied. Quantities like the first two moments of the buffer filling, or the mean first access time to the boundaries of the stock also carry also an important amount of information. Indeed, it is precisely the empty or full states of the buffer which induce the blockages or the starvations of the machines in these systems. The lack of studies especially devoted to the transcient behaviours originates from the usually untreatable form of the equations governing the dynamics. Let us now study of one of the simplest configuration for which dynamics is of the form we have so far considered..Let us introduce the modelization. We consider the following as a simple realization of a production line L (see Fig. 4.1).

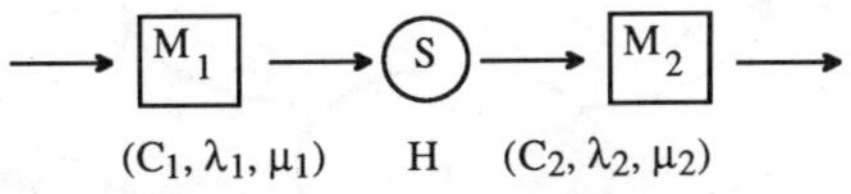

Fig. 4.1. The production line L.

The production line L is composed by two machines M_1 and M_2 with respective constant production rates C_1 and C_2.respectively. Between M_1 and M_2, we equip our system with a buffer stock S whose maximum capacity is H parts. We assume that the filling of S is described by a continuous variable $x(t) \in [0, H]$, t being the (continuous) time. The reader is referred to [6] for a discussion on the validity of this "hydrodynamic" description of a production line. The study of a single arrangement like L is important as it is the basis and the cornerstone of the aggregation techniques available to discuss more complex devices [6].

As usual, we say that M_1 is blocked when S is full and M_2 is starved when S is empty. Assume that we have fictive infinite buffers at the extremities of L such that M_1 is never starved and M_2 never blocked. In addition, a lost of productivity is induced by the fact that generally both M_1 and M_2 are stochastically failure prone. We denote by $\Gamma_i = \{0, 1\}$, $i = 1, 2$; the state of the machine M_i, where 0 stands for M_i out of working order and 1 in working order. Therefore, the state of M_i can be represented by two alternating renewal processes as introduced in section 2.2 . Consequently, the population of the buffer x(t) is governed by the stochastic differential equation :

$$\frac{d}{dt} x(t) = v(t) \qquad v(t) \in \{C_1 - C_2, C_1, 0, -C_2\} \text{ with } C_1 \geq 0 \text{ and } C_2 \geq 0 \tag{4.1}$$

where C_1 and C_2 denote respectively the production rates of M_1 and M_2. The transitions between the different states of v(t) are sketched in the Markovian case in Fig 4.2.

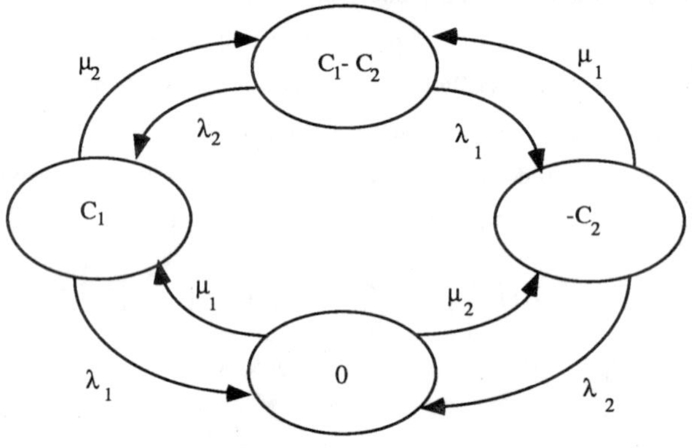

Fig. 4.2. Sketch of the probability transitions between the stages characterizing two Markovian failure prone machines.

While the first access time to the boudaries of S will be discussed in the next section, we first restrict our attention to shorter time intervals i.e., those times for which we have :

$x(t) \in \;]0, H[$ (4.2)

Hence, we can directly use the general theory introduced in section 2.1 to define an operator $A(\xi, \gamma)$ which will be, for the simplest situation sketched in Fig. 4.2, a fourth order differential operator. To simplify this discussion, we restrict from now on, our attention to the special case where the first machine M_1 is not failure prone, in other wordσ, $\lambda_1 = 0$ while M_2 fails with an exponential law and ithe reparation time is described by an Erlang law of the type Eq. 2.6. The situation is a particular case of Fig. 2.1.and is represented in Fig. 4.3.

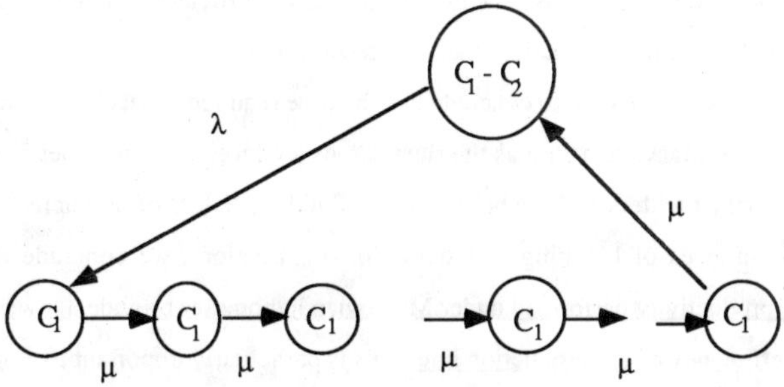

Fig. 4.3 . Sketch of a typical situation describing a production line. The state $C_1 - C_2$ is occupied when both M_1 and M_2 are operating. In the state C_1 M_2 is under reparation. Each stage C_1 has been chosen to have the same mean sojourn time μ. The number of stages in C_1 is N.

In the production lines, it is obviously clear that the buffers are efficient only when the two machines have approximately balanced production rates. Indeed, if this is not realized, the resulting drift immediately drives the population of the buffer to one of its boundaries (i.e. full or empty), hence the presence of the buffer becomes rapidly useless. A balanced production line is characterized by the condition :

$$\langle v \rangle = (C_1 - C_2)\,\tau_\lambda + C_1\,\tau_\mu = 0 \tag{4.3}$$

With Eq. 4.1, Eq. 2.18 immediately gives :

$$\langle x(t)^2 \rangle = \frac{C_1^2 \, \tau_\mu^2 \left[1 + CV_\mu^2\right]}{\tau_\lambda + \tau_\mu} t = \frac{C_1^2 \, I \left(1 + \frac{1}{N}\right)}{\mu\,(1 + I)} t \tag{4.4}$$

where in Eq. 4.4, we have introduced an indisponibility factor, $I = \lambda/\mu$ of the machine M_2. Worded, I is the proportion of the mean time that M_2 is under reparation compared with the mean time it is operating. Remember that the number of stage v_0 is N which leads to $CV_\mu = N^{-1/2}$.

Eq. 4.4 exhibits explicitly the influence of non-Markovian effects on the time evolution of the variance of the population in the buffer S. Indeed, the Markov situation occurs for N = 1. So far, we have assume the size of S to be sufficiently large to ensure the validity of Eq. 4.4. In actual production lines it is obvious that the buffer capacity is finite. From Eq. 4.4, it however possible to conclude that the time required to reach the boundaries of S is shorter in the Markovian case as the fluctuations are larger. On the other hand, it is clear that each contact with one of the boundaries of S induces a loss of productivity of the system (i.e. occurences of blocking and starvations). Therefore, we conclude that the calculations of productivity performed under Markovian hypotheses provide a lower bound on the overall efficiency of the production line. This is particularly important to emphasize as it is precisely for the Markovian cases that calculations are possible. Accordingly, we henceforth pursue the discussion under the Markovian hypothesis, (i.e. N = 1). In view of Eqs. 2.2 and 2.13, we have :

$$\left(\frac{\partial^2}{\partial t^2} + (\mu + \lambda)\frac{\partial}{\partial t}\right)\Sigma(x,t) = (2C_1 - C_2)\frac{\partial^2}{\partial x\,\partial t}\Sigma(x,t) + \left\{C_1\,(C_2 - C_1)\frac{\partial^2}{\partial x^2} + \frac{\langle v\rangle}{\lambda\,\mu}\frac{\partial}{\partial x}\right\}\Sigma(x,t) \tag{4.5}$$

Using Eq. 4.5, we can calculate the time evolution of the moments in the general situation but for time prior to the boundaries access of S. We obtain :

$$\langle x(t)\rangle = x_0 - \lambda\,\mu\,\langle v\rangle\left[t + \frac{1}{\lambda + \mu}\left(\exp\left[-(\lambda + \mu)\,t\right] - 1\right)\right] \tag{4.6}$$

and

$$\langle x^2(t)\rangle = x_0^2 + \frac{2\,C_1\,(C_2 - C_1)}{\lambda + \mu}\left[t + \frac{1}{\lambda + \mu}\left(\exp\left[-\left(\lambda + \mu\right)t\right] - 1\right)\right] + \psi(V,t) \tag{4.7}$$

where Ψ (<v>, t) is a function such that $\Psi(<v> = 0, t) = 0$. Let us again concentrate on the balanced case <v> = 0. In this case Eqs. 4.6 and 4.7 yield :

$$\langle x^2(t)\rangle \cong x_0^2 + \frac{C_1\,(C_2 - C_1)}{\lambda + \mu}\,t^2 = x_0^2 + C_1^2 I\,t^2 \qquad \text{for } t \ll (\lambda + \mu)^{-1} \tag{4.8a}$$

and

$$\langle x^2(t)\rangle \cong \frac{2C_1\,(C_2 - C_1)}{\lambda + \mu}\,t = \frac{2C_1^2 I}{\lambda + \mu}\,t \qquad \text{for } t \gg (\lambda + \mu)^{-1} \tag{4.8b}$$

From Eq. (4.8b), we deduce that for asymptotically large times, a diffusive behaviour of the buffer filling process. This behaviour is naturally reminiscent of the central limit theorem invoked in section 2.2. On the contrary, for short intervals, the behaviour of the second moment is quadratic in time ; we call this the ballistic regime Eq. 4.8a.

Ideally, the buffer capacity should be infinite which is obviously impossible in the practical cases. In actual situations, the aim is to improve the efficiency obtained with a finite buffer size. This will be achieved by dimensioning the buffer so as to mostly absorb the fluctuations of the ballistic regime. Indeed, the rapid growth of $<x^2(t)>$ in the ballistic regime, generates frequent accesses to the boundaries of S. When one of the boundaries of S is reached, one of the machines is idle and therefore the performance of the production line L is reduced. Once the ballistic fluctuations are absorbed, i.e. when the diffusive regime prevails, the efficacity is poorly improved by an additional increase of the buffer capacity . Indeed, due to the slow evolution of the diffusive fluctuations, the frequency of the accesses to the boundary of the stocks is only slighty affected by changes in the capacity of S.

Invoking the same argument, it is also interesting to note that even when the line is balanced in the mean (i.e. <v> = 0), a large divergence on the cycle times of M_1 and M_2 is likely to drop the productivity of the production line This observation is easily confirmed by the permanent regime analysis [9-11] which for our configuration gives an effective indisponibility factor of the form :

$$I_{\text{effective}} = \frac{\lambda}{\mu}\,\frac{C_1(C_2 - C_1)}{C_2\lambda H + C_1(C_2 - C_1)} = I\,\frac{C_1^2\,I}{C_1\,(1 + I)\,\lambda\,H + C_1^2\,I} \tag{4.9}$$

Clearly in order to reduce $I_{effective}$, it is necessary either to reduce the difference $(C_2 - C_1)$ or to enlarge the buffer capacity H as it is of course expected.

4.2 First acces time to the boundaries of the buffer.

Let us now calculate the mean first passage time $T(x_0, \Delta)$ to the boundaries. To perform this calculation we assume the line to be balanced in the mean ($\langle v \rangle = 0$) which is the relevant situation.

Generally, the calculation of $T(x_0, \Delta)$ for a non-Markovian process is a tedious task. However, for a simple Markovian alternating process x(t) , the calculation has been performed. Using Eq. 17 of [26], with $z = H/2$, $a = C_1$, $b = C_2 - C_1$, $\lambda_a = \mu$, $\lambda_b = \lambda$ and under the assumption that at time $t = 0$ M_2 has just failed, we have :

$$T(\tilde{x}_0, \tilde{\Delta}) = \frac{H/2 - \tilde{x}_0}{C_1} + \frac{H/2 - \tilde{x}_0}{C_1} \frac{\mu H \frac{(1-I)}{2I}}{C_1 + \mu H} + \left\{ \frac{(H^2/4 - \tilde{x}_0^2)\mu(1+I)}{2C_1^2 I} \right\}$$

$$= \frac{H/2 - \tilde{x}_0}{C_1} + \frac{H/2 - \tilde{x}_0}{C_1} \frac{\mu H \frac{(1-I)}{2I}}{C_1 + \mu H} + t^* \tag{4.10a}$$

where we have introduced the notation :

$$\tilde{x}_0 \in \left[\frac{-H}{2}, \frac{H}{2}\right] = \tilde{\Delta} \quad \text{and} \quad \tilde{x}_0 = x_0 - \frac{H}{2} \tag{4.10b}$$

The right-hand side of Eq. 4.10a can be separated into three components. The first term $(H/2 - x_0)/C_1$ corresponds to the deterministic time to reach the boundary when M_2 is broken down. The second and curly bracket terms correspond to stochastic corrections.

Remark that t* in Eq.4.10a, describes a diffusive contribution. By using Eq.4.8b, t* is directly interpreted as the time required, in the diffusive regime, for $\langle x^2 t) \rangle$ to attain the size of the buffer capacity. From Eq. 4.8b, we have :

$$\frac{2C_1^2 I_2}{\lambda + \mu} t^* = \left(\frac{H^2}{4} - x_0^2\right) \tag{4.11}$$

We explicitly see that the smaller the indisponibility facto I is, the higher $T(x_0, \Delta)$ and hence the larger the efficiency is.

Instead of a single initial condition x_0, we can consider x_0 to be drawn at random from a probability density $f(x_0)$. Hence, we can write:

$$T(\tilde{\Delta}) = \int_0^H f(x_0)\, T(x_0, \Delta)\, dx_0 = \int_{-H/2}^{H/2} f(\tilde{x}_0)\, T(\tilde{x}_0, \tilde{\Delta})\, d\tilde{x}_0 \tag{4.12}$$

The density $f(x_0)$ can for instance be chosen as the stationary density leading to the efficiency Eq.4.9. We have [11]:

$$f(x_0) = N\left[\frac{C_2}{C_2 - C_1} + 2\left(\frac{C_1}{\mu_2}\,\delta(H - x_0) + \frac{C_1}{\lambda_2}\,\delta(x_0)\right)\right] \tag{4.13}$$

with

$$N = \frac{I}{(1 + I)[H + C_1/\mu]} \tag{4.14}$$

We introduce the dimensionless parameter:

$$F = \frac{\mu H}{C_1} \tag{4.15}$$

In terms of F, we obtain:

$$\int_0^H f(x_0)\, x_0^n\, dx_0 = \frac{H^n}{(1 + F)(1 + I)}\left[\frac{F(1 + I) + (n + 1)I}{n + 1}\right] \qquad n \geq 1 \tag{4.16}$$

Using Eqs. 4.12, 4.13 and 4.15, we get:

$$T(\tilde{\Delta}) = \frac{F}{4\lambda}\left[\frac{1}{1+I} - \frac{1-I}{(F + 1)^2}\right] + \left\{\frac{F^3}{12\,\lambda(1 + F)}\right\} \tag{4.17}$$

Using the definition of F, Eq. 4.9 takes the suggestive form:

$$I_{eff} = I\left\{\frac{1}{1 + (1 + I)\,F}\right\} \tag{4.18}$$

Again in Eq. 4.17, we can identify two contributions characterized by the leading order in the parameter F. The first term is of leading order linear in F, whereas the curly bracket (the diffusive contribution to T) behaves as F^2. Hence, for large F, the diffusive contribution dominates. An efficient choice of the buffer size should absorb the fluctuations of production in the ballistic regime. From Eq.4.8, the ballistic regime clearly dominantes for small F, while for F chosen of the order 1 or 2, both contributions in Eq. 4.17 become comparable. This range of F values enables us to select an efficient buffer size.

By using the permanent analysis results Eq. 4.9, it is interesting to observe that the selection of the buffer size given by F=1 is indeed judicious. For this choice, the effective indisponibility factor of the system already drops by a factor of at least two. Let us note that based on purely heuristic grounds, the value F = 1 has been proposed proposed in [27] on heuristic grounds.

Finally, we mention that our conclusions do not crucially depend on the choice of the distribution Eq. 4.13. In fact, we can draw similar conclusions by simply considering Eq. 4.10a with the single initial condition $x_0 \approx 0$

5.Conclusions.

The buffer population in automated production lines with failure prone machines and the transport of sediments in porous medium find a natural means of modelling in terms of stochastic differential equations. The noise sources are generated by continuous time Markov chains. The transition probability densities obey Chapman- Kolmogorov equations expressible as high order hyperbolic partial differential equations. The order of the hyperbolic differential operator is directly related to the number of stages of the chain which generate the noise. For asymptotically large times, a central limit theorem holds and enables us to calculate explicitly the relevant moments of the dynamics of these systems. Using the multistage Markov chains, it is moreover possible, by a projection technique, to calculate the non-Markovian contributions to the dynamics. More precisely, the marginal probability densities obtained by summing on the noise states describe non-Markovian evolutions of the dynamical quantity of interest. An illustration, which plays the central role in the applications proposed here, is given by the dichotomous, alternating renewal processes with non-exponential sojourn times in each of its two stages. The transition probability densities are calculated for arbitrary sojourn times. We further specialize our calculations to situations where the sojourn times are distributed according to the generalized Erlang densities. In this case, we calculate the non-Markovian contribution to the growth of the variance of the process. The central role is observed to be played by the coefficients of variation CV of the sojourn times of the process. Explicitly, the variance is proportional to CV^2t when the time $t \to \infty$, (the diffusive nature results from the central limit theorem just invoked). The coefficient of variation CV of the Erlang probability densities are smaller than unity. The limiting cases CV = 1 holds for the exponential density which gives the Markov dynamics while CV = 0 corresponds to the purely deterministic dynamics). Hence, the non-Markovian effects tend to reduce the growth of the variance. This has, in particular, the important consequence that calculations of the throughput of production lines performed with exponentially distributed "times to failure" and "times to repair" furnish lower estimates, thereby making it precious to the engineer.

Acknowledgements. We thank Prof. J. Figour and Prof. Dr. C. W. Burckhardt for their stimulating interest.

References.

1. M.S. Klamkin, Math. Modelling,**1** (1980),63.

2. J. Gani and T .Todorovic, Stoch. Hydrol. Hydraul. **1**, (1987), 209.

3. P. Todorovic, Math. Scientist. **5**, (1980), 45.

4. P. Todorovic, Water Resources Res. **11**, (975),348.

5. A. Gut and P. Ahlberg, Chemica scripta. **18**, (1981), 248

6.Y. Dallery and S. Gerschwin, *Manufacturing flow line systems. A review of models and analytical results* . Rapport MASI 91-18, (1991). Institut Blaise Pascal, Université Paris VI, Paris VII and CNRS.

7. B. Zimmern, Rev. Stat. Appliquées, 4 (1956), 85.

8. J. P. Forestier, R.A.I.R.O. Automatique, 14 (1980), 127.

9. D. Dubois and J. P. Forestier, R.A.I.R.O. Automatique, 16 (1982), 105.

10. J. Winjgaard, A.I.E.E. 11 (1979), 42.

11. P. Coillard and J. M. Proth, Revue Belge de Stat., Inform. et Rech. Oper., 24 (1983), 1

12. D. Mitra, Adv. Appl. Probab. 20, (1988), 646.

13. R. David, X.Xie and Y. Dallery, IMA J. of Math. Appl. in Business and Industry, **6**, (1990), 281.

14. Y. Dallery, R David and X.-L. Xie, IEEE Trans. on Autom. Control. **34**, (1990), 943.

15. S. Goldstein, Quart. J. Mech and Appl. Math. **4**, (1951), 130.

16. M. Kac, Rocky Mountain J. of Math. **4**, (1974), 497.

17. D. C. Heath, *Probabilistic analysis of hyperbolic systems of partial differential equations* . PhD Thesis, (1969), University of Illinois.

18. R. Hersch, Rocky Mountain J. of Math. **4**, (1974), 443.

19. M. Pinsky, Z. Warscheinlichkeitstheorie verw. Geb. **9**, (1968), 101.

20. V. Cane, Bull. Intern. Stat. Inst. **42**, (1967), 622.

21. E. Orsingher, Adv. Appl. Probab. **22**, (1990), 915.

22. P. Chr. Hemmer, Physica. **27**, (1961), 79.

23. D. R. Cox, *Renewal theory.* (Methuen , 1962).

24. S. Karlin and H. M. Taylor, *A first course in stochastic processes (* Acad. Press 1975).

25. I.S. Gradshteyn and M. Ryzhik,*Tables of Integrals and Products* (Academic Press, 1963).

26. J. Masoliver, K. Lindenberg and B. J. West , Phys. Rev. **A33**, (1986), 2177.

27. J. Figour, *Cours d'Assemblage automatisé,* Institut de microtechnique EPFL (1989).